P9-DOG-511

Signaling System #7

Other McGraw-Hill Telecommunication Books of Interest

Signaling System #7

Travis Russell

Second Edition

McGraw-Hill

New York San Francisco Washington, D.C. Auckland Bogotá
Caracas Lisbon London Madrid Mexico City Milan
Montreal New Delhi San Juan Singapore
Sydney Tokyo Toronto

Library of Congress Cataloging-in-Publication Data

Russell, Travis.
Signaling system #7 / Travis Russell.—2nd ed.
p. cm.
Includes bibliographical references and index.
ISBN 0-07-058032-4 (hardback)
1. Telecommunication—Switching systems. 2. Signaling system 7.
I. Title.
TK5103.8.R88 1998
621.382—dc21 98-4157
CIP

McGraw-Hill

A Division of The McGraw-Hill Companies

2 3 4 5 6 7 8 9 0 DOC/DOC 9 0 3 2 1 0 9 8

ISBN 0-07-058032-4

The sponsoring editor for this book was Steve Chapman, the editing supervisor was Bernard Onken, and the production supervisor was Claire Stanley. It was set in Century Schoolbook by Estelita F. Green of McGraw-Hill's Professional Book Group composition unit.

Printed and bound by R. R. Donnelley & Sons Company.

This book is printed on recycled, acid-free paper containing a minimum of 50% recycled, de-inked fiber.

Contents

Preface

It has been several years since I wrote the first edition of *Signaling System #7*. Since then there have been many changes in our industry. The most sweeping change to affect all of telecommunications was the Telecommunications Act of 1996. With this new legislation came many new changes that have and will continue to have an impact on SS7.

In this second edition, I have captured many new changes made to the SS7 protocol. You will see these changes in virtually every chapter. New parameters in TCAP, ISUP, and SCCP have been defined. I expanded on the ISUP chapter as both ANSI and Bellcore have further defined BISUP.

I have also made some corrections from readers who have written to me over the last few years. Some of these readers were involved in development of SS7 products, some involved in standards bodies defining various portions of the SS7 protocols and their functions

I added a new chapter in this edition, titled "Local Number Portability (LNP)." This chapter provides an overview of what LNP is and how SS7 is impacted. It is my hope in future editions to add other applications to this chapter, and provide explanations of various real world uses for SS7 and Intelligent Networks.

Each chapter in this book has been arranged to provide a three-step approach to learning. In the beginning of the chapters discussing SS7 protocol, I provide an overview of the topic, getting you familar with the topic before discussing specifics. This is ideal for students learning SS7 for the first time, or those just wanting to understand more about what SS7 is.

This is followed by a more detailed discussion about the functions of the various parts of the protocol, identifying the various procedures and operations of the SS7 network as they relate to the protocol being discussed. This is useful to those who need to understand the functions of SS7 and its various protocols.

The last section of these chapters provides a bit level description of the protocol, identifying the message formats, parameter values, and parameter descriptions. This section of the chapter is useful to those involved in the development of SS7 products.

The audience for this book is broad. Students studying telecommunications should learn SS7, because they will be involved with this technology in some form or fashion if they choose a career in telecommunications. Engineers will find this book useful as a reference. I have seen this book on the shelves of many engineering firms developing SS7 products. Sales, marketing, management; all will benefit from this book.

Acknowledgments

First, thank you to my beautiful wife. So many nights were spent typing away while she sat alone watching movies, or reading books. Thank you Deb for being so understanding and endearing throughout this project.

Once again, I have my friends at Tekelec to thank for all of their support for this book. Special thanks to Mr. Allan Toomer, retired CEO of Tekelec. Alan has promoted this book all along, and his encouragement has been much appreciated. Also thanks to Mr. Gord Werner, GM of the Network Switching Division at Tekelec. Being part of a cutting edge company like Tekelec has been a great benefit to me, learning about new advances in SS7 before they are implemented in our nation's networks.

I also must thank my family for their patience while I worked on this second edition. For those interested, I am finally taking the family to Disney World as promised in the first edition. A little late, but a well-deserved vacation for all of us.

I also want to thank the many readers who took the time to send e-mail and provide comments and inputs, and for buying the first edition. I never thought this book would sell as well as it has, only because SS7 is such an obscure and specialized subject matter. You the readers proved me wrong and provided confirmation that SS7 is a technology that everyone is gaining interest in.

Travis Russell

Introduction

First there was ISDN. Then came "portable" 800 numbers. Soon, cellular providers were talking of tying their networks together so that subscribers could roam from one cellular network to another without special roaming numbers. And, of course, we now have the Information Highway and all of the talk behind its many features and intrinsic value to the public.

But during all of these discussions, nothing (or at least very little) has been mentioned about how all of this is possible. How is it possible that many networks can communicate with one another, and what control mechanism is used to maintain the connections and data communications?

Behind the scenes is a quiet, highly reliable, fault-tolerant data communications network that links the world's telephone networks together and allows them to share vital signaling and control information. This one network, which controls telecommunications networks around the world, is being prepared to control the broadband networks that will form the framework for tomorrow's "Information Highway."

This data communications network is so robust, so sophisticated, that few really know much about it. Only recently has anything been written about this network, and what information does exist has been sparse and ambiguous. The very standards that define the network are written to answer the needs of so many different audiences that they remain ambiguous and reliant on self-interpretation.

This network, often dubbed the world's largest data communications network, is known as Signaling System #7 (SS7). It started as a way to access 800 databases here in the United States, although its functions provide much more. Soon SS7 was being used to send signaling information from exhange to exchange. Today, the same network is being used to control central office switching equipment from remote locations.

SS7 is really a *control* network, as well as a signaling network. This is important to understand, because as the Information Highway rolls out, and as the Advanced Intelligent Network (AIN) is implemented, SS7 will be relied on almost exclusively as a means for telephone companies and other service providers to share database information and switching control without human intervention.

Already we are seeing SS7 play an important role in Local Number Portability (LNP), a new requirement for all telephone service providers (both wireline and wireless) defined in the Telecommunications Act of 1996. SS7 continues to grow in use and function, as the industry discovers new uses for this powerful technology.

Without SS7, AIN is not possible. Cellular roaming is not seamless. 800 numbers are not portable. And the many features and services we take for granted today (such as caller ID) would not be feasible.

When someone asks me how to break into the telecommunications field, I often provide one suggestion. Learn SS7. Every telephone company needs it. Every manufacturer develops around it. Yet there is a lack of expertise in SS7 in this country. That is what this book is all about.

As I began studying SS7 protocols, and began providing seminars around the country, I came to the conclusion that there is no reference material available that provides a comprehensive view of SS7. Nothing to explain the various acronyms and the true applications of this network.

Hence the reason for this book. If you are looking for more than a reference book—a tutorial on SS7, a text book which provides real-world applications, and a futuristic look at the telephone networks—read on.

The intent of this book is to provide a comprehensive introduction to this fascinating network and its protocols, as well as a reference for those already familar with SS7. While the various standards are always the best source for specific details, this book provides something the standards do not: explanations as to when and why procedures are used and what they mean to the network, the subscriber, and the service provider.

Although it is based on ANSI and Bellcore standards, the text is of value to anyone seeking knowledge about SS7 networks. While the message types and the protocol parameters may be different between countries, the principles and applications are universal.

I hope you find this to be as fascinating as I did while writing this book. Truly, we are at an exciting period for the telecommunications industry, as new services and applications are defined almost monthly. Keeping up with all of the new technologies and advances can be a dizzying experience. Yet the future is today, and this technology is a key player in making it all happen.

Travis Russell

Acronym List

A-links	Access links
AAL	ATM adaptation layer
ABOM	A-bis Operations and Maintenance
AC	authentication center
ACD	automatic call distribution
ACG	Automatic Code Gap
ACM	address complete message (message)
AE	application entity
AERM	alignment error rate monitor
AIN	Advanced Intelligent Networks
AK	data acknowledgment
AMI	alternate mark inversion
ANI	automatic number identification
ANM	answer message (message)
ANSI	American National Standards Institute
ASE	application service element
ASN-1	Abstract Syntax Notation One
ASP	Application Service Part
ATM	asynchronous transfer mode
B-links	bridge links
BIB	backward indicator bit
BISDN	broadband ISDN
BISUP	broadband ISUP
BITS	building integrated timing system
BLA	blocking acknowledgment (message)
BLO	blocking (message)
BOP	bit-oriented protocol
BRI	basic rate ISDN
BSC	base station controller
BSDB	business service database
BSN	backward sequence number

BSS	base station subsystem
BSSMAP	Base Station Subsystem Mobile Application Part
BTA	basic trading area
BTS	base transceiver station
BTSM	Base Transceiver Station Management
C-links	cross-links
CAE	Communications Applications Environment
CBA	changeback acknowledgment (message)
CBD	changeback declaration (message)
CC	connection confirmed (message)
CCE	consistency check end (message)
CCEA	consistency check end acknowledgment (message)
CCR	continuity check request (message)
CCRA	consistency check request acknowledgment (message)
CCS	common channel signaling
CD	carrier detect
CESID	Callers Emergency Service Identification
CFN	confusion (message)
CGB	circuit group blocking (message)
CGBA	circuit group blocking acknowledgment (message)
CGU	circuit group unblocking (message)
CGUA	circuit group unblocking acknowledgment (message)
CIC	circuit identification code
CLLI	COMMON LANGUAGE™ Location Identifier
CMC	call modification completed (message)
CMR	call modification request (message)
CMRJ	call modification reject (message)
CMSDB	call management services database (message)
CON	connect (message)
COO	changeover order (message)
COT	continuity test (message)
CPG	call progress (message)
CPU	central processor unit
CQM	circuit query message (message)
CQR	circuit query response (message)
CR	connection request (message)
CRA	circuit reservation acknowledgment (message) *also* consistency check request (message)
CRC	cyclic redundancy check
CREF	connection refusal (message)
CRG	charge information (message)
CRM	circuit reservation message (message)

CRST	cluster-route-set-test (message)
CSA	Canadian Standards Association
CTIA	Cellular Telecommunication Industry Association
CTS	clear to send
CVM	circuit validation test (message)
DCE	data communications equipment
DN	directory number
DPC	destination point code
DPNSS	Digital Private Network Signaling System
DRS	delayed release (message)
DSR	data set ready
DSU	data service unit
DTAP	Direct Transfer Application Part
DTE	data terminal equipment
DTMF	dual-tone multifrequency
DTR	data terminal ready
DT1	data form 1
DT2	data form 2
DUP	Data User Part
EA	expedited data acknowledgment (message)
EAS	exchange access signaling
ED	expedited data (message)
EIA	Electronic Industries Association
EIR	equipment identification register
ERR	error (message)
FA	Framework Advisory
FAA	facility accepted (message)
FAR	facility request (message)
FCC	Federal Communications Commission
FDDI	Fiber Distributed Data Interface
FIB	forward indicator bit
FISU	fill-in signal unit
FOT	forward transfer (message)
FR	Family of Requirement
FRJ	facility rejected (message)
FSN	forward sequence number
GMT	Greenwich Mean Time
GPS	Global Positioning System
GRA	circuit group reset acknowledgment (message)

GRS	circuit group reset (message)
GSM	Groupe Special Mobile
GUI	graphical user interface
HLR	home location register
IAA	IAM acknowledgment (message)
IAM	initial address message (message)
IAR	IAM reject (message)
IN	Intelligent Network
INA	information network architecture
INF	information (message)
INR	information request (message)
IP	intelligent peripheral
ISDN	integrated services digital network
ISDNUP	Integrated Services Digital Network User Part
ISNI	Intermediate Signaling Network Identification
ISO	International Standards Organization *or* International Organization for Standardization
ISUP	ISDN User Part
IS-41	Interim Standard-41
IT	inactivity test (message)
ITU	International Telecommunications Union
ITU-D	ITU Telecommunication Development Sector
ITU-RS	ITU Radiocommunication Sector
ITU-TS	International Telecommunications Union Telecommunications Standardization Sector
LAN	local area network
LAPD	Link Access Procedure on the D Channel
LATA	Local Access Transport Area
LI	length indicator
LIDB	Line Information Database
LLI	logical link identifier
LPA	loopback acknowledgment (message)
LSB	least significant bit
LSSU	link status signal unit
MAP	Mobile Application Part
MF	multifrequency
MIN	mobile identification number
MLPP	Multi-Level Precedence and Preemption
MSC	mobile switching center

MSU	message signal unit
MTA	major trading area
MTP	Message Transfer Part
NISDN	narrowband ISDN
NNI	network-to-network interface
NOF	Network Operations Forum
NRC	Network Reliability Council
NRM	network resource management
NRZ	nonreturn to zero
OAMP	Operations, Administration, Maintenance, and Provisioning
OLM	overload message (message)
OMAP	Operations, Maintenance, and Administration Part
OMC	Operations and Maintenance Center
OOS	out of service
OPC	origination point code
OPDU	Operations Protocol Data Unit
OS	operations system
OSI	Open Systems Interconnections
OSS	Operations Support System
PAM	pass-along message
PBX	private branch exchange
PCR	preventive cyclic retransmission
PCS	Personal Communications Services
PIN	personal identification number
POP	point-of-presence
POTS	Plain Old Telephone Service
PRI	primary rate ISDN
PRS	primary reference source
PSC	PCS switching center
PSTN	Public Switched Telephone Network
PVC	permanent virtual circuit
QoS	Quality of Service
RAO	regional accounting office
RBOC	Regional Bell Operating Company
RD	receive data
REL	release (message)
RES	resume (message)

RJ registered jack
RLC release complete (message)
RLSD released (message)
RSA regional service area
RSC reset confirmation (message) *also* reset circuit (message)
RSR reset request (message)
RTS request to send, ready to send

SAM subsequent address message
SAP service access point
SAT supervisory audio tone
SBR subsystem-backup-routing (message)
SCCP Signaling Connection Control Part
SCE service creation environment
SCLC SCCP connectionless control
SCMG SCCP management
SCOC SCCP connection-oriented control
SCP service control point
SCRC SCCP routing control
SDLC signaling-data-link-connection
SDU service data unit
SF status field, single frequency
SGM segmentation message
SIF service information field
SIO service indicator octet
SLC signaling link code
SLS signaling link selection
SLTA signaling-link-test-acknowledgment
SLTM signaling-link-test-message
SMDR station detailed message recording
SMS service management system
SNR subsystem-normal-routing
SOG subsystem-out-of-service-grant
SONET Synchronous Optical NETwork
SOR subsystem-out-of-service-request
SR Special Reports
SRCT signaling-route-set-congestion-test
SRST signaling-route-set-test
SRT subsystem-routing-status test
SSA subsystem-allowed
SSP service switching point *also* subsystem-prohibited
SST subsystem-status-test
SS7 Signaling System #7
ST *Science and Technology* (publication)
STP signal transfer point

SUERM	signal unit error rate monitor
SUS	suspend message
SIB	status indicator—busy
SIE	status indicator—emergency
SIN	status indicator—normal
SIO	status indicator—out of alignment
SIOS	status indicator—out of service
SIPO	status indicator—processor outage
TA	Technical Advisory
TCA	transfer-cluster-allowed
TCAP	Transaction Capabilities Application Part
TCP	transfer-cluster-prohibited
TCR	transfer-cluster-restricted
TD	transmitted data
TFA	transfer-allowed
TFC	transfer-controlled
TFP	transfer-prohibited
TFR	transfer-restricted
TIA	Telecommunication Industry Association
TR	Technical Reference
TRA	traffic-restart-allowed
TRW	traffic-restart-waiting
TUP	Telephone User Part
T1A6	T1 Advisory Group
UCIC	unequipped circuit identification code
UBA	unblocking acknowledgment (message)
UDT	Unitdata
UDTS	Unitdata Service
UL	Underwriters Laboratories
UNI	user-to-network interface
UPA	user part available
UPT	user part test
USIS	user-to-user indicators
USR	user-to-user information (message)
VPCI	virtual path connection identifiers
VLR	visitor location register
WATS	Wide Area Telephone Service
XUDT	Extended Unitdata
XUDTS	Extended Unitdata Service

Signaling System #7

Chapter

1

Signaling System #7

Early telephone networks were the result of years of evolution, with little thought about future technology. Based around analog equipment, the telephone network of the early telephone company was not well suited for services such as data and video. Many individual technology service providers began popping up during the 1960s, providing packet-switching networks and data communications services the telephone companies were just not equipped to provide.

The international telephone network was facing the same problems. In many countries, just getting telephone service was a feat in itself. As international bodies began investigating alternative technologies for providing telephone service to the masses (such as cellular), the need for an all-digital network became apparent. Thus arose the beginnings of an all-digital network with intelligence.

The International Telecommunications Union (ITU) commissioned the then CCITT to study the possibility of an all-digital intelligent network. The result was a series of standards known now as Signaling System #7 (SS7). These standards have paved the way for the *Intelligent Network* (IN) and, with it, a variety of services, many yet to be unveiled.

This book outlines the technologies related to the SS7 protocols and details how the protocols work within the Intelligent Network (IN).

Introduction to SS7

The ITU-TS (once known as the CCITT) developed a digital signaling standard in the mid-60s called Signaling System #6 that would revolutionize the telephone industry. Based upon a proprietary, high-speed data communications network, SS6 later evolved into SS7, which has now become the signaling standard for the entire world.

The secret to its success lies in the message structure of the protocol and the network topology. The protocol uses *messages,* much like X.25 and other message-based protocols, to request services from other entities. These messages travel from one network entity to another, independent of the actual voice and data they pertain to, in an envelope called a *packet.*

Common Channel Signaling (CCS) was first introduced in the United States in the 1960s as Common Channel Interoffice Signaling System #6 (SS6). Developed by the International Telecommunications Union—Telecommunications Standards Society (ITU-TS), SS6 used a separate facility for sending signaling information to distant telephone offices.

The first deployment of SS6 in the United States used 2.4-kbps data links. These were later upgraded to 4.8 kbps. Messages were sent in the form of data packets and were used to request connections on voice trunks between two central offices. This became the first use of packet switching in the Public Switched Telephone Network (PSTN). The packets were assembled by placing 12 *signal units* of 28 bits each into a data block. This is similar to the method used in SS7 today.

Signaling System #7 (SS7) was derived from the earlier SS6, which explains the similarities. SS7 provides much more capability than SS6. Where SS6 used fixed-length signal units, SS7 uses variable-length signal units (with a maximum sized length), providing more versatility and flexibility. SS7 also uses high-speed data links (56 kbps). This makes the signaling network much faster than SS6. In international networks, the data links operate at 64 kbps. Study is under way to increase this in the United States to 1.544 Mbps, and internationally to 2.048 Mbps.

As of 1983, SS6 was still being deployed throughout the United States telephone network, even though SS7 was being introduced. As SS7 began deployment in the mid-1980s, SS6 was phased out of the network. SS7 was used in the interoffice network and was not immediately deployed in the local offices until many years later.

In fact, the first usage of SS7 was not for call setup and teardown, but for accessing remote databases. In the 1980s, the telephone companies offered a new service called Wide Area Telephone Service (WATS), which used a common 800 area code regardless of the destination of the call. This posed a problem for telephone-switching equipment, which uses the area code to determine how to route a call through the Public Switched Telephone Network (PSTN).

To overcome this problem, a second number was assigned to every 800 number. This second number is used by the switching equipment to actually route the call through the voice network. But the number had to be placed in a centralized database where all central offices could access it. This database became a popular commodity for all telephone companies and still exists today.

When an 800 number is dialed, the telephone company switching equipment uses a data communications link to access this remote database and look up the actual routing number. The access is in the form of a message packet, which queries the network for the number. The database then responds with a response message packet, providing the routing telephone number as well as billing information for the 800 number. The switching equipment can then route the call using conventional signaling methods.

SS7 provides that data communications link between switching equipment and telephone company databases. Shortly after the 800 number implementation, the SS7 network was expanded to provide other services, including call setup and teardown. Still, the database access capability has proven to be the biggest advantage behind SS7 and is widely used today to provide routing and billing information for all telephone services including 800 numbers, 900 numbers, 911 services, custom calling features, caller identification, and many new services yet to be offered.

800 numbers at one time belonged to one service provider. If subscribers wanted to change service providers, they had to surrender their 800 number. This was due to the location of the routing information. All routing information for 800 numbers is located in a central database and accessed via the SS7 network. SS7 is now used to allow 800 numbers to become transportable and to provide subscribers the option of keeping their 800 numbers even when they change service providers.

Without SS7, number portability would be impossible. Local Number Portability (LNP) is a service mandated by the FCC in 1996 which requires telephone companies to support the *porting* of a telephone number. If customers wish to change their service from Plain Old Telephone Service (POTS) to ISDN, they would normally be forced to change telephone numbers. This is because of the way telephone numbers are assigned in switching equipment, with switches assigned ranges of numbers.

With LNP, the telephone number does not change. This requires the use of a database to determine which switch in the network is assigned the number, very similar to the way 800 numbers are routed. Future implementations of LNP will support subscribers moving from one location to another without changing their telephone number (even if they move to a new area code). This obsoletes the former numbering plan and the way calls are routed through the telephone network.

In addition to database access, the SS7 protocol provides the means for switching equipment to communicate with other switching equipment at remote sites. For example, if a caller dials a number which is busy, the caller may elect to invoke a feature such as automatic callback. When the called party becomes available, the network will ring the caller's phone. When the caller answers, the called party phone is then rung. This feature relies on the capabilities of SS7 to send messages from one switch to another switch, allowing the two systems to invoke features within each switch without setting up a circuit between the two systems.

Cellular networks use many features requiring switching equipment to communicate with each other over a data communications network. Seamless roaming is one such feature of the cellular network that relies on the SS7 protocol.

Cellular providers use the SS7 network to share subscriber information from their Home Location Registers (HLRs), so cellular subscribers no longer have to register with other service providers when they travel to other areas. Cellular providers can access each other's databases and share the subscriber information so that subscribers can roam seamlessly from one network to another.

Before deploying SS7, cellular providers were dependent on X.25 networks to carry IS-41 signaling information through their network. This did not allow them to interconnect through the Public Switched Telephone Network (PSTN) because the X.25 network was not compatible with the PSTN signaling network (SS7). The cellular providers are aggressively changing this situation today, deploying their own SS7 networks.

Today, SS7 has been deployed throughout the Bell Operating Companies network, and is being deployed by almost all independent telephone companies and interexchange carriers as well. This makes SS7 the world's largest data communications network, linking telephone companies, cellular service providers, and long distance carriers together into one large information-sharing network.

SS7 supports many new features and applications. Because of its ability to transfer all types of digital information, this new network is being used to deliver many sophisticated services to the customer premises such as custom calling features, ATM, ISDN, and cellular. Many new applications are still under development. The SS7 network interconnects thousands of telephone company providers all over the world into one common signaling network.

New technology will continue to place demands on the signaling network. SS7 continues to evolve and become more sophisticated as new features are added. While the network is sophisticated enough to work on its own with very little interaction from maintenance personnel, when problems do arise, knowledge of the protocols and the processes that take place between network entities is critical.

Yet to fully understand what SS7 is about, one must understand the conventional signaling methods used prior to SS7 in telephone networks. The following discussion explains the signaling methods used prior to SS7.

Introduction to Telephony Signaling

Ever since the beginning of the telephone, signaling has been an integral part of telephone communications. The first telephone devices depended on the receiving party standing next to the receiver. Early telephones did not have ringers like today's telephones, and used crude speakers to project the caller's voice into the room. If the party being called was not within close proximity of the speaker, he or she would have no indication of an incoming call.

Later, after the formation of the Bell Telephone Company, Alexander Graham Bell's faithful assistant Watson invented the telephone ringer. This new signaling method served one purpose: to alert the called party of an incoming call. When the called party lifted the receiver, another form of signaling used DC battery and ground to indicate the called party had answered the telephone and completed the circuit. Although not having an immediate impact, this method became important when the first telephone exchange was created. By lifting the receiver and allowing DC current to flow through the phone and back through the return of the circuit, a lamp would be lit on the exchange operator's *switchboard.* This signaled the operator when someone needed telephone service, and was often accompanied by a buzzer.

Signaling has evolved over the decades to include significantly more information than these early methods could. Consider the typical long distance telephone call today. When a caller dials the area code and prefix of the telephone number, the local exchange must determine how to route the call. In addition, billing information must be passed to a central database. If the caller is using a contemporary digital facility (such as T-1 or ISDN), information regarding the digitization of the line must also be provided.

Early signaling methods were limited because they used the same circuit for both signaling and voice. They were also analog and had a limited number of *states,* or values, which could be represented. The circuit would be busy from the time the caller started dialing until the caller went "on-hook." To compound the problem, the telephone companies were quickly running out of facilities and were in desperate need of additional facilities.

Many telephone companies in metropolitan areas such as Los Angeles were facing substantial investments to add new facilities to support the millions of customers that were creating an enormous amount of traffic. The telephone companies had to find a way to consolidate their facilities, making more economical use of what they had. In addition, they needed a service that would vastly improve their network's capability and support the many new services being demanded by subscribers.

Europe had already begun the process of digitizing the network in the early '60s. One of the first steps was to remove signaling from the voice network and place the signaling on a network all its own. This way, the call setup and teardown procedures required with every call could be faster than the previous methods, and voice and data circuits could be reserved for use when a connection was possible, rather than maintaining the connection even when the destination was busy. Common Channel Signaling (CCS) paved the way for services the early pioneers of signaling never dreamed of. CCS is the technology that makes ISDN and SS7 possible.

The concept behind Common Channel Signaling (CCS) SS7 is simple. Rather than use voice trunks for signaling, they are used only when a connection is established. For instance, when a call is placed to a distant party using conventional signaling, the signaling for that call begins from the time the caller lifts the receiver and goes off-hook until the caller goes back on-hook. After the end office has received the dialed digits, an outgoing trunk to the destination end office is seized, based on a routing table entry and the digits dialed.

The voice circuit remains busy even if the distant party never answers the call until the calling party hangs up. Meanwhile, other subscribers are tying up other voice circuits by placing calls of their own. This is not good utilization of voice circuits and it placed immediate limitations on the networks. But if the signaling could be placed over a different network and the voice circuit used only when the called party answered, the voice circuit would remain available for a longer period of time. This meant the availability of voice circuits would be higher and the need for additional circuits would decrease.

When a caller is to receive an intercept recording ("all circuits are busy"), the same trunk used for the voice is also used for the recording. The recording is sent by the distant office. Busy tones and other service tones are sent over the trunk by the distant office to the caller. With SS7, the caller's local office can provide these tones and recordings at the command of the distant office. These commands are received via the signaling network. The voice trunk is left unconnected (although in some implementations, one side of the trunk is connected for transmitting tones and recordings).

The procedure for tearing down a circuit is much faster in Common Channel Signaling (CCS) than in conventional signaling, and is not as error prone. Even if voice circuits do get connected, with the speed of the signaling network, circuits can be disconnected and quickly connected again for a new call. While a call is in progress, information regarding the call can be sent through the SS7 network (for instance, information from a database requested during an interactive multimedia call).

CCS uses existing telephone company resources, so it does not require additional facilities to be installed. When signaling information is placed on existing digital transmission facilities (such as DS0 or DS1), it uses a fraction of the circuits required for in-band signaling (discussed later in this chapter). One digital data link can carry the signaling information for thousands of trunks and maintain thousands of telephone calls.

CCS is in wide use today, even though many in the telecommunications business do not understand it. SS7 is the protocol and architecture used in this new network and is the topic of this book.

There are many methods used for signaling, with only a few used in the telephone network. None of the methods described in this section can support network management functions or control information between switches and operations systems. The exception is SS7. Because SS7 consists of a data network using data messages, SS7 can meet the demands both now and in the future of the evolving telephone network.

Signaling takes place in two parts of the telephone network between the subscriber and the local end office, and from switching office to switching office within the telephone company network. The signaling requirements are similar, though interoffice signaling can be more demanding.

There are two basic functions of signaling: addressing and supervision. With the earlier methods of signaling, supervision was simple. If current existed from one end to the other, the circuit was good. For addressing, dialed digits would be passed through the network in the same fashion as they were originated, either in pulses or tones. Only the destination address could be provided.

But as the telephone network grew more sophisticated, the signaling methods grew as well. Signaling between the subscriber and the central office now includes the calling party number, which is forwarded to the called party and displayed before the phone is even answered. Interoffice signaling now includes information obtained from regional databases, pertaining to the type of service a subscriber may have or billing information.

Calling card validation is another important function of these databases, and provides security against telephone fraud. Personal identification numbers are kept in a subscriber database and verified every time a call is placed using a calling card.

Previous to SS7, signaling was accomplished over the same facility as the voice call. This was accomplished in many cases using DC current. There are many disadvantages to DC signaling, which is what led to the development of SS7.

In addition to DC signaling, many companies used *Single Frequency* (SF) signaling. This was also accomplished over the same facility as the voice. This method of signaling used tones above the voice frequencies, but still within the 4-kHz bandwidth of the facility to set up and tear down circuits. Other methods have been used in addition to DC and SF signaling, depending on the type of facility. But with all of the various signaling methods, none can offer the features and versatility of SS7.

Conventional signaling

Conventional signaling relies on many different types of mechanisms, depending mostly on the location within the network. *Dual-Tone Multi-Frequency* (DTMF) is used between the subscriber and the end office. Single frequency (SF) is used between telephone company offices. Following is an example of how conventional signaling is used to process a call.

DC signaling relies on DC current to signal the distant end. The simplest example of DC signaling is used in *Plain Old Telephone Service* (POTS) between the subscriber and the local end office. When a subscriber goes off-hook, DC current from the central office is allowed to flow through the telephone (the switch-hook provides the contact closure between the two-wire interface) and back to the central office. The central office switch uses a DC current detector to determine when a connection is being requested.

The central office acknowledges receipt of the loop current by sending a dialtone. A dialtone signals the subscriber to begin dialing the telephone number. This can be done using a rotary dial or a Dual-Tone Multi-Frequency (DTMF) dial. Rotary dials use a relay to interrupt the current creating pulses (10 pulses per second). The central office switch counts each series of pulse "bursts" to determine the number dialed.

When DTMF is used, the dial creates a frequency tone generated by mixing two frequencies together (hence the name *dual-tone*). The central office switch "hears" these tones and translates them into dialed digits.

After the telephone number has been dialed, the central office switch must determine how to connect to the destination. This may involve more than two central offices. A facility (or circuit) must be connected between every telephone company office involved in the call. This circuit must remain connected until either party hangs up. The originating office determines which circuits to use by searching its routing tables to see which office it must route the call through to reach the final destination. That office, in turn, will search its routing tables to determine the next office to be added to the call.

Once the circuits are all connected, the distant party can be alerted by sending a generator (80 V AC at 20 Hz) out to the telephone. This activates a ringer inside the telephone. At the same time, the distant telephone company switch sends a ringback tone to the originator to alert the caller that the called party's phone is being rung. When the distant party answers, the ringback tone is interrupted and the circuits now carry the voice of both callers.

If the called party is busy, the same facilities are used so that the far end office can send a busy tone back to the originator. This means those facilities cannot be used for any other calls and are being tied up to send the busy tone.

The limitations of DC signaling are somewhat obvious. For example, the telephone number of the originator cannot be sent to the called party (at least not without long delays in setup). Signaling is limited to seizing circuits, call supervision, and disconnect. Because DC signaling uses the voice trunk, the trunks are kept busy even when the two parties are never connected.

DC signaling

As mentioned previously, DC signaling relies on direct current to signal distant offices. This is a very limited signaling method, because of the minimum number of states that can be represented by voltages and current. When a subscriber lifts the receiver of a phone, current is allowed to flow through the phone and back to the central office. Current detectors on the line cards in the central office switching equipment detect the current and provide the subscriber with a dialtone.

Other types of trunks use similar techniques. E&M signaling is another form of DC signaling. These trunks use a separate pair of wires for signaling. The two wires are labeled as E and M, or "ear" and "mouth." These, of course, are not what the letters really stand for, but this designation is often used to describe their function.

The M lead is used to send 248 V DC or ground to the distant switch (implementation dependent). The M lead of one switch must be connected to the E lead of the distant switch. When the distant switch detects current on its E lead, it closes a relay contact and allows the current to flow back to the sending switch through its M lead.

When the sending switch detects the current flow on its E lead, the connection is considered established and transmission can begin on the separate voice pairs. This type of trunk is often used between two PBXs, and is often referred to as tie lines.

In-band signaling

In-band signaling is used when DC signaling is not possible, for example, in tandem offices. In-band signaling uses tones in place of DC current. These tones may be Single Frequency (SF) tones, Multi-Frequency (MF) tones, or Dual-Tone Multi-Frequency (DTMF). The tones are transmitted with the voice. Because these tones must be transmitted over the same facility as the voice, they must be within the voice band (0 to 4 kHz). There is the possibility of false signaling when voice frequencies duplicate signaling tones. The tones are designed for minimal occurrence of this, but this is not 100 percent fault tolerant. Signal delays and other mechanisms are used to prevent the possibility of voice frequencies from imitating SF signals.

Single Frequency (SF) signaling is used for interoffice trunks. Two possible states exist: on-hook (idle line) or off-hook (busy line). To maintain a connection, no tone is sent while the circuit is up. When either party hangs up, a disconnect is signaled to all interconnecting offices by sending a tone (2.6 kHz) over the circuit. Detectors at each end of the circuits detect the tone and drop the circuit.

SF signaling has become the most popular of all the in-band methods, and the most widely used of all signaling methods. SF is still in use today in some parts of the telephone network. However, as deployment of the SS7 network spreads, SF is no longer needed.

Multi-Frequency (MF) is much like Dual Tone Multi-Frequency (DTMF), and is used to send dialed digits through the telephone network to the destination end office. Because voice transmission is blocked until a connection to the called party is established, there is no need for mechanisms that prevent the possibility of voice imitating signaling tones.

MF is also an interoffice signaling method used to send the dialed digits from the originating office to the destination office.

Out-of-band signaling

Out-of-band signaling has not shared the popularity and widespread usage of SF signaling. Out-of-band signaling was designed for analog carrier systems, which do not use the full 4-kHz bandwidth of the voice circuit. These carriers use up to 3.5 kHz, and can send tones in the 3.7-kHz band without worrying about false signaling. Out-of-band signaling is an analog technology, and is of no advantage today.

Digital signaling

As digital trunks became more popular, signaling methods evolved that greatly enhanced the reliability of the network. One technique used in digital trunks (such as DS1) is the use of signaling bits. A signaling bit can be inserted into the digital voice bit stream, without sacrificing voice quality. One bit is "robbed" out of designated frames and dedicated to signaling (robbed-bit signaling). The digitized voice does not suffer from this technique since the loss of one bit does not alter the voice signal enough to be detectable by the human ear.

Because of its digital nature, digital signaling is much more cost effective than Single Frequency (SF). SF requires expensive tone equipment both for sending and detection, whereas digital signaling can be detected by any digital device loaded in the switching equipment and can create any kind of signaling information. This has fueled the efforts to make carriers digital rather than analog.

Digital signaling has another fundamental difference. It does not use messages as SS7 and other message-based protocols do. This limits the type of signaling it can provide.

Common channel signaling

As discussed earlier, Common Channel Signaling (CCS) uses a digital facility, but places the signaling information in a time slot or *channel* separate from the voice and data it is related to. This allows signaling information to be consolidated and sent through its own network apart from the voice network. It is this method that is used in ISDN and SS7 today.

In addition, this method of signaling is capable of sending and receiving messages, supporting an unlimited number of signaling values. Even information retrieved from a remote database can be transferred from one entity to another using CCS.

Introduction to Telephony

The telecommunications industry has undergone many organizational changes as well as service changes. With the current trend to merge data communications with telephony, many new professionals have entered the industry without the opportunity to learn the nuances of the industry and its players.

This section provides a fundamental outline of the telephony industry and its players. Of major importance is understanding how the telephone network is structured, and how it has evolved over the years.

Bell System hierarchy

The Bell System network really can be divided into two distinct functions: signaling and switching. The signaling network is the SS7 network. The switching network is the portion used for the transfer of data and voice from one subscriber to another, through the telephone service providers' network.

There exists a hierarchy within the switching network, to ensure efficient use of trunk facilities and to provide alternate circuits in the event of failures or congestion. Before the divestiture of the Bell System in 1984, the hierarchy was much different than today. The hierarchy provided five levels of switching offices, with the class five office being the end office or local office (Fig. 1.1).

The class five office was capable of connecting to other end offices within its calling area, but relied on the class four office to connect to offices outside of its calling area. The calling area was not defined by area codes, but was a geographical service area drawn by the Bell System. Service access areas have since been reallocated as *Local Access Transport Areas* (LATAs) by the Justice Department as a result of the Bell System divestiture.

The class four office allowed the Bell System to aggregate its facilities and used high-capacity trunks to interconnect to other class four offices. In this way, the class five office did not require high-capacity facilities and handed off the bulk of its calls to the class four office. This also prevented the necessity for the class five office to have trunks to every other class five office in the service area. The class four office was also known as the toll center.

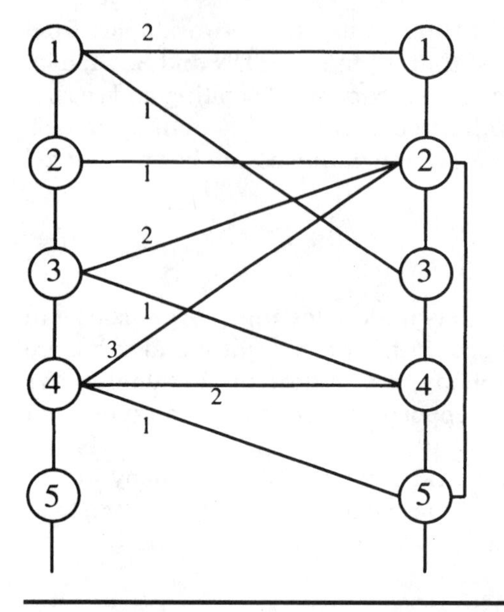

Figure 1.1 This figure illustrates the Bell System Switching Hierarchy prior to the divestiture of the Bell System. The priority of the routes is indicated by the numbers.

As seen in Fig. 1.1, the class four office provided two paths for a telephone connection. The interconnection of these various offices depended mostly on distance. There were many occurrences of class five offices connecting directly to class one offices.

The toll office searched its trunks for an available trunk as low in the hierarchy as possible. If one was not available, it would search for a trunk to a primary center in the destination calling area. If there were no available trunks to the primary center, then the last choice would be an overflow trunk to its own primary center.

The class three office, or primary center, was part of the toll network. This office connected to class two offices, or sectional offices, but also provided a path to other class three offices and class four offices. This office served as an overflow switching center in the event that other routes lower in the hierarchy were not available.

The class two office was also known as the sectional center, and provided access to the regional center. Only two routes were available at this level, one to its peer in the destination calling area and one to the regional switching center, or class one office.

The class one office was known as the regional center, and was used for toll calls. The regional center also provided access to the long distance network. A typical toll call required an average of three trunks. The maximum number of trunks allowed in a connection was nine.

Postdivestiture switching hierarchy. In the mid-1980s, technology allowed many of the functions just described to be combined. As switching equipment was improved, systems were given the capability to act as local switches, tandems, and even toll switches. In addition to better routing functionality, these switches were also given the ability to record billing records and perform alternate routing in the event of congestion or failures.

The new hierarchy consists of fewer levels, consolidating many of the functions of the previous hierarchy into two layers. Long distance access is accomplished through a Point-of-Presence (POP) office. The long distance carrier will also have its own multilevel hierarchy, which may be several layers as well.

After divestiture of the Bell System, the calling areas were also redefined and changed to Local Access Transport Areas (LATAs). Within each LATA is a simple hierarchy, with three levels. With newer advanced switching equipment, many of the functions once found in the higher layers of the hierarchy can now be combined and located in the end office (Fig. 1.2).

Local Access Transport Areas (LATAs)

After divestiture, the telephone companies' service areas (or *exchanges,* as they were sometimes called) were redrawn by the Justice Department so that telephone companies would have evenly divided

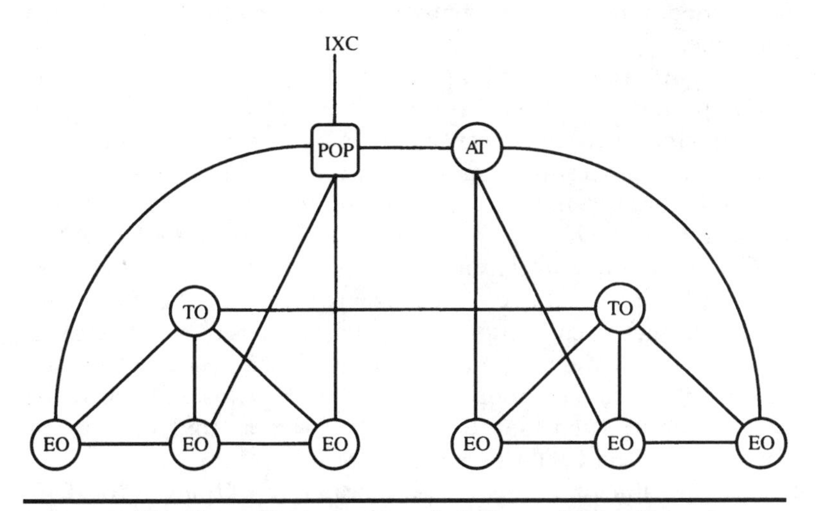

Figure 1.2 This figure illustrates a much flatter switching hierarchy, used after the divestiture of the Bell System.

service areas with equal revenue potential. These areas, called Local Access Transport Areas (LATAs), were divided according to census information regarding the demographics of each LATA. Each Regional Bell Operating Company (RBOC) and each independent telephone company received a service area that would provide it a fair and equal market. There were originally 146 LATAs, but as changes take place in the networks, the number of LATAs is growing.

Other considerations had to be made regarding LATAs. Telephone companies already had a significant amount of equipment and capital invested in the previous service areas, which meant reassigning service areas could have a major financial impact. The service areas were divided into LATAs, which are much smaller than the original service areas defined by the telephone companies themselves. Each telephone company could have more than one LATA for which it provides service. This allowed telephone companies to maintain their original investment, but forced them to divide areas into smaller chunks.

The difference was made between LATAs. The local operating companies and independents were not allowed to carry traffic from one LATA to another. Instead, they were required to use a long distance carrier such as AT&T or MCI to provide them that service. This ensured that the local telephone companies did not interfere with long distance competition, and provided the long distance companies a fair and profitable boundary for which they could compete with other carriers.

In the mid-1980s, the telephone companies were forced to provide the long distance carriers equal access into the telephone network. This was accomplished in the current switching hierarchy by establishing a point-of-presence (POP), which serves as an interface to all interexchange carriers into the LATA. Every LATA must have one POP. The telephone companies collected access fees from the long distance carriers for this interface into their network, to offset the cost of equipment and ensure a revenue stream from long distance traffic.

The local telephone companies further divided each LATA into a *local market* and a *toll market.* The toll market is within the LATA but considered by the telephone company as a long distance call, because of the distance from one city to another or the distance between central offices handling the call. These toll calls are currently very expensive, and have recently been open for competition. These are the only long distance calls local telephone companies are allowed to provide. Many states have allowed long distance carriers to compete in this market, opening up the LATA to competition.

In 1996, the government approved the Telecommunications Act of 1996. This piece of legislation has changed the face of the telecommunications industry in many ways, and impacts both users of the telephone network and the telephone companies themselves. One part of this legislation allows long distance companies to provide local telephone service in their markets. They must pass specific criteria defined by the Telecommunications Act before the FCC can grant them permission to provide local service. This reverses the legislation put into place by Judge Green when divestiture of the Bell System reshaped the nation's telephone industry and limited long distance companies from offering local service.

At the same time, the Telecommunications Act of 1996 also allows local telephone companies to offer long distance (interLATA) service in their market areas. They must demonstrate that they have allowed competition in their market areas, and pass criteria set in the Telecom-munications Act before offering such service. The result of this sweeping new legislation is yet to be seen, but many anticipate local telephone companies will partner and merge with long distance companies taking advantage of one another's markets.

There are over 300 LATAs presently defined throughout the United States. Throughout these LATAs are hundreds of telephone companies competing for revenues. Many of the smaller independent companies are investing capital to get connected to the SS7 network so that they can offer the same types of advanced calling features the larger service providers offer. These independent companies have joined forces forming telephone associations across the nation. This allows them to represent their industry in standards committees, and voice their concerns to the government as a collective body rather than a lone voice. These associations also pool their resources and build their own networks using monies from the member companies to pay for the cost of the network. This is how many small independent companies are getting involved in SS7.

Hierarchy of the synchronization network

All digital networks rely on timing mechanisms to maintain integrity of the data transmission. Since all digital transmission is multiplexed and based on time division, accurate timing is critical. This is especially true when DSO links are used in SS7. DSO links must have accurate timing sources in order for them to synchronize and carry signaling traffic.

Digital facilities must have reliable, accurate clock sources to determine proper bit timing. These clocks must be synchronized with the same source, and are deployed throughout the telephone network. To maintain timing in the telephone network, a separate synchronization network has been defined (Fig. 1.3).

The source for a clock signal is referred to as the *Primary Reference Source* (PRS). These clock sources reside in the various regions of the telephone network. They are highly accurate clocks, usually cesium beam or rubidium-based clocks. These clocks must be resynchronized and continuously verified using a universal time source. Loran-C and the Global Positioning System (GPS) are currently used by many companies to check

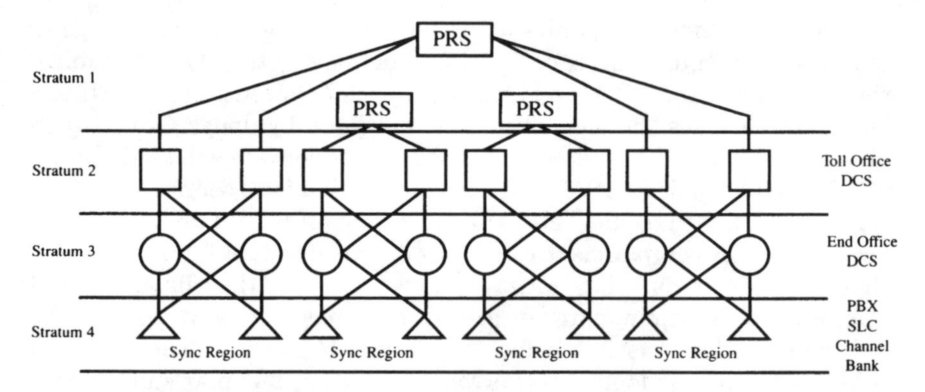

Figure 1.3 To maintain synchronized timing throughout the digital network, the Bell System Operating Companies (BOCs) use this timing network.

the accuracy of their clocks. The distribution of clock signals is implemented at different levels, referred to as *strata.* The highest stratum obtainable is stratum 1, which is the primary clock source.

Clock signals are distributed in a primary/secondary relationship to all other levels. This means that central office switching equipment at stratum 2 distributes its clock signal to equipment at stratum 3.

The PRS distributes clocking signals to toll offices. The toll office is considered stratum 2, and must redistribute the clock to end offices within its LATA. Each LATA must have at least one stratum 2 clock. Whenever a clock is redistributed, it loses some accuracy. Yet the clock signal is accurate enough to operate throughout the network reliably, despite the loss of accuracy. End offices are considered stratum 3.

The end office will distribute clock signals to other users of digital transmission facilities, such as private branch exchanges (PBXs) and channel banks. These are considered to be at stratum 4. In some cases, devices at this level can redistribute the clock signal to other adjunct equipment. These devices are considered to be at the lowest level of the hierarchy, stratum 5.

Within a central office, clocks are distributed through a *building integrated timing system* (BITS). BITS is a distribution system for clock signals, and is distributed throughout the office to switching equipment in that office. BITS is critical to the proper operation of DS0A links in the SS7 network. Failure of this clock signal will result in the failure of the signaling links.

Digital signaling hierarchy

The telephone network also has a digital hierarchy for all digital transmission facilities. This digital hierarchy is a means for expressing the

TABLE 1.1 North American Hierarchy

Digital signal destination	Bandwidth	Channels (DSOs)	Carrier designation
DS0	64 kbps	1 channel	None
DS1	1.544 Mbps	24 channels	T-1
DS1C	3.152 Mbps	48 channels	T-1c
DS2	6.312 Mbps	96 channels	T-2
DS3	44.736 Mbps	672 channels	T-3
DS4	274.176 Mbps	4032 channels	T-4

capacity of these various facilities. Typically, the highest level of the hierarchy is an aggregate of the levels below it. Thus, the lowest digital signal in the hierarchy is multiplied to establish the next level, which is multiplied to obtain the next level, and so forth. Table 1.1 illustrates the correlations between levels in the hierarchy.

The SS7 network uses DS0s for signaling links. This is a 56- or 64-kbps data link, capable of sending voice or data. The DS0 in the United States is always 56 kbps, because the telephone company uses 8 kbps for control information. This control information is used by the transmission equipment to maintain the integrity of the data link. Some studies have been under way regarding the use of 64 kbps, but the multiplexers used throughout the telephone network today do not support 64-kbps links.

In order for equipment to use the DS0, special digital interfaces that are capable of sending and receiving signals at this level must be installed. If a DS1 is used, a device called a *channel bank* must be used as the interface. The channel bank divides the 24 time slots of the DS1 into 24 separate DS0s, which can then be distributed to their proper destinations. Other types of multiplexers/demultiplexers exist that perform the same task, depending on the location in the network.

Currently, only the DS0 level is supported in the SS7 network. This is currently under study, however, as new high-speed networks are deployed. *Asynchronous Transfer Mode* (ATM) and *Broadband ISDN* (BISDN) may require signaling speeds beyond that of DS0. As ATM is deployed in the public networks, signaling traffic will migrate to these new facilities. The current digital facilities will be replaced and the signaling network will become integrated with the new broadband network. The first step is to support DS1 speeds over an ATM interface. This is currently under development and will be deployed in many networks by the end of 1999. ANSI and ITU-TS are currently studying the use of DS1 in the SS7 network.

As seen in Table 1.1, the DS1 facility provides a total bandwidth of 1.544 Mbps. This should be more than adequate for almost any signaling requirements. The impact on the existing network will require new hardware as well as software upgrades. The SS7 protocol will undergo many changes to support these new high-speed data links.

The digital hierarchy is quickly being replaced with a newer and faster technology. Fiber optics is quickly replacing copper facilities throughout the telephone network. Fiber optics has the capability to transmit at much higher data rates than copper, and is critical to the success of technologies such as broadband and ATM.

Synchronous Optical NETwork (SONET) is currently found in telephone company networks worldwide and has become the transmission medium of choice. SONET provides data rates up to 2.4 Gbps, and will support broadband ISDN and ATM. SONET is also being used to link local area networks (LANs) through the Public Switched Telephone Network (PSTN).

As seen in Table 1.2, SONET is also divided into different levels of service, each level being an aggregate of the levels below it. There are two designations used for these levels: the electrical signal itself and the optical signal. They are terms used for different reasons. Electrical signals are directly related to the optical signals and, therefore, can almost be used synonymously in most discussions. In this book, we will always refer to the optical signal.

When compared to the digital signal hierarchy, there is a stark difference (Table 1.3). Even at the lowest level of the optical hierarchy, 28 DS1s can be supported on one facility. This represents a significant cost savings to telephone companies. At OC-1, 672 time slots are supported for voice, data, or even signaling.

A single SONET facility is not dedicated entirely to one application. One channel may be used for signaling, while the remainder carry voice, data, and video. This practice allows telephone companies to use existing transmission facilities between offices, rather than deploying a special link just for SS7.

TABLE 1.2 SONET Digital Hierarchy

Electrical signal	Optical signal	Data rate (Mbps)	ITU designation
STS-1	OC-1	51.84	
STS-3	OC-3	155.52	STM-1
STS-9	OC-9	466.56	STM-3
STS-12	OC-12	622.08	STM-4
STS-18	OC-18	933.12	STM-6
STS-24	OC-24	1244.16	STM-8
STS-36	OC-36	1866.24	STM-12
STS-48	OC-48	2488.32	STM-16

TABLE 1.3 Optical and Digital Compared

Digital signal	Optical signal
DS0 (64 Kbps)	OC-1 (51.84 Mbps)
DS1 (1.544 Mbps)	OC-3 (155.52 Mbps)
DS1C (3.152 Mbps)	OC-9 (466.56 Mbps)
DS2 (6.312 Mbps)	OC-12 (622.08 Mbps)
DS3 (44.736 Mbps)	OC-18 (933.12 Mbps)
DS4 (274.176 Mbps)	OC-24 (1244.16 Mbps)
	OC-36 (1866.24 Mbps)
	OC-48 (2488.32 Mbps)

Current Trends in Telecommunications Technology

Today's telecommunications industry has changed dramatically. Data communications and voice networking have been merged to provide a variety of services that leave even the most educated somewhat confused and baffled. These services all revolve around the backbone of the new Intelligent Network (IN), SS7.

Because of SS7, these new technologies can be supported in the Public Switched Telephone Network (PSTN) rather than having to have a separate network for each type of service (as previously done to support packet switching in the '70s and '80s). In fact, all data and voice communications will be simplified to the point that the subscriber does nothing but dial a number and get connected. The signaling network will handle the rest.

The goal of the telephone network is to provide seamless service to all subscribers, regardless of the information being sent through the network. As previously discussed, the IN will provide this capability. But before the IN is fully deployed, there are many different pieces that must first be put into place.

The telephone network of today will not support the types of services that subscribers are asking for. If there is a need for high-speed data, a special circuit must be installed from the customer premises to the other end of the circuit. If video is to be transmitted through the telephone network, special high-capacity circuits must be installed from the studio through the telephone network to another high-capacity circuit at the transmitter.

The ultimate goal is to provide one network capable of transferring all kinds of information regardless of the bandwidth necessary and sending it through the network just as if placing a telephone call. To support this level of service, the network must be changed.

Technology is changing the network in anticipation of the IN. To understand the role SS7 plays in each of these technologies, the rest of this chapter will provide some overview of current services being offered by telephone companies and independents. Some of these services are still under development and may not be offered for general availability for some years to come.

Introduction to the intelligent network

The 1990s will become known as the years of the Clinton administration's information highway. The fact is that the information highway was already under way before this time. The Intelligent Network (IN) has been under development for many years, with the goal of allowing all types of information to pass through the telephone network without special circuits or long installation cycles.

The concept of being able to access all kinds of information is not new. The network to support this type of service is new. The IN provides the backbone to support and define these services.

As the need for new features and services becomes more important to customers, the need to deliver those services and features in an economical way becomes equally important. The problem facing telephone companies today is being able to provide these features and services quickly and efficiently. Ordering an 800 line for two weeks' usage is now easy and can be implemented within hours instead of days.

The IN makes it easier, because now, when subscribers order new services, technicians do not have to be dispatched to add programming to the switching equipment and to cross-connect the circuits.

In the IN, everything is controlled or configured by workstations with user-friendly software interfaces. Telephone service representatives can create new services and tailor a subscriber's service from the terminal while talking with the customer. The changes are immediately implemented in the switches. Circuits are cross-connected using digital cross-connect systems, which are also controlled by the workstation. Customers today can order high-speed communications, video, audio, and digital voice facilities on an as-needed basis.

Networks were not always equipped to handle such demand. But switch manufacturers have added new features to switching equipment that allow services to be added to subscriber lines by simple commands at a terminal. Some new products even allow customers to order services and features by dialing a sequence of codes on their telephones. Soon, ordering an 800 line will be as simple and as fast as ordering a pizza.

Welcome to the age of the Intelligent Network. The IN is just what its name implies: intelligent. Services and features can be changed or deployed using simple procedures through a terminal, rather than through expensive programming changes made by certified technicians. All the customer needs is the facility (trunks) to utilize the new services.

As more and more customers line up to deploy T-1 and ISDN, the IN will become as commonplace as Touch Tone™ dials. Yet, little is understood about the IN. To understand what the IN is about, let us first examine the architecture of such a network. The Intelligent Network consists of a series of intelligent nodes, each capable of processing at various levels, and each capable of communicating with one another over data links.

The Intelligent Network relies on the SS7 network, which forms its backbone. SS7 provides the basic infrastructure needed for the *Service Switching Point* (SSP), which provides the local access as well as an ISDN interface, for the *Signaling Transfer Point* (STP), which provides packet switching of message-based signaling protocols for use in the IN, and for the *Service Control Point* (SCP), which provides access to the IN database. The SCP is connected to a *Service Management System* (SMS), which provides a human interface to the database, as well as the capability to update the database when needed. The SMS uses a command-line interface and a man-to-machine language to build services and manage the network. The SMS can also be used in some applications as a central control point for updating multiple databases and controlling the updates to those databases from a central authority.

One additional node used in the Intelligent Network that is not seen in the SS7 architecture is the *Intelligent Peripheral* (IP). The IP provides resource management of devices such as voice response units, voice announcers, and DTMF sensors for caller-activated services. The IP is accessed by the Service Control Point (SCP) when services demand its interaction. IPs provide the Intelligent Network with the functionality to allow customers to define their network needs themselves, without the use of telephone company personnel.

When a call is placed in the IN, a request for call-handling instructions is sent to the SCP using the Transaction Capabilities Application Part (TCAP) protocol. The database provides the instructions for handling the call based upon the customized service instructions the subscriber has programmed, and sends them to the end office switch. The end office switch then communicates to the Intelligent Peripheral using the ISDN protocol to attain the use of resources such as recordings and other devices. The call setup and teardown is handled using conventional SS7 protocols.

Advanced Intelligent Networks (AINs—Fig. 1.4) provide many components not found in the earlier versions of Intelligent Networks (INs). One of the key components is the Service Creation Environment (SCE). In the AIN standard, SCE defines the look and feel of the software used to program end office switches to provide a new service. This look and

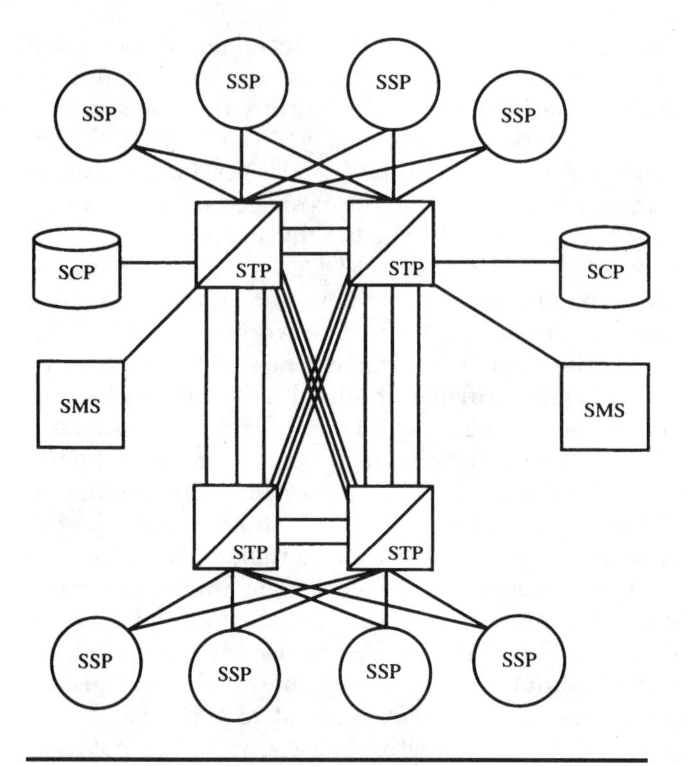

Figure 1.4 The Advanced Intelligent Network (AIN) relies on SS7 for interconnection of network switches.

feel defined in the AIN standard provides a graphical user interface (GUI), which uses icons, for building customized services. AIN administrators can then tailor services to meet the customers' specific needs by clicking on network capability icons rather than programming via commands on a command line.

Eventually, this terminal and the Service Creation Environment (SCE) will be extended to the customer premises, allowing large and small companies with special network needs to tailor their services on an as-needed basis, without telephone company assistance. This concept has already intrigued many large companies with large volumes of inbound and outbound calls, especially those in the telemarketing industry whose network needs vary from week to week.

The first offerings of an Intelligent Network (IN) were in the early '80s when AT&T announced a centralized database for all 800 numbers. End offices wishing to handle 800 calls would have to access this database through the SS7 network. Control of the network was provided by *operations systems* (OSs), which used intelligent workstations to provide maintenance and administration for the network databases and other network nodes.

In 1988, AT&T asked vendors to participate in defining the standards for the evolving IN. The end result was the release of an evolutionary path defined in a series of releases, now known as the Advanced Intelligent Network (AIN). The AIN defines how features will be invoked by the telephone company or the subscriber.

The AIN deployed in today's network is known as AIN Release 0.1, and is presently under trial with several local exchanges. AIN Release 0.1 offers a variety of services to the consumer, such as redirecting the destination of calls on a per-line basis. This can be altered by dialing a code through a phone and entering in the destination telephone number. This feature would be of huge value to marketing firms with large inbound call volumes. Eventually, the feature will be deployed by the subscriber through a terminal located at the customer premises and connected to the telephone company's signaling network.

Other features currently available include call screening (Do Not Disturb), selective call acceptance, calling name delivery, and spoken caller identification. The many services and features offered to customers are based on databases linked to the SS7 network through nodes called Service Control Points. Local end offices and other networks can access these databases by sending database query messages through the SS7 network to the SCP. The SCP replies to the query after accessing the information from the appropriate database and sending the requested information in an SS7 message format through the SS7 network to the requesting end office. Based on the information received, the end office (or service node) is then able to create the requested services. Intelligent Networks (INs) use the Transaction Capabilities Application Part (TCAP) protocol for sending database queries to the SCP.

The SCP provides the call-handling instructions and service instructions to the end office so that it knows how to handle calls for a specific subscriber. The following features are just a sampling of the features of an IN. As the IN evolves, new services will become available. Features include:

- Find Me Service
- Follow Me Service
- Computer Security Service
- Call Pickup Service
- Store Locator Service
- Call Routing Service
- Multilocation Extension Dialing
- Name Delivery
- Outgoing Call Restriction

Find Me Service. This service allows calls to be forwarded to another location. The difference between this feature and today's call forwarding feature is the ability to screen unwanted calls from forwarding. Only authorized callers are forwarded to the new location.

Follow Me Service. Similar to call forwarding, this allows a number to be forwarded on a time schedule. The subscriber determines the time forwarding is to take place when the feature is invoked. Destinations can include cellular telephones or *Personal Communications Services* (PCS) handsets.

Computer Security Service. This feature prevents unauthorized callers from accessing a computer via modem. Only callers with the authorized access code or calling from an authorized number can access the computer. The SS7 network delivers the calling party number to the destination end office. This number is then checked in a database located with a Service Control Point (SCP), and, if authorized, is allowed to connect with the modem.

Call Pickup Service. When a call is placed to a number and is unanswered, the called party can be paged via radio pager. The called party can then dial a code from any telephone at any location and immediately be connected with the waiting caller. Some manufacturers of Personal Communications Services (PCS) devices have already developed two-way pagers that connect the caller with the party being paged. The pager is a two-way transceiver capable of receiving calls (pages) and connecting the caller with the paged party (similar to the voice pagers used a few years back, but this pager allows two-way conversation).

Store Locator Service. Businesses can advertise one number, and callers are automatically transferred to the nearest location based on their own telephone number. The telephone company provides the routing service based on the prefix of the calling party number. This allows businesses to advertise nationwide for all locations without special ads based on geography. The calling party number is matched in a routing database located at a Service Control Point (SCP). The SCP provides the end office with the routing instructions based on the calling party number.

Call Routing Service. This allows businesses to reroute calls when congestion occurs or after business hours. It is an excellent feature for telemarketing and reservation centers with multiple locations. ACD switches can be interfaced to the end office by SS7 data links. This allows the ACD to send network management messages via SS7 protocols to a database, located by a Service Control Point (SCP). Calls are then rerouted around the call center based on the routing instructions in the database.

Multilocation Extension Dialing. This allows the usage of abbreviated extension numbers to reach personnel regardless of their location and without PBX equipment. Very similar to PCS offerings, the subscribers receive personal numbers which can be used to reach them no matter where they go.

Name Delivery. As a call rings the telephone, the caller's name is displayed on a digital display. This is offered to residential and business customers alike. The digital display is built into a digital phone or is an adjunct to any standard telephone. This is somewhat different than the controversial *Automatic Number Identification* (ANI) feature, which displays the caller's telephone number. ANI allows telemarketing companies to store calling party numbers in their databases, which are later sold to other telemarketing companies. Name delivery delivers the name only, retrieved from a line subscriber database.

Outgoing Call Restriction. This feature allows the restriction of specific numbers or prefixes and area codes, allowing customers to restrict long distance calls and service numbers such as 900 and 976 numbers from being dialed on their phones. Presently, the telephone companies can offer restriction of 900 and 976 numbers, but they cannot provide restriction of specific area codes, prefixes, or individual telephone numbers. This feature allows those restrictions to be programmed by the subscriber.

It is important to note that AIN does not define the features and services, but how those features and services will be deployed by the customer. The features and services are defined by the service providers themselves, and may or may not be consistent from one telephone company to another. They are limited, however, to the capabilities of the switching equipment used in their end offices.

AIN is currently offered in two releases, Release 0.1 and Release 0.2, with new releases on the way. Currently, Release 0.2 has been deployed, with plans for future releases to follow.

Many telephone companies are reluctant to deploy Advanced Intelligent Network (AIN) because of talk of a newer network called *Information Network Architecture* (INA). This new technology is still in development, but many feel it will succeed Advanced Intelligent Network. Yet others view INA as a subset of AIN. At any rate, INA provides better utilities for managing the new broadband services being deployed by major telephone companies. It is very likely that there will be two architectures: AIN for the voice network and INA for managing the broadband network.

The Advanced Intelligent Network (AIN) is certainly not a new concept. As we have already seen, this concept evolved from an earlier introduction to make 800 services easier to customize and maintain. The ability to create new services quickly and efficiently was one of the early driving forces behind the Intelligent Network. In fact, 800 services today can be customized by the customer according to their immediate needs with a simple phone call. AIN has not been deployed for local telephone service, however, and that is the undertaking today.

Personal Communications Services (PCS) and Centrex also benefit from the Intelligent Network. PCS will be based on the ability to tailor subscriber services in real time. This ability cannot be supported without the Intelligent Network (IN). Centrex popularity is on the rise. With the IN, customers can tailor their specific service requirements within hours instead of days. With features such as networked voicemail and Automatic Call Distribution (ACD), Centrex will become a powerful competitor to the PBX.

Presently, Centrex is difficult to market against the PBX because of the limited feature set offered with Centrex. With PBX systems, proprietary phones can offer a multitude of high-tech features that greatly enhance the systems' capabilities. Centrex does not use electronic telephones, but standard analog telephones. While Centrex can offer simple features such as forwarding, conferencing, and speed dialing, it cannot match the features of a PBX.

When Centrex can be offered with Intelligent Network (IN) support, subscribers can take advantage of the telephone network to provide them with seamless end-to-end call-handling capability. Callers can be routed to noncongested calling centers dependent on traffic and/or time-of-day. Voice-mail can be linked so that users can reach their voicemail from any phone at any company location, without dialing into the system from an outside line. Even station detailed message recording (SMDR), the feature that provides the calling records of every extension in the system, can be bridged to include all extensions in the corporation, rather than just those within a specific office handled by one carrier. The IN will allow Centrex to provide many of the same features previously seen only in the PBX environment.

The Intelligent Network (IN) is under implementation even as this book is being completed. Yet, there is a lot of work to be defined before it can be called complete. It may take many years before the IN reaches its full potential, but the framework has been defined and the infrastructure is being laid.

The Integrated Services Digital Network (ISDN)

The *Integrated Services Digital Network* (ISDN—Fig. 1.5) was first offered to the public in the 1980s, as the telephone service providers began deploying their SS7 networks. With ISDN, many new services could be extended to the customer premises. By using ISDN, subscribers can consolidate all their trunks to one DS1 facility. The ISDN protocols provide circuit allocation within the ISDN.

When additional bandwidth is needed for high-speed data communications, the protocol is capable of allocating additional channels with-

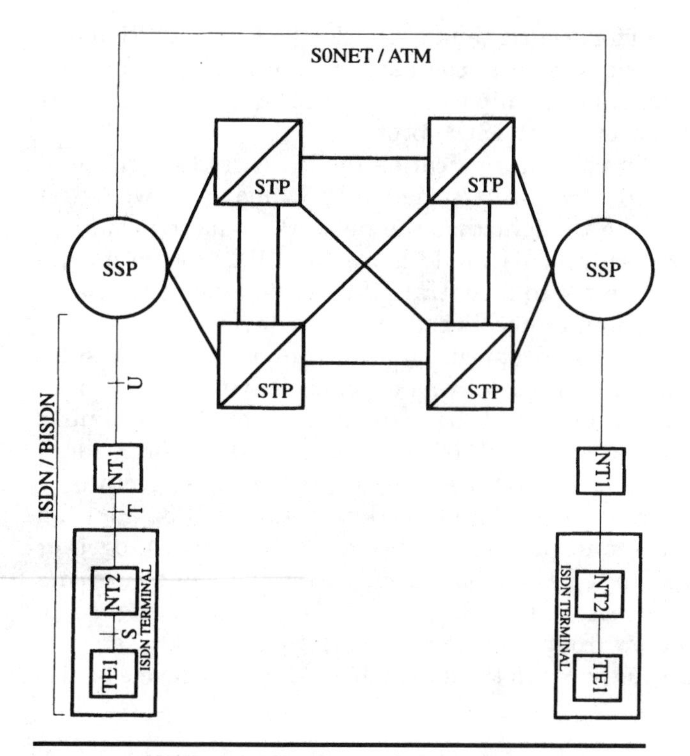

Figure 1.5 ISDN and BISDN extend the services of the Intelligent Network out to the subscriber premises. They both act as interfaces to the signaling network (SS7) without providing direct links.

in the DS1 to carry the call. When the call is terminated, the channels are released and made available for other calls. This is often referred to as dynamic bandwidth allocation and is one of the principal features of ISDN. Besides transmitting voice, ISDN is also capable of transmitting data using the same facilities as the voice.

ISDN signaling uses a separate channel and is compatible with SS7. The signaling information is handed off to the SS7 network and transferred to the distant end using the SS7 ISDN User Part (ISUP) protocol. The ISUP was developed for all call setup and teardown, and replaced the SS7 Telephone User Part (TUP) protocol in ANSI networks.

The term ISDN was originally used to refer to the entire Intelligent Network, including SS7. This later evolved to reference only the subscriber interface. Originally, the creators of SS7 thought of extending the SS7 network all the way to the subscriber. This was abandoned, however, over concern for security and network fraud. The solution was to create an intelligent interface, compatible with SS7, which could offer the same services and intelligence as the SS7 network. It was this that spurred the creation of the ISDN protocol.

Perhaps the most important application for ISDN is the concept of connecting PBXs within a private network. When SS6 was first deployed, there was the thought that the network signaling could be extended to the local PBX. This would allow the PBX to send its signaling information directly to a central office switch using the same message packet-switching protocol used by SS7 today.

This concept was quickly dropped, however, due to security issues. Instead, a separate access protocol was developed. With a specialized access protocol, signaling could be extended through the Public Switched Telephone Network (PSTN) to distant PBXs, without sacrificing security of the PSTN. ISDN was created as the access protocol to deliver PBX signaling through the network to distant PBXs, allowing large companies with multiple PBXs to bridge their switches together transparently. The United Kingdom already uses the *Digital Private Network Signaling System* (DPNSS), an ISDN protocol designed to extend PBX signaling through the SS7 network to distant PBXs.

ISDN offers many services to the subscriber. The basic levels of service are defined as:

- *Transport elements.* Allow information to be transported through the telephone service provider's network and its switches, routers, multiplexers, and other network equipment transparently, without alteration to the original data.
- *Control elements.* Support real-time operations of transport capabilities (connection establishment and database queries).
- *Network management elements.* Provide procedures and capabilities to administer, maintain, and operate the communications infrastructure. Include provisioning of transmission facilities, fault management, congestion control, and administration of databases and routing tables.
- *Communications applications environment.* Provides a development environment for programmers from which applications can be developed, using the other three elements.
- *Transport.* Provides the lower three layers of OSI, providing allocation of bandwidth, routing, relaying, and error detection/correction.

To understand how ISDN can be of significant benefit to PBX networks, consider this example. Many large corporations own several PBXs at different locations. Tie lines are often used to tie the PBXs together. This allows users to access extensions in any other company location by dialing an access number (to access the tie line) and dialing the extension. Many digital PBXs allow callers to dial extension numbers of remote extensions without dialing an access code. Automatic routing features provide software to determine which trunk the call must be routed to.

Another advantage of tying PBXs together is that long distance calls can be routed over tie lines to a PBX in the calling area of the dialed number. The call is then routed to a trunk in the remote PBX, as a local call. This can save corporations thousands of dollars in long distance charges. The problem is that the remote PBX does not know what class of service the calling phone has been assigned. The class of service is a software feature that determines what numbers a phone is allowed to dial. This programming takes place in the PBX terminating the telephone.

ISDN allows this information to be passed along to the remote PBX. In addition, other information regarding the privileges of a telephone and even features can be passed from one PBX to another. This allows corporations to create their own proprietary network, without expensive facilities between PBX locations.

ISDN never got a good start and has not shared the popularity of other service offerings, mostly because it was somewhat premature in its release. Many telephone service providers tried to market this service to residence subscribers, but failed because of the prohibitive cost and lack of intrinsic value. To truly take advantage of ISDN capabilities, the signaling information must be able to travel from the originating end to the distant end. Without SS7, this is not possible. When ISDN was first introduced, SS7 was not yet fully deployed, leaving "islands" of ISDN service that could not be connected with other ISDN networks.

Now with SS7 deployed in all the Bell Operating Companies' networks as well as those of many of the independent telephone companies, ISDN can finally be used to its full potential. Unfortunately, the telephone companies still do not understand how to market ISDN and often cite the many features ISDN provides, rather than real applications.

One feature often used in marketing ISDN is automatic number identification. For a short period there was a lot of interest in ANI, as telemarketing companies scrambled for databases that could provide names and addresses of callers triggered by the calling party information supplied by the ISDN network. All interest in ANI was quickly lost as the states began legislation blocking ANI, or at least forcing telephone companies to provide ANI blocking at no charge. Without the guarantee that at least a majority of the calls would provide ANI information, and without any other substantial advantages in the way of features, ISDN quickly faded into the background.

The problem was that most areas could not provide the calling party information. This was due to the fact that many central offices had not yet been upgraded to digital switches and were not capable of passing this information through the SS7 network. By the time networks caught up with ISDN and the SS7 network bridged all the "islands," state legislation in many states had already begun blocking the usage of the feature on the grounds that it was an invasion of privacy.

Another plague of ISDN has been the failure of manufacturers to follow a standard. Unfortunately, many manufacturers created ISDN terminals and devices that were proprietary and did not share the concept of open interconnection, making interoperability difficult, if not impossible.

This was later addressed by the North American ISDN User's Forum in June 1988. The NIUF comprises industry vendors and service providers. This forum agreed on a set of features and how they would be deployed in the ISDN network and ratified a new ISDN standard, now known as National ISDN #1. This new standard is an agreement among ISDN vendors here in the United States to standardize the equipment interfaces, interoperability, and features, guaranteeing ISDN's success in North America.

Still, ISDN continues to be marketed for its features rather than for true applications. What ISDN really offers to any business with a PBX is the ability to consolidate its trunking requirements to one or more digital spans, or T-1s. Using the ISDN protocols, they can take advantage of end-to-end digital communications for both voice and data. This allows owners of PBX equipment to rid themselves of costly dedicated lines for data communications, facsimile, and video conferencing. ISDN can provide all these services and more, with a common facility, eliminating the need for special circuits. In addition, these same facilities can be assigned dynamically, on an as-needed basis.

ISDN may also become popular as an access to the new broadband network. Small and medium-sized companies wishing to connect their local area networks (LANs) can use *narrowband ISDN* to connect to the ATM network. This interface is being considered by many service providers as the midpriced interface to the broadband network.

Larger corporations with higher bandwidth requirements will find *broadband ISDN* (BISDN) more attractive, with enough bandwidth to support Ethernet, Token Ring, and even Fiber Distributed Data Interface (FDDI). BISDN will provide bandwidth up to 155 Mbps and higher, and will most likely be the interface of choice to the new ATM network currently under development.

It is this dynamic capability that makes ISDN attractive to businesses. But ISDN can be used in residential applications as well. This is a tougher market, since consumers look for obvious advantages and features to offset cost. This is the market ISDN has had a tough time breaking into, primarily because of the prohibitive cost for the average consumer. But as the cost of ISDN decreases and becomes comparable to Plain Old Telephone Service (POTS), ISDN will become as commonplace as the telephone itself. This is especially true in metropolitan areas where telecommuting has become a way of life for many living in congested areas. ISDN provides high-speed data communications and seamless connectivity to the office local area network (LAN). As the information highway evolves, ISDN should gain in popularity.

ISDN is also an excellent medium for telemetering, something many utilities do not like to talk about because of its impact on the workforce. Telemetering has already been defined and is waiting on the digital capability of ISDN before it becomes a reality. With ISDN at the residence, utilities can attach interfaces to the ISDN circuit and use its channels during idle periods to send meter information to a collection center. This allows utilities to read meters in real time, rather than waiting until the end of a billing cycle. Telemetering also allows them to monitor usage in geographic areas during peak periods, providing invaluable information for studies and preparation for heavy loads.

There are two classes of ISDN service: *Basic Rate* (BRI) and *Primary Rate* (PRI). BRI service provides two 64-kbps bearer channels (B channels) and one 16-kbps signaling channel (D channel). This service is designed for residential and small business usage. PRI offers 23 64-kbps B channels and one 64-kbps D channel. PRI is designed for larger businesses with large call volumes. Many PBX manufacturers already provide ISDN-compatible trunking interfaces for their equipment, making ISDN a good choice for companies who need end-to-end voice and data communications.

ISDN cannot be successful by itself. As we have already seen, without SS7, ISDN remains a local digital service providing a limited number of features and applications. With the addition of SS7, ISDN can become an extension of the telephone network to the customer premises, offering true end-to-end voice and data communications with no boundaries.

The ISDN standards can be found in the ITU-TS "I" series. The signaling standards are defined in publication Q.921 (defines the Link Access Procedure—D Channel) and the Q.931 publication (defines the ISDN call control procedures at layer 3).

The cellular network and SS7

Cellular networks have evolved into two different networks. The European cellular network uses Groupe Special Mobile (GSM). The GSM network is described below and in Fig. 1.6. In the United States, cellular networks are still primarily analog (although cellular companies are quickly deploying all-digital technology to replace the analog network), but the addition of IS-41 for signaling has increased the effi-

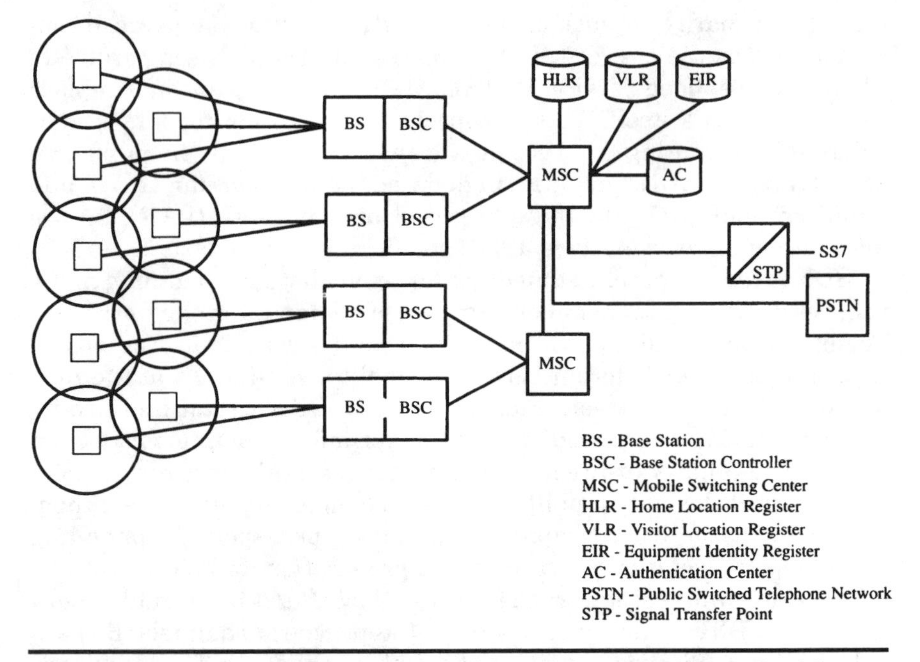

Figure 1.6 The GSM network is the standard for cellular networks outside of the United States. Many U.S. cellular providers are now looking at GSM for new PCS networks.

ciency of these networks. Both networks are very similar, the difference being in the procedures used for call handling and the entities which exist in these networks.

Previous to the deployment of SS7, cellular networks in North America used X.25 as the transport for signaling information. The signaling information in itself is somewhat new and is made possible by an application called IS-41. The IS-41 application is not particular as to what transport it uses, so X.25 was sufficient for carrying this information within one network.

Many North American cellular networks have already deployed SS7, replacing their X.25 networks. Cellular networks use the Transaction Capabilities Application Part (TCAP), which supports the access of remote databases. The ISDN User Part (ISUP) is now being implemented in cellular networks for the connection to the wireline network.

European cellular networks have always relied on SS7 for their signaling requirements and have enjoyed a much more robust and feature-rich network because of SS7. GSM is currently under trial in North America, and in many networks has become the technology of choice for supporting the new Personal Communications Services networks.

The GSM network. The GSM cellular network comprises two different segments: the radio segment and the switching segment. The radio segment consists of the cellular telephone itself, or transceiver, and the antenna system used to receive and aggregate signals within a geographical location. This antenna is referred to as the cell site.

Cell sites use different types of antennas, depending on the coverage required. Omnidirectional antennas provide coverage in a circular pattern, radiating in all directions from the cell site. Directional antennas can be used to cover a specific sector, and normally cover an area of 120 degrees.

All cell sites consist of the antenna and the radio transceiver, as well as interface equipment. This equipment is referred to as the *base transceiver station* (BTS) and the *base station controller* (BSC). The combination of both the BTS and the BSC is called the *base station subsystem* (BSS).

The purpose of the base transceiver station is to communicate with the cellular telephone. This is actually the radio transceiver, capable of transmitting and receiving within the 900-Mhz frequency band. When a caller is connected with a cell site, the BTS measures the strength of a *supervisory audio tone* (SAT), which is sent at a much higher frequency than the actual voice transmission. The SAT is sent at regular intervals by cellular telephones and is received by multiple cell sites. Each cell site reports the strength of the signal it receives to the *Mobile Switching Center* (MSC), which then determines which cell site will take possession of the call. If a call is in progress and the cellular phone moves into a new cell site coverage area, the call is handed off by signaling messages through the MSC to the new cell site.

The hand-off is controlled by the Mobile Switching Center (MSC), which communicates with all cell sites within its geographic area. The MSC does not communicate directly with the base transceiver station (BTS) but communicates with the base station controller (BSC), which serves as an interface between the radio segment and the switching segment. The BSC uses digital facilities to communicate with the MSC.

The interface between the MSC and the BSC is called the A-bis interface. The A-bis interface is a 64-kbps digital link, and uses three protocols to transport signaling information to the MSC:

- Link Access Procedure on the D Channel (LAPD)
- Base Transceiver Station Management (BTSM)
- A-bis Operations and Maintenance (ABOM)
- Direct Transfer Application Part (DTAP)

The LAPD protocol is used as the layer 2 transport protocol and provides the node-to-node communications necessary to send packets through the network. The Base Transceiver Station Management (BTSM) protocol is used for managing the radio equipment resident at the base station, as well as the interface between the base station and the MSC. Data and other signaling information are sent from the base station equipment via the Direct Transfer Application Part (DTAP).

In addition to the Public Switched Telephone Network (PSTN), the Mobile Switching Center (MSC) must also interface to other entities within the cellular network. These entities include:

- Home Location Register (HLR)
- Visitor Location Register (VLR)
- Operation and Maintenance Center (OMC)

The Home Location Register (HLR) is a database used to store the subscriber information for all subscribers within the home service area of the service provider. This database and its relationship to the rest of the cellular network prevent seamless roaming in American networks. In European GSM networks, the HLR is linked to other service areas so that subscriber information may be shared between networks, a concept not yet used throughout the United States.

The Visitor Location Register (VLR) is used to store information about visiting subscribers who are not in their home service area. This is where the roaming number information gets stored, so that subscribers may use their cellular phones while in another city. This is also linked to other networks, so that this information may be shared. While subscribers are in this network, the information regarding their service will remain in the VLR. This information is retrieved from the home HLR by the network.

SS7 protocols are used throughout the cellular network to provide the signaling information required to establish circuit connections and disconnect circuit connections, as well as share database information from one entity to another. In addition to the Message Transfer Part (MTP) and the Signaling Connection Control Part (SCCP), the following protocols are used from the Mobile Switching Center (MSC) to other entities within the network:

- Mobile Application Part (MAP)
- Base Station Subsystem Mobile Application Part (BSSMAP)
- Direct Transfer Application Part (DTAP)
- Transaction Capabilities Application Part (TCAP)

Cellular entities. The Home Location Register (HLR) is a database used to store information regarding the users of the cellular network. When a cellular telephone is purchased, the phone must be activated before it can be used in the network. The purpose of the activation is to program its serial number into the HLR database.

Whenever the cellular telephone is activated (powered on), the serial number is transmitted to the closest cell site. This information is used to identify the location of a cellular station, so that incoming calls can be routed to the appropriate cell site for reception by the destination cellular phone. The cellular phone continues to transmit its identification at regular intervals, so the network always knows the whereabouts of all active cellular phones. This is in addition to the supervisory audio tone (SAT), used by the Mobile Switching Center to measure the signal strength of cellular telephones as they move from one cell area to another.

The Visitor Location Register is a database used to store information regarding cellular telephones being used in the coverage area that are not normally registered to operate in this home area. For example, if cellular users travel to another state and wish to use their cellular phones while visiting that state, they must contact the cellular provider and obtain a *roaming* number. The roaming number is a temporary number to which all their calls will be forwarded in the area they are visiting.

The cellular provider in the area being visited must store information regarding the visiting user in its Visitor Location Register (VLR). Unfortunately, roaming is not seamless, and requires users to make arrangements prior to their travels. However, with the advent of IS-41 [a new protocol used by the Mobile Switching Center (MSC) to connect to other cellular providers], roaming can now be seamless. No advance provisions are necessary—the user just carries the cellular telephone from one area to the next. This technology is still being deployed in many cellular networks today. IS-41 provides the signaling protocols necessary for cellular providers to share database information.

The Operation and Maintenance Center is used to access the *Equipment Identification Register* (EIR) and the *Authentication Center* (AC). The EIR stores the identification serial number of all cellular telephones activated within the coverage area. The AC stores a security key embedded into all cellular phones. This code is transmitted along with the serial number when the phone is activated, and prevents unauthorized phones from being used in the network.

So far, we have discussed only the cellular network. This network must connect to the Public Switched Telephone Network (PSTN) at some point. The signaling information used to request service and connect calls within the public network is sent through the SS7 network. The Mobile Switching Center (MSC) connects to the SS7 network via a Signaling Transfer Point (STP) and the ISUP, TCAP, and MTP protocols. The SS7 network is instrumental in connecting all the cellular providers together and allowing their various databases to be shared with one another.

The IS-41 network. In the United States, moving from one calling area to another requires prior negotiations with the service providers. If the same service provider covers the calling areas being traveled through, there is no problem and no notice must be given. When the calling area belongs to another service provider, access to subscriber records is not available, and the subscriber must negotiate with the other service provider to get service in the visited area.

To provide for seamless roaming between calling areas, the EIA/TIA developed the IS-41 protocol. IS-41 is really an application entity, which relies on the Transaction Capabilities Application Part (TCAP) and the Signaling Connection Control Part (SCCP) protocols to travel through the network. Both of these protocols are particular to SS7 networks and are not commonly found in the cellular network.

Before SS7 was deployed in the cellular network, X.25 has provided the carrier services for data messages traveling from one Home Location Register (HLR) to another, or from a Visitor Location Register (VLR) to an HLR. With SS7 services and the new application entity, registration of a cellular subscriber in a new area is automatic and transparent to the user. The IS-41 provides the messages and transactions necessary to register and cancel registration in various databases, while TCAP and SCCP provide the routing and transport of these messages.

IS-41 was developed by the Electronic Industries Association (EIA) and endorsed by the Cellular Telecommunication Industry Association (CTIA). The standard is divided into a series of recommendations presented by the Telecommunication Industry Association (TIA) 45 subcommittee (TR45). The same subcommittee is responsible for Personal Communications Services (PCS) standards.

The principal differences between IS-41 and GSM lie in the protocols used to communicate between the various entities and the frequencies of the telephone units themselves. The network topology is virtually the same (Fig. 1.7).

IS-41 aligns with the ANSI version of SS7, using the Transaction Capabilities Application Part (TCAP) protocol from the SS7 protocol stack to communicate with databases and other network entities. These are the same databases found in the GSM network, the Home Location Register (HLR), Visitor Location Register (VLR), and Equipment Identification Register (EIR). The ISDN User Part (ISUP) and Message Transfer Part (MTP) are also used in the IS-41 network to connect cellular calls to the public switched telephone network, and to connect cellular circuits from the Mobile Switching Center (MSC) to the base stations.

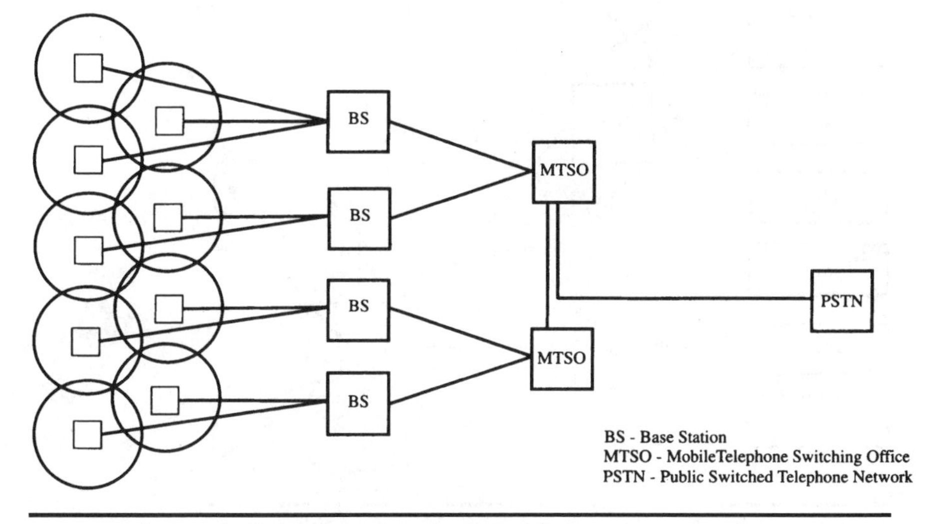

Figure 1.7 A typical cellular network in the United States.

The IS-41 network uses an MSC, as does the GSM network, for connecting to other networks and databases. Little has changed functionally within the MSC other than the fact that the data stored in the Home Location Register (HLR) can now be shared with other MSCs across the SS7 network. This requires a transport protocol (TCAP) to move the data through the SS7 network. The HLR and VLR can be colocated with the MSC.

The biggest advantage to IS-41 is the passing of cellular telephone information needed when cellular subscribers wander from one service provider's calling area to another calling area. Previously, users had to call ahead and arrange for a special roaming number. Callers then had to call the roaming number when the user was in a different area. With IS-41, the information stored in each service provider's network is shared by passing signaling information between cellular networks.

In essence, the cellular networks are much like ISDN was a few years ago. Every service provider can provide service within its own network, but is not connected with other service providers. This creates a bunch of individual "islands" of service. IS-41 is the bridge that allows service providers to share database information with other networks and eliminate the need for setting up roaming in advance. Subscribers can now move from calling area to calling area without worrying about coverage.

The PCS network. Personal Communications Services (PCS) is a new type of wireless communication based on the same philosophy as cellu-

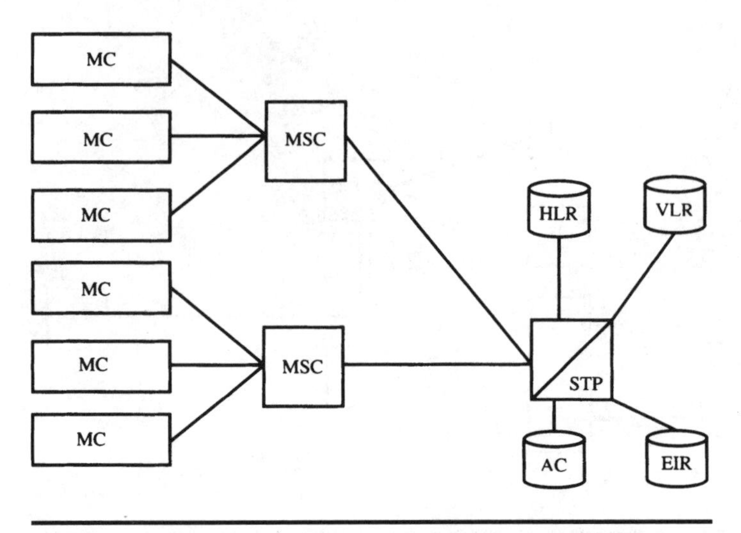

Figure 1.8 The PCS network being proposed by many cellular providers today.

lar, but with significant differences (Fig. 1.8). The most apparent difference is in the distance between the base station and the handset.

PCS is really a combination of the Intelligent Network and cellular networks, with a different topology. The PCS network relies heavily on the IN for delivery of custom features on a demand basis. The PCS network offers many new features not currently available in the cellular network. These features include:

- Available mode
- Screen
- Private
- Unavailable

Available mode. This allows all calls through except for a minimal number of telephone numbers, which can be blocked from reaching the PCS subscriber. Relies on delivery of the calling party number, which is checked against a database. The subscriber updates the database through the dialpad of a PCS handset.

Screen. The name of the caller appears on the display of the PCS handset, allowing PCS subscribers to screen their own calls. Unanswered calls are forwarded to voice-mail or another number. The forwarding destination is determined by the subscriber.

Private. All calls can be forwarded, except for a limited list of numbers that are allowed to ring through. The list is maintained by the subscriber and can be changed at any time through a special set of codes, which are input by the subscriber through the dialpad of the handset.

Unavailable. No calls are allowed to get through to the subscriber.

In the cellular network, a cell site is deployed to cover a wide area (typically four to eight miles). This distance does not permit calls to be of the same quality as terrestrial lines. In PCS networks, the area of coverage is much smaller (only a radius of about a quarter of a mile). This allows calls to be of much higher quality than in the cellular network. In addition, PCS handsets use digital technology, which is quieter than alalog.

The disadvantage to this type of deployment is in the cost of the network. To cover such a small area in a metropolitan area requires many more antennas and transceivers than in the cellular network. The positioning of these antennas also is critical. Large towers, such as those currently used for cellular, are not acceptable in most neighborhoods, which is where many of these PCS antennas have to be located.

The base station communicates to other networks (whether cellular or the Public Switched Telephone Network) through the PCS Switching Center (PSC). The PSC connects directly to the SS7 network with a link to a Signaling Transfer Point. PCS networks rely heavily on the SS7 network for internetworking, as well as for database access.

The most important feature will be the database access. The problem in cellular networks is that the database access takes place through the mobile switching center, and the databases are a noncompatible mixture of mainframes and minicomputers deployed in proprietary networks. In the case of PCS, the database is located in the SS7 network itself. The SS7 Service Control Point acts as the interface point to these databases. These databases also store data for the Public Switched Telephone Network, such as calling card information and subscriber data, and also are capable of providing information to all networks regarding PCS telephones.

This is an improvement over cellular, in which each network has its own database. In theory, the database in PCS will be centrally located, so all service providers can share it equally, but, in practice, each service provider will deploy its own database services and charge other service providers an access fee for sending queries to their database.

The purpose of having quick access to the telephone network databases is to allow subscribers of PCS services to define their own call-handling instructions and tailor their services to meet their immediate needs. For example, a subscriber may want to be left undisturbed during an important meeting, yet is expecting an important phone call from the president of a new client company. The subscriber wants that call to get through during the meeting. The telephone number of the person calling could be entered into the database with a service that would route all calls to a recording and voice-mail, except that from the president of the new client company, whose call would be routed to the PCS subscriber immediately.

This type of tailored communications is not possible today, for two reasons. First, the network must know the number the called party is going to be at, which may change when the meeting is over. Users must invoke a forwarding feature and provide a new number every time they move to another location. With PCS, the network will know where subscribers are at any time, because the network uses a mechanism much like that in cellular cell sites to track the movements of subscribers.

Second, subscribers do not as yet have the kind of access to the telephone network databases needed to tailor their services. This requires the deployment of the Advanced Intelligent Network and Personal Communications Services.

The PCS radio spectrum is much higher than the 900-MHz cellular network. In 1993, the Federal Communications Commission (FCC) allocated 160 MHz of bandwidth in the 1850- to 2200-MHz range. The FCC divided this range into licensed frequencies and unlicensed frequencies. The licensed frequencies have been further divided into seven separate bands.

The service areas have been drawn according to Rand McNally's boundaries for *Major Trading Areas* (MTAs) and *Basic Trading Areas* (BTAs), although this is under much debate. There are 51 MTAs and 492 BTAs in the United States. Much of the debate is related to the fairness of trading areas versus the Local Access Transport Areas (LATAs) created by the Justice Department during the divestiture of the Bell System. Many service providers argue that it would be unfair to use different boundaries for Personal Communications Services (PCS) than those issued for local telephone companies. Others argue that the LATAs are a better approach because they represent a more fair and equitable market.

To add to the subscribers' confusion in this market, consumers are now faced with the decision of whether to buy 49-MHz cordless telephones, 900-MHz cellular telephones, or the new PCS communications devices. Many vendors are confusing the issue even further by describing new cellular features under the auspices of PCS. Many features are being offered today that emulate PCS in the cellular market, such as one-number dialing (the ability to use one telephone number to reach a subscriber no matter where that subscriber is). However, there is more to PCS than just simple features. It represents a whole new network architecture, very different from the current cellular network.

Many vendors have already alluded to the possibility of GSM technology for use in their new PCS networks. The only difference between other digital technologies lies in the digitizing method used at the air interface. GSM may prove to be the best solution for cellular providers here in the United States who are looking for a proven technology for their new wireless networks.

Video and the telephone network

During the early to mid 90s, there was a lot of acitivty between cable television companies and telephone companies. That interest has faded somewhat due to the lack of interest from subscribers. There is much speculation surrounding the mergers taking place between the cable television companies and the Bell Operating Companies. Some speculate that the cable providers will begin competing for dialtone in other market areas. Others believe this provides yet another service that the Bell companies can provide to their subscribers: video on demand. But many have missed the underlying motivation for making these mergers today. The cable companies own lots of coaxial cable that is already installed to nearly every American home. In fact, statistics show cabling companies already pass 90 percent of the homes in the United States.

Along with that coaxial cable is a healthy fiber optic backbone, capable of handling lots of high-speed data. While the cable network may be inadequate for broadband data services, it certainly can handle the demands of most telecommuters today. In fact, with a bandwidth of at least 10 Mbps, coaxial could be the answer for many of those who must reach local area networks from their homes.

But whether or not the telephone companies want to use this installed base of coaxial for data or video remains to be seen. One thing is certain: the merger of these networks into the Public Switched Telephone Network (PSTN) will place new requirements on SS7. This will be especially true if video dialtone becomes a reality.

Video dialtone will be offered by the local telephone companies within this decade. The Bellcore philosophy is that subscribers want convenience. Most subscribers will be willing to pay for that convenience. (Domino's Pizza proved that theory.) If the local telephone companies can provide pay-per-view movies over the telephone line for the cost of a long distance telephone call, they are guaranteed success—or so the theory goes. In order for this to be successful, the network must be able to support it.

The telephone companies will also find it necessary to add new entities to their networks. Cable service providers have already begun defining the specifications for video servers. A video server is nothing more than a large disk array, providing digital storage for scores of movies. These video servers can then be located anywhere in the telephone company network and, through signaling, triggered to download any movie to the appropriate telephone circuit.

That circuit may very well be a coaxial service or digital service such as ISDN to a subscriber's home. Once again, the key to this process will be signaling to the video server to download the movies to the proper circuit and to initiate billing.

On the voice networking side, little is required. With the advent of SONET between end offices, there is enough bandwidth to pump video signals through the voice network. However, a higher-layer protocol must manage the video transmission over the SONET network. Currently, work is being completed on standards that will allow Asynchronous Transfer Mode (ATM) to deliver high-speed data and video over the SONET network.

The ATM protocol provides the capability of packetizing video and any other form of information and moving it through the network in a switched fashion, just like placing a telephone call. ATM also supports the bandwidth requirements of video dialtone, currently offering 155 Mbps of bandwidth and migrating to a stealthy 600 Mbps by the end of this decade.

At the subscriber level, there will need to be some additional hardware. Vendors are already releasing prototypes of television set top boxes which will connect to the telephone line and allow movies downloaded from the telephone company to be stored in memory and played back at the viewer's convenience. These playback boxes will also provide rewind, fast-forward, and pause functions just as a VCR does, but will only allow a movie to be played in its entirety one time.

Video dialtone will also bring home shopping options a little closer to convenience than they already are. While viewing the shop-at-home channels, purchases can be made by choosing a selection key on the television set top box remote control. This will allow the subscriber's information, such as credit card number and shipping address, to be downloaded to the shopping network without picking up the telephone. At the touch of a button, one will be able to purchase groceries, appliances, automobiles, and jewelry without ever leaving the armchair.

As far as SS7 is concerned, this new addition to the telephone network will be nothing more than one more facility. For multimedia applications, voice and video are combined over the same facility. The interactive portion (subscriber data downloading to home shopping networks) will rely on the signaling network for database access and control information.

Another application is multimedia. Multimedia allows video to be combined with graphics and a user interface that allows the user to make selections by choosing icons on the television screen. When an icon is selected, a database is accessed to retrieve specific information. The medical profession is a big supporter of this type of service. Distance learning and teleconferencing are also likely candidates.

However, the medical profession has already begun using this technology for patient diagnosis.

For example, a doctor is sitting in his office in Fargo, North Dakota, diagnosing a patient with possible heart disease. For a second opinion, the doctor dials a specialist in New York City. By using a video camera connected to the telephone and a computer connected to the same interface, the doctor can video the entire office visit live to the specialist in New York City. The specialist can ask pertinent questions regarding the patient and view the patient through video as if the specialist were in the office with the patient.

The patient's medical records can be transmitted in seconds to the specialist, while the office examination is under way using the computer. The specialist can even add comments to the file and send it back to the patient's doctor. Sound futuristic? This is the impetus behind the Information Highway and it is already under trial in several major cities today.

SS7 is in the background during all of this activity, quietly setting up connections and accessing databases. Information needed by the telephone companies to complete the video conference is provided by SS7 by accessing a Service Control Point where all the information is stored and sending it to the requesting end office. SS7 will provide the connection control and billing services needed for such services.

Broadband data communications

Broadband (Fig. 1.9) is formally defined as any data communications with a data rate from 45 up to 600 Mbps. Broadband standards are still being defined today, with ATM receiving the bulk of attention. SS7 will not carry the actual user data, but will provide the services necessary to connect the end-to-end telephone company facilities required for data transfer between two end points. The specific requirements of SS7 are still being defined. The trial networks using ATM do not require SS7, because the trials being performed with broadband ISDN and ATM are currently permanent virtual circuits (PVCs) and are not switched. However, as the new services are deployed for public access, SS7 will provide the signaling for ATM.

ATM is a transport protocol that relies on upper-layer protocols for functions above and beyond layer 3. The ATM technology was created to support broadband ISDN, but has found favor with a number of other networking protocols as well. Frame relay, with speeds up to 155 Mbps, will most likely serve as a good solution for those requiring a PVC service. These work much like a dedicated special circuit and are not typically switched. However, there are plans to provide switched frame relay, making it a viable service for switched data networks.

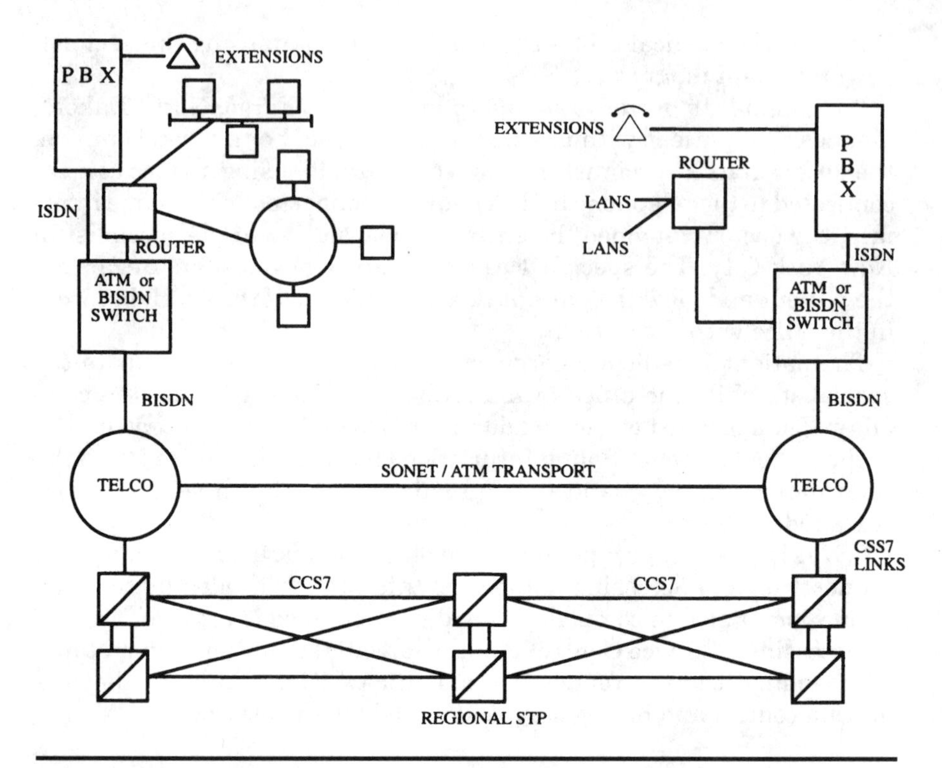

Figure 1.9 This drawing depicts a typical broadband network. The subscriber side of the network may consist of these components or others. The SS7 network is used to control the connection of circuits between exchanges (end-to-end).

Narrowband ISDN (1.554 Mbps) is being suggested by many service providers as a cheaper access to broadband services, while broadband ISDN (155 Mbps) is being targeted at the meganetworks that need much more bandwidth.

Broadband ISDN (BISDN) will provide fast data transmission for services such as video dialtone. With the bandwidth of BISDN, video and audio can be transmitted simultaneously on the same facility. This will prove to be an important feature for the information highway.

A point that often gets lost among all the marketing hype over ATM is the purpose of this technology. Talk of a new telephone network is nothing new. Telephone company officials have been planning for many years to upgrade their old analog networks to support the services and applications under much demand. Subscribers are no longer complacent with slow data transmission and expensive special circuits for sending data and other information through the network.

ATM was developed to support the broadband ISDN interface being defined for the customer premises. Many vendors have placed ATM switches in the hands of the consumer (philosophically), thinking that this will meet the needs of every business and become a necessity in office buildings everywhere. Some have even been so bold as to claim that ATM will replace the local area network.

In reality, ATM will only be found at customer premises where nothing else will deliver the bandwidth required for video services. ATM will be much too expensive for the average business to deploy in place of a LAN, and certainly constitutes overkill for most daily business activities. But for the university needing the 600 Mbps of bandwidth and the hospital using the telephone network to send high-resolution medical images, ATM and broadband ISDN are likely to be an integral part of their networks.

Narrowband ISDN will serve as the customer interface for the majority of subscribers, while ATM and broadband ISDN will become the transport mechanism within the telephone network to the destination. ATM requires a transmission facility capable of carrying the bandwidth (600 Mbps and higher). Synchronous Optical NETwork (SONET) is being deployed as the physical medium for ATM.

The Information Highway

No book on telecommunications networking would be complete without some discussion on the information highway. The information highway is a coined phrase from the Clinton administration describing the technologies required to provide information services to all citizens. The requirements for the information highway are unclear. The only criterion with which politicians seem to agree is the ability to access a wealth of information using a variety of media, all accessible through the Public Switched Telephone Network (PSTN).

This information highway in itself will require a major investment in the telephone network infrastructure. Certainly, this task could be achieved today if private networks were acceptable and if information was somewhat centralized. This is not the philosophy of the present administration, and they have set their agendas to transform the telephone network into an all-purpose information network, capable of providing access to telephones, databases, and video sources anywhere in the world.

Just recently, this message was carried to the United Nations in an effort to encourage all nations to jump onto the information highway, resulting in a worldwide network. This is certainly achievable—maybe not within the time frame the Clinton administration would like, but certainly within the next decade.

The challenge is not how to access information from a variety of locations. This can be done today through packet switching. The challenge is how to provide this access through the PSTN, to anyone anytime, and to support all forms of media, whether it be audio, video, high-resolution graphics, or data.

Today's telephone network is not equipped to handle much more than voice and some data transmission. The telephone companies have embarked on a major upgrade of the existing infrastructure. This upgrade is being implemented in phases.

The first phase is to upgrade the carrier facilities. Telephone switching offices typically use DS1 or DS3 circuits for interoffice trunks. These facilities carry voice, data, and signaling traffic between exchanges. Video and high-speed data require more bandwidth than what any of these facilities provide and, for that reason, fiber optic facilities are now replacing the previous DS1s and DS3s.

This new fiber optic technology, called Synchronous Optical NETwork (SONET), provides the bandwidth necessary to meet almost all applications. Yet another technology will be necessary for the switching and routing functions to carry the information from originator to destination. Asynchronous Transfer Mode (ATM) has been developed for this reason.

ATM will provide the mechanism for transporting information in all forms of media from one exchange to another, and eventually to the subscriber. ATM development has been slow, as telephone companies battle with the cost of existing infrastructure and the cost of replacing that infrastructure with new technology. Until then, many point-to-point networks will begin offering this service to a selected number of subscribers on a trial basis.

Already there has been much talk about the information highway, ATM, video on demand, and high-speed data available to every household by the end of the Clinton administration. But this development is not happening as fast as most would like. There are many issues to be resolved and much work to be completed on the standards that will deliver these services.

As many in the industry can attest, technology does not happen overnight. SS7 took almost 20 years before telephone companies began rapid deployment. ATM will take 10 years for full development, and at least another five years to replace the existing infrastructure with the new switching equipment required to support ATM.

With the cost of replacing the existing infrastructure with new switching equipment, new cabling, and new subscriber interfaces, the general public is not likely to see the results of ATM for some time to come. Nonetheless, development is well under way, and the industry has embraced this wholeheartedly. The telephone companies are aggressively pursuing this technology as the technology of the next decade.

Now that we have discussed the technologies used today to form the world's most advanced communications network, let us look at the various organizations responsible for developing these standards.

Standards Organizations

Standards such as SS7 and ISDN do not happen quickly. They are the result of years of research and development conducted by standards committees. These organizations are usually composed of government agencies or industry representatives from manufacturers and service providers.

To understand the various standards available today, one must first understand the purpose of new standards and how they are developed. There are two different types of standards: *de jure* and *de facto.*

The de jure standard is formed by committee. These standards take many years to develop because the processes used in committees are long and bureaucratic. Nonetheless, many of the standards used today are the result of standards committees.

A de facto standard is the result of a manufacturer or service provider monopolizing a market. A good example of a de facto standard is the DOS personal computer. IBM was instrumental in saturating the market with their PCs, but, more importantly, with encouraging third-party vendors to use their architecture and build IBM PC clones, using the same disk operating system. The result of this IBM marketing strategy is still felt by its competitors today. There are so many DOS PC computers in the market that introducing a new platform requiring a different operating system is a high risk. Yet there are no standards in existence that define the use of DOS in all personal computers.

De jure standards and de facto standards can be voluntary or regulatory standards. Voluntary standards are adopted by companies on a voluntary basis. There are no rules that say all manufacturers and service providers must comply with a voluntary standard. However, the advantages are many. Voluntary standards help ensure that everyone developing networking products builds their products for interconnectivity. Without this interconnectivity and interoperability, only a few equipment manufacturers would win the market—those with the largest install base.

Interoperability is another issue in data communications. Interoperability is the ability for equipment to communicate with equipment from different vendors in a network environment. Often, vendors will implement varying protocol versions in their equipment and are noncompliant with the standards. When this occurs, other equipment cannot communicate with the noncompliant system because it uses a proprietary interface, forcing subscribers to purchase all their equipment from the same manufacturer.

Voluntary standards help ensure the interoperability of all networking equipment. With voluntary standards, all those participating in the technology can have a voice in the final "product." The organizations responsible for creating these voluntary standards are usually made up of industry representatives. These representatives work for the same companies who build the equipment. As the technologies evolve, companies participating in the development of the standards can also get a sneak preview of what the final standards will consist of, and can be first to market with product that is compliant with the standards.

Regulatory standards are created by government agencies and must be conformed to by the industry. These standards do not hold any major advantage to the service provider or the manufacturer, but are in place in most cases to protect the consumer.

Regulatory standards are monitored by government agencies such as the Federal Communications Commission (FCC). These agencies ensure the protection of the public and other network users by enforcing standards covering safety, interconnectivity, and, in some cases, health (i.e., radiation emission from computer terminals and cellular phones).

SS7 networks use standards from a variety of organizations and standards committees. Some of the standards used in SS7 networks were developed for other applications as well, not specifically for SS7. The following organizations have written standards directly related to SS7:

- International Telecommunications Union—Telecommunications Standardization Sector (ITU-TS)
- American National Standards Institute (ANSI)
- Bell Communications Research (Bellcore)

In addition to these organizations, many other standards have been written that affect SS7. The following standards organizations have contributed to the SS7 network with standards not written specifically for SS7 but used by equipment in the network:

- Electronic Industries Association (EIA)
- ATM Forum
- Federal Communications Commission (FCC)
- Underwriters Laboratories (UL)
- Canadian Standards Association (CSA)
- International Organization for Standardization (ISO)

These organizations are responsible for the standards which govern the quality of cables, quality standards for manufacturing practices, electrical specifications, and interfaces used to interconnect telecommunications equipment. In addition to these standards organizations, many new forums have evolved. As the industry begins to understand the importance of standards bodies and compliance to these standards, new industry forums evolve composed of vendors and service providers in the industry with a vested interest in the technology.

These forums are often commissioned by the ITU-TS and ANSI to develop new standards in their behalf (such as ATM) or work on issues with existing standards (such as ISDN). In many cases, these forums can develop standards much faster than the standards committees themselves, saving the committees years in development time.

Currently, there are just a few of these forums which are relevant to the telecommunications industry. They are described here because their work involves portions of the SS7 network. They are:

- Network Operations Forum (NOF)
- ATM Forum

The rest of this section will look at all of these organizations in greater detail. The purpose of describing these organizations is to provide a better understanding of who the players are and what significance they carry. The major organizations are described in greater detail than some of the less significant ones.

International Telecommunications Union (ITU-TS)

Formerly known as the CCITT, this organization is a part of the ITU, which is a United Nations (UN) Treaty organization. The purpose of the ITU-TS is to provide standards that will allow end-to-end compatibility between international networks, regardless of the countries of origin. The standards are voluntary standards, but many countries require full compliance to connect to their networks.

The members of the ITU-TS are government representatives from the various nations. The representative for the United States is the Department of State. In addition to government agencies, manufacturers and service providers carry some influence as well. Membership is limited to four categories:

- Administrations of a country's public telephone and telegraph companies
- Recognized private operating agencies
- Scientific and industrial organizations
- Standards organizations

The ITU was reorganized into three sectors: the Radiocommunication Sector (ITU-RS), the Telecommunication Development Sector (ITU-D), and the Telecommunication Standardization Sector (ITU-TS). The ITU-TS is the sector responsible for defining SS7 standards and other related standards.

The ITU-TS standards for SS7 have been embraced by every country that is deploying SS7, yet not every country's network is the same. One would think that, with international standards in place, interconnectivity would not be an issue. Yet every country creates its own standards to meet the requirements within its own networks.

Because of this independence within individual countries, the SS7 network hierarchy consists of an international network and many national networks. The national networks are based on the ITU-TS standards, but modified for usage within individual countries. The United States uses the ANSI standards. The differences between the ANSI standards and the ITU-TS standards are mainly in the addressing (point codes) and in network management procedures. Bellcore has also modified and published a set of standards, endorsed by ANSI. The ANSI standards and the Bellcore standards are virtually the same.

The SS7 ITU-TS standards were first defined in 1980. This is referred to as the Yellow Book. ITU-TS standards are published every four years in a set of documents, which are color-coded to indicate the year in which they were published. The color code is as follows:

- 1980—Yellow Book
- 1984—Red Book
- 1988—Blue Book
- 1992—White Book

The SS7 standards can be found in the ITU-TS documents numbered Q.701 through Q.741. The following list identifies all those documents which are SS7 standards or related to SS7:

- Q.700–Q.709—Message Transfer Part (MTP)
- Q.710—PBX Application
- Q.711–Q.716—Signaling Connection Control Part (SCCP)
- Q.721–Q.725—Telephone User Part (TUP)
- Q.730—ISDN Supplementary Services
- Q.741—Data User Part (DUP)
- Q.761–Q.766—ISDN User Part (ISUP)
- Q.771–Q.775—Transaction Capabilities Application Part (TCAP)

- Q.791–Q.795—Monitoring, Operations, and Maintenance
- Q.780–Q.783—Test Specifications

The ITU-TS recently announced a change in the documentation structure, which will affect all new publications from the ITU-TS. There should be no effect, however, on existing SS7 standard publications. These will not be changed.

American National Standards Institute (ANSI)

The American National Standards Institute (ANSI—Fig. 1.10) is responsible for approving standards from other standards organizations for use in the United States. There are many organizations considered accredited standards bodies by ANSI, including Bellcore and the EIA. ANSI is divided into committees. The ANSI Accredited Standards Committee T1 is responsible for standards associated within the telecommunications industry.

The T1 committee's responsibilities include developing standards for interconnection and interoperability of telecommunications networks.

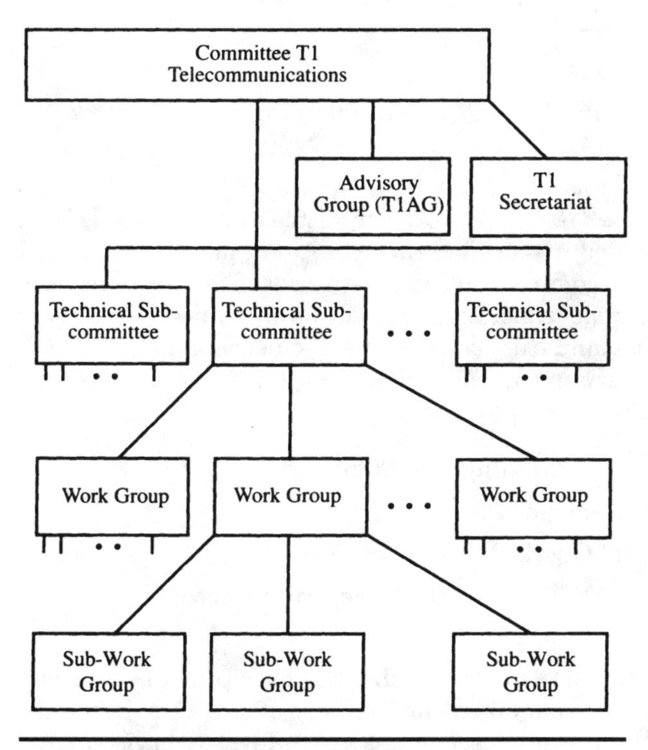

Figure 1.10 The ANSI organizational chart.

The T1 committee is divided into seven Technical Subcommittees which receive their direction from the T1 Advisory Group (T1AG).

The T1 Advisory Group meets on a bimonthly basis and is responsible for the establishment and administration of procedures for the committee's activities. As illustrated in Fig. 1.10, the Advisory Group reports to the T1 committee.

The secretariat provides support functions to the subcommittees and interacts with ANSI to get approval and publication of the standards created by the subcommittees. Duties of the secretariat include scheduling committee meetings and approving memberships.

Following is a description of each of the seven subcommittees along with their missions, according to the *T1 Committee Procedures Manual.*

Technical Subcommittee T1E1. This subcommittee defines standards for network interfaces, concentrating on physical layer interfaces. This includes the electrical, optical, and magnetic specifications of data communications interfaces, as well as telephony interfaces.

It consists of four working groups, all which work closely with each other and with other groups within the T1 Committee:

- T1E1.1 Analog Access
- T1E1.2 Wideband Access
- T1E1.3 Connectors and Wiring Arrangements
- T1E1.4 DSL Access

Technical Subcommittee T1M1. This subcommittee's primary focus is on the processes and procedures used in the operations, maintenance, and administration of telecommunications networks. This includes the testing of facilities, measurements, routine maintenance, and traffic routing plans. Their standards focus on the engineering and planning functions, network resources, and support systems. There are four working groups in this subcommittee:

- T1M1.1 Internetwork Planning and Engineering
- T1M1.2 Internetwork Operations
- T1M1.3 Testing and Operations Support Systems and Equipment
- T1M1.5 OAM&P Architecture, Interfaces, and Protocols

Technical Subcommittee T1P1. This subcommittee provides support services and program management for the rest of the T1 subcommittees. It provides high-level descriptions, high-level overviews and architectures, scheduling for interactive sessions between subcommittees, publishing support of standards, and reference models, and it determines T1 endorsement of programs.

Technical Subcommittee T1Q1. This subcommittee focuses on performance issues of network traffic, switching, transmission, maintenance, availability, reliability, and restoration. Its performance specifications are standards used from carrier to carrier and from carrier to customer interfaces. There are four working groups:

- T1Q1.1 4-kHz Voice and Voiceband Data
- T1Q1.2 Survivability
- T1Q1.3 Digital Packet and ISDN
- T1Q1.5 Wideband Program

Technical Subcommittee T1S1. The T1S1 subcommittee is directly involved with SS7 signaling standards as well as ISDN and other related services, architectures, and signaling. It reviews international standards and makes decisions on how those standards will be implemented in the United States. In addition, it works closely with the ITU-TS in developing standards for the international community. There are four working groups in this subcommittee:

- T1S1.1 Architecture and Services
- T1S1.2 Switching and Signaling Protocols
- T1S1.3 Common Channel Signaling
- T1S1.5 Broadband ISDN

Technical Subcommittee T1X1. The members of this subcommittee define the standards used to define the hierarchy of digital networks and synchronization networks. This subcommittee defines standards used for internetworking, focusing on the functions necessary to interconnect at the network transport level. There are four working groups:

- T1X1.1 Synchronization Interfaces
- T1X1.4 Metallic Hierarchical Interfaces
- T1X1.5 Optical Hierarchical Interfaces
- T1X1.6 Tributary Analysis

Technical Subcommittee T1Y1. This subcommittee defines standards not covered by any of the other subcommittees and serves as a sort of miscellaneous standards committee. Specialized video and audio services, including broadcast services, teleconferencing, and graphics, as well as specialized voice and data processing, are within this group's jurisdiction. There are three working groups:

- T1Y1.1 Specialized Video and Audio Services
- T1Y1.2 Specialized Voice and Data Processing
- T1Y1.4 Environmental Standards for Exchange and Interexchange Carrier Networks

The ANSI publications regarding SS7 define the function of the protocols. Bellcore has published numerous other documents detailing the specific requirements of all SS7 entities and management procedures. Bellcore protocol chapters and ANSI publications for SS7 are numbered as follows:

- T1.110—Signaling System 7, General
- T1.111—Message Transfer Part (MTP)
- T1.112—Signaling Connection Control Part (SCCP)
- T1.113—ISDN User Part (ISUP)
- T1.114—Transaction Capabilities Application Part (TCAP)
- T1.115—Monitoring and Measurements
- T1.116—Operations, Maintenance, and Administration Part (OMAP)

An ANSI catalog is available for all ANSI publications by contacting the ANSI organization in New York City.

Bell Communications Research (Bellcore)

Bellcore is the research and development arm of the seven Bell Operating Companies. These seven companies were divided from the Bell System in 1984 as part of the divestiture. AT&T was separated from the local exchanges, taking with it Bell Laboratories. The new Bell Operating Companies (BOCs) then founded Bellcore to replace the services once provided them by Bell Laboratories. The seven operating companies are:

- Ameritech
- Bell Atlantic
- BellSouth Telecommunications
- NYNEX
- Pacific Telesis
- Southwestern Bell
- US West

In addition to these seven regional operating companies, Bellcore also provides services to the following "nonowners":

- Cincinnati Bell, Inc.
- Southern New England Telephone Co.
- Centel Corporation
- General Telephone Company and its local telephone companies
- Sprint and its local telephone companies
- Canadian local telephone companies

Bellcore publishes standards and recommendations for all kinds of telecommunications services, maintenance and operations procedures, and network architecture. These documents are available for a fee to any individual through the Bellcore organization.

Recently, Bellcore changed their documentation numbering system. What is described below is the former identity used to communicate document phases. The three levels described below have since been collapsed into one identity, known as the Generic Requirement (GR). All new Bellcore documents will use this new convention. The former identity is defined below because there are still many documents in circulation using the older identification.

Bellcore documents are developed and published in three phases. The first phase of publication is the *Framework Advisory* (FA). Any document beginning with "FA" is a Framework Advisory, and is still in draft format. These documents are published and submitted to the industry for comments and suggestions.

The second phase is known as the *Technical Advisory* (TA). These documents are preliminary publications and are subject to minor changes. They are submitted to the industry for final comment and suggestions before reaching final publication.

The third phase of publication is the *Technical Reference* (TR), which is the final phase of the document. Technical References represent final released versions of a document. All of these publications are numbered with the following convention:

TR-NWT-000082

The preceding document number is just one example of the document numbering of Bellcore publications. The "TR" indicates that this is a technical reference, the "NWT" indicates this is a network-related publication, and the six-digit number identifies the unique publication. As mentioned above, Bellcore has recently implemented a new document numbering scheme, which alleviates the three levels of numbers and identifies all new documents as Generic Requirements (GR).

In addition to these types, Bellcore also publishes documents on technology findings and services offered through their regional companies. These are referred to as *Science and Technology* (ST) publications and *Special Reports* (SRs).

A publication regarding Bellcore requirements for interconnectivity within Bell networks is called a *Family of Requirements* (FR) and specifies design requirements, as well as the functionality of specific network elements. These are classified as:

- LATA Switching Systems Generic Requirements (LSSGR) (FR-NWT-000064)
- Operator Services Systems Generic Requirements (OSSGR) (FR-NWT-000271)
- Operations Technology Generic Requirements (OTGR) (FR-NWT-000439)
- Reliability and Quality Generic Requirements (RQGR) (FR-NWT-000796)
- Transport Systems Generic Requirements (TSGR) (FR-NWT-000440)

The Bellcore publication TR-NWT-000246GR-246-CORE is a three-volume series, which matches the ANSI publications almost word for word and defines the SS7 protocol. The chapter numbering is the same as in ANSI publications, allowing cross-referencing between publications. The Bellcore version identifies the Bell System implementation of SS7 and identifies specific procedures and functions required in the Bellcore networks. The protocol descriptions are identical to the ANSI publications, with the exception of some Bellcore implementations. A catalog of Bellcore documents is available through Bell Communications Research.

Electronic Industries Association (EIA)

The EIA develops standards focused on the physical interfaces used in the data communications industry. Probably the best example of an EIA standard is the RS-232C. This standard was originally created for interfacing modems to computer systems, but it proved so simple and versatile that it quickly became the de facto standard for any application requiring a serial interface.

The EIA has developed many other standards besides the RS-232C, including some faster interfaces designed to replace the RS-232C. Recently, the EIA released a new cellular standard called the Interim Standard-41 (IS-41). This standard defines the interfaces between cellular network entities as well as the communications protocols used at these interfaces. The IS-41 standard was developed by the EIA/TIA Subcommittee TR-45.2, Cellular System Operation.

ATM Forum

The ATM Forum is a voluntary organization consisting of industry and public sector representatives. Their mission is to assist the ITU-TS in developing a standard for international use in ATM networks. The ITU-TS has been actively working on this standard, but was not expected to complete the standards until the year 2000. The ATM Forum was formed to help expedite the process of development, in hopes of finishing this standard much sooner.

Much of the work accomplished by the ATM Forum has been submitted and accepted (with modification) by the ITU for inclusion in the final ATM standards. It is important to understand, however, that the ATM Forum does not write standards. They write implementation agreements, which allow vendors to agree on certain aspects of the technology and begin development on products prior to the final standards being completed.

Formed in November 1991, the ATM Forum is now over 400 members strong, with representatives from a wide spectrum of interests. Data communications and telephone companies, service providers, and private networking types have joined together in this consortium to ensure a standard which meets the needs of all interested industries.

Completion of standards is now expected by 1998, although this may prove to be difficult in light of all the issues surrounding ATM. Transmission of high-speed data is not an issue in ATM, but making ATM switchable through the public switched telephone network is not an easy task and will require much work.

The ATM Forum has defined many implementation agreements to encompass the LAN aspects of ATM, as well as the *user-to-network interface* (UNI), but there is still much work to be finished on the *network-to-network interface* (NNI) implementation agreements and the signaling standards.

Federal Communications Commission (FCC)

The FCC was created as part of the Communications Act of 1934 and is responsible for the regulation of the airwaves, as well as for the regulation of the telecommunications industry. There are five commissioners in the FCC, all appointed by the president of the United States. Each commissioner serves a term of five years. One commissioner is appointed by the president to serve as chairperson of the FCC.

There are four operating bureaus within the FCC organization, with a total staff of 1700 personnel. Each operating bureau serves a specific function: mass media, common carrier, field operations, and private radio. The common carrier bureau regulates all aspects of the telecommunications industry, including paging, electronic message service, point-to-point microwave, cellular radio, and satellite communications.

The FCC regulates the access of interconnect companies and determines the type of interfaces to be used. They do not necessarily create standards, but they enforce regulations regarding the interconnection of networks and network devices. They also issue licenses for radiotelephone circuits and assign frequencies for their operation. Recently, the FCC allocated new frequencies for the Personal Communications Services being deployed in the United States by many major service providers. Rather than create new frequencies, the FCC reallocated frequencies previously used by microwave.

The FCC also supervises charges and practices of the common carriers, approves applications for mergers, and determines how the carriers will maintain accounting for their operations. The FCC also governs the type of services that a service provider can provide.

A good example of how the FCC regulates the telecommunications industry is the registered jack program. All manufacturers of telephone equipment, or any equipment which connects to the telephone network, must have their equipment registered with the FCC. The FCC then determines which type of interface (*registered jack,* or RJ) the equipment must use to connect to the network.

The Bell System sometimes gets credit for these interfaces, since they usually are the ones who install and maintain them. But the FCC determines how these interfaces will be used and who must use them. The RJ-11, used to connect single-line telephones to the central office line, is found in every home across the United States today. Most of us refer to them as *modular jacks.*

The Network Reliability Council (NRC) was formed to monitor network outages and seek resolutions through the industry vendors and service providers. This council works closely with the key corporations in resolving key issues which can be related to service outages in the nation's SS7 networks. This information is published and shared with all service providers to prevent future outages from occurring due to common failures. The NRC was recently formed by the FCC after several network outages in the SS7 network shut down telephone service in several major cities.

Today, all outages must be reported by the service providers to the NRC. This information is then disseminated among service providers as an information-sharing mechanism. The FCC hopes that sharing information regarding software deficiencies from the various vendors can help deter many network outages.

Underwriters Laboratories (UL)

The Underwriters Laboratories is a not-for-profit organization which began in 1894. The UL uses a suite of tests and gives approval to any equipment which passes the minimal requirements set by the test suite.

UL approval is not required here in the United States, but is sought after by most manufacturers of electrical or electronic equipment. The UL label certainly influences buying decisions, and in most building codes it is a requirement for electrical equipment.

Canadian Standards Association (CSA)

The CSA is Canada's equivalent of the Underwriters Laboratory. CSA approves all electrical and electronic products for sale in Canada. Manufacturers in the United States who plan to sell their equipment in Canada usually have the equipment approved by the CSA. If the equipment is sold in the United States as well, the equipment must be UL and CSA approved.

As with the UL, CSA approval is not a requirement. However, in the telecommunications industry, most buyers will require CSA approval before purchasing equipment. This is their guarantee that the equipment they are buying went through some level of testing for electrical compliance.

International Standards Organization (ISO)

The ISO is an international standards organization responsible for many data communications standards, including the Open Systems Interconnection (OSI) model. The OSI model was developed after SS7 and was not adopted as a standard until 1984. However, layering was well understood and practiced during the early to late '70s, which is when the work was being done on SS7 protocols. The OSI model is still used today to define the functions of the various levels within a protocol stack.

The ISO is composed of other standards bodies from various countries, mostly government agencies responsible for setting communications standards within their own governments. The United States representative is ANSI.

The ISO does not limit itself to just data communications standards. They have created many other types of standards as well. One of their most recent contributions to industry is the ISO 9000 quality standards. The ISO 9000 defines processes to be used in manufacturing to ensure quality production. Again, these are not mandatory standards, but are essential for companies selling products in Europe because most European buyers now require ISO 9000 compliance.

Other agencies

In addition to the standards organizations, there are other agencies that have made a significant impact on the telecommunications network. These agencies are responsible for ensuring reliability in our telecommunications network. Following are the most prominent agencies:

Network Reliability Council (NRC). First commissioned by the FCC in 1992, this council was chartered to investigate network outages and report them to all network providers, as well as to vendors. They were placed into existence only after numerous network outages caused telephone service to be out for extended periods of time, costing many corporations millions of dollars (including Wall Street in New York) and even necessitating the closing of an airport.

Their charter was to have expired in 1994, but they received a new charter by the FCC chairman, Reed Hundt, and are still tasked with the investigation of network outages. The commission is made up of CEOs from leading carriers and manufacturers, and provides reports regarding the reliability of the nation's network, as well as explanations for outages and how they can be prevented in the future.

Network Operations Forum (NOF). The NOF was originally formed in 1984, over concerns as to who would track and clear trouble reports that crossed network boundaries. They have since expanded their operations to include definitions for interoperability testing and interworking issues. They also meet with manufacturing companies to resolve issues regarding the reliability of network equipment.

Chapter

2

SS7 Network

The SS7 network is separate from the voice network, and is used solely for the purpose of switching data messages pertaining to the business of connecting telephone calls and maintaining the signaling network. Packet switching is the method used for transferring messages through the network.

The telephone switches used in many exchange offices today perform a dual function. Besides connecting voice circuits to other exchanges and switching voice circuits from one exchange to another (as well as from one subscriber to another) they must also perform signaling functions. Often this is accomplished through adjunct computers, which are connected through digital links to other computers in the network.

These computers are referred to as *signaling points.* There are three functions required of signaling points. The originator and receiver of all messages in the network is located at the end office. All messages are switched through the network using transfer points. These transfer points do not originate messages and are seldom the receivers of messages; they are used to through-switch the packets which are received from end offices.

Another function is to provide access to databases. The front-end system receives packets destined to an addressed database and converts from the SS7 protocol to either an X.25 protocol or some other transport protocol (such as TCP/IP) carrying as its payload the primitives which the database can read directly. This front end must be capable of receiving messages, routing to the appropriate database (based on an address), and maintaining reliable transfer of messages from the SS7 network into the database environment.

All nodes in the SS7 network are called signaling points. A signaling point has the ability to perform message discrimination (read the address and determine if the message is for that node), as well as to route SS7 messages to another signaling point. There are three different types of signaling points:

- Service Switching Point (SSP)
- Signal Transfer Point (STP)
- Service Control Point (SCP)

Signaling points provide access to the SS7 network, provide access to databases used by switches inside and outside of the network, and transfer SS7 messages to other signaling points within the network. Signaling points are deployed in pairs for redundancy and diversity. The secret to the SS7 network and to making sure the network is always operational is to provide alternate paths in the event of failures. These alternate paths provide the reliability needed in a network of this nature, and ensure that SS7 messages can always reach their destinations (Fig. 2.1).

The facilities which link the signaling points to one another are also deployed in pairs. Data links are used between signaling points to provide the speed necessary for SS7 message delivery (56 or 64 kbps, or 1.536 Mbps). These data links are typically DS0s from existing DS1/DS3 facilities used for interexchange trunking, for the exception of the 1.536-Mbps links. These are ATM facilities now being deployed for voice, data, and signaling transmission.

Only 56-kbps or 1.536-Mbps links are supported in the ANSI network here in the United States, due mostly to the fact that the infrastructure already in place will support them.

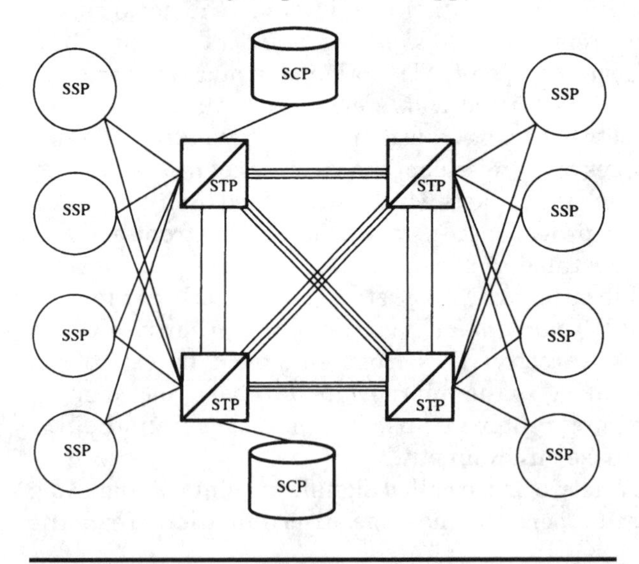

Figure 2.1 This figure depicts a typical SS7 network, with multiple B links and A links.

The network is deployed as two distinct levels, or *planes*. There is an international plane, using the ITU-TS standard of the SS7 protocol, and there is the national plane. The national plane uses whatever standard exists within the country in which it is deployed. For example, in the United States, ANSI is the standard for the national plane. Bellcore is an extension of the ANSI protocol, and ensures the reliability required to interwork with Bell Operating Companies' networks.

In other nations, there may be one or several different versions of national protocols for SS7. Yet all countries are capable of communicating with one another through gateways which convert the national version of the SS7 protocol to the international version of the SS7 protocol. This ensures that all nations can interwork with one another, while still addressing the requirements of their own distinct networks.

Using these two planes, communications are possible among all nations in the world, allowing automatic circuit connections from one network to the next. Yet when SS7 was first deployed here in the United States, it was used for a different purpose. The emphasis then was placed on database access. In fact, early implementation provided nothing more than just that: access to network databases. These databases could be accessed directly by end offices using data links connecting them directly to the database entity or to a signaling point (such as an STP) which would route their database query to the appropriate database (as is in today's network).

Even today, the SS7 network provides two types of services circuit: circuit-related and non-circuit-related. Circuit-related signaling is used for the setup and teardown of central office trunks. Non-circuit-related services are all other services provided by the network, such as database access for translations and subscriber information and network management.

This book explores the network as it is used here in the United States. International networks may have different objectives and, while similar, do have fundamental differences. However, an understanding of the requirements of Bellcore and ANSI will provide great insight to anyone working with SS7, regardless of the country and the version being used. The Bellcore standard provides an excellent model of what a data communications network should look like for high reliability.

Service Switching Point (SSP)

The Service Switching Point (SSP) is the local exchange in the telephone network. An SSP can be a combination voice switch and SS7 switch, or an adjunct computer connected to the local exchange's voice switch. The SSP provides the functionality of communicating with the voice switch via the use of primitives and creating the packets, or signal units, needed for transmission in the SS7 network (Fig. 2.2).

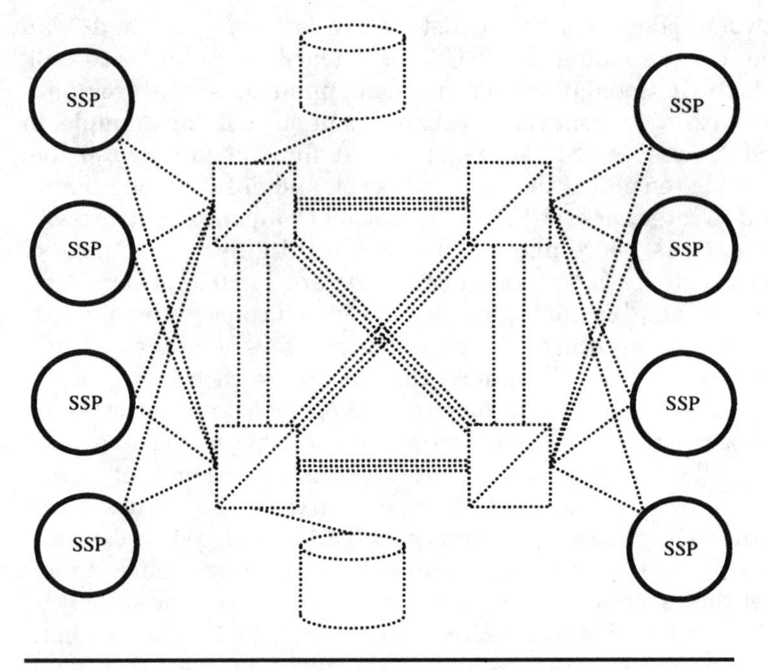

Figure 2.2 This figure shows the relationship of the SSP to the SS7 network.

The Service Switching Point (SSP) must convert signaling from the voice switch into SS7 signaling messages, which can then be sent to other exchanges through the SS7 network. The exchange will typically send messages related to its voice circuits to the exchanges with a direct connection to it.

In the case of database access, the SSP will be sending database queries through the SS7 network to computer systems located centrally to the network. This was the first usage of the SS7 network, as the need for 800 number lookup became necessary.

The traffic mix found in most SS7 networks is still primarily circuit-related messages. With the implementation of applications such as Local Number Portability (LNP), the traffic mix is changing significantly, becoming predominantly non-circuit-related messages. These messages originate from SSPs and are used to connect voice circuits from one exchange to another exchange. The SSP does not use circuit-related messages exclusively, however.

Before a switch can route a call, it must first be able to access information regarding the destination of the call. For most Plain Old Telephone Service (POTS) systems, the telephone number dialed is sufficient for routing. However, with 800 and 900 numbers, routing is impossible, because the dialed digits do not provide enough information about the destination.

Even POTS may now require a database query. If the switch determines that the called number has been "ported," it will generate a primitive to the SSP requesting the SSP send a query to an LNP database to determine to which exchange the ported number has been reassigned. Porting is the movement of a number from one switch to another (regardless of geography) allowing subscribers to move or change primary telephone companies without changing their telephone numbers.

For this reason, the Service Switching Point (SSP) must access a remote database to learn the routing number assigned to the 800 or 900 number, or the new location of a ported number. Once this information has been retrieved, the SSP can then begin circuit connections based on new the routing number information.

The SSP function is to use the information provided by the calling party (such as dialed digits) and determine how to connect the call. A routing table will identify which trunk circuit to use to connect the call, and which exchange this trunk terminates at. An SS7 message must be sent to this adjacent exchange requesting a circuit connection on the specified trunk.

The adjacent exchange grants permission to connect this trunk by sending back an acknowledgment to the originating exchange. Using the called party information in the setup message, the adjacent exchange can determine how to connect the call to its final destination. This may require several trunk connections between several adjacent exchanges. The SSP function manages these connections until the final destination is reached.

Many SSP functions are accomplished by adding a computer adjunct to existing switches. This computer receives signals from the voice switch which are used to trigger the transmission of specific SS7 messages. The called and calling party address must be passed from the voice switch to the SSP for transfer across the network.

Using adjuncts allows telephone companies to upgrade their SS7 signaling points without replacing expensive switches, providing a modular approach to networking. Upgrades are typically limited to software loads, since these computers require very little hardware.

There are very few features required of an SSP. The ability to send messages using the ISDN User Part (ISUP) protocol and the Transaction Capabilities Application Part (TCAP) protocol is the only requirement, other than the network management requirements defined in the Bellcore publications. Specific Bellcore requirements for an SSP can be found in Bellcore publication TR-TSY-000024, *Service Switching Points (SSPs) Generic Requirements* (this has since been updated, and the new document is known as GR-024-CORE).

Signal Transfer Point (STP)

All SS7 packets travel from one SSP to another through the services of a Signal Transfer Point (STP). The STP serves as a router in the SS7

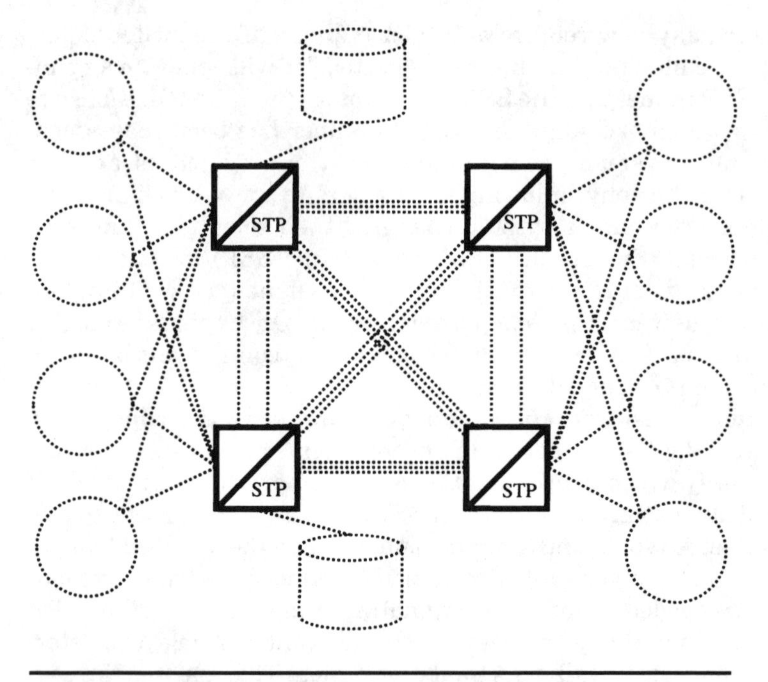

Figure 2.3 This figure shows the relationship of the STP to the SS7 network.

network. Messages are not usually originated by an STP. The STP switches SS7 messages as received from the various SSPs through the network to their appropriate destinations (Fig. 2.3).

The STP is also typically an adjunct to a voice switch. Many tandem switches provide the capability of voice switching through the switch and STP functionality through the use of an adjunct computer. There are very few manufacturers of these devices, and rarely is an STP a stand-alone system built for the sole purpose of STP functionality.

There are three levels of STPs:

- National Signal Transfer Point
- International Signal Transfer Point
- Gateway Signal Transfer Point

The national Signal Transfer Point exists within a national network and is capable of transferring messages using the same national standard of protocol. Messages may be passed to another level of STP, the international Signal Transfer Point, but the national STP has no capability of converting messages into another version or format. Protocol converters are often used to interconnect the national STP with an international STP, providing the conversion from a national standard (such as ANSI) to a international standard (ITU-TS).

The international Signal Transfer Point (STP) functions the same as the national STP, but is used in the international network. The international network provides interconnection of all countries using the SS7 protocol, using the ITU-TS standards. This ensures interconnectivity between worldwide networks, despite the use of different point code structures and network management. All nodes connecting to the international STP must use the ITU-TS protocol standard.

The gateway Signal Transfer Point provides protocol conversion from a national standard to the ITU-TS standard, or some other standard. A gateway STP is often used as an access to the international network, providing access and conversion of messages to the ITU-TS protocol standard. This eliminates the need for adjunct protocol converters in the network. Gateway STPs must be able to work using both the international standard of protocol and the national standard, depending on the location of the STP.

Gateway STPs are also finding their way into the cellular network. Cellular networks presently use X.25 as a transport protocol between their Mobile Switching Centers and databases. The X.25 networks are private networks and do not provide the support for accessing other cellular providers' networks.

The X.25 network has other limitations that do not lend themselves well to the applications of the cellular network. All information transfers in X.25 networks require connection-oriented services. When trying to connect to multiple networks and entities in various locations, this can become somewhat cumbersome. Many messages originated in the cellular network or the SS7 network do not require connection-oriented services.

For this reason, many cellular providers are slowly changing their networks to include an STP and data links for the use of SS7 protocols within the cellular network. The MSCs can then exchange information about the location of mobile phones and update their databases using the Transaction Capabilities Application Part (TCAP) protocol, which is much better suited to this task.

The gateway Signal Transfer Point (STP) serves as the interface into another network. Long distance service providers may have access into the local telephone company's database for subscriber information, or the local service provider may need access into the long distance service provider's database. In any event, this access is accomplished through a gateway STP.

Gateway STPs use screening features to maintain network security. Screening is the capability to examine all incoming and outgoing packets and allow only those which are authorized. This is determined through a series of gateway screening tables that must be configured by the service provider. Gateway screening also prevents messages from unstable networks that have not been approved by the service provider from entering into the network and causing service conflicts.

In the international network, gateway STPs may provide an additional function. International SS7 is based on ITU-TS standards, yet every country uses a national version that is not 100 percent compliant with the ITU-TS standard, For example, in the United States, we are ITU compliant, yet the ITU-TS standards have been modified for usage here. The major difference between the two standards is in addressing and network management functions of the protocol.

The gateway STPs that connect us to other countries do not deviate from the ITU-TS standards. They must be 100 percent compliant. They can perform a protocol conversion, allowing ITU-TS messages into the network, converting the messages into the national format before transferring them into the network. This is true gateway functionality, yet not all gateway STPs must perform protocol conversion.

Yet another feature of the Signal Transfer Point is measurements. There are many types of measurements defined by the Bellcore and ITU-TS standards. Measurements can be divided into two basic functions: traffic measurements and usage measurements.

Traffic measurements provide peg counts and statistical information regarding the type of messages entering and leaving the network. For maintenance purposes, network events are also recorded (such as link out-of-service duration, local processor outage, etc.). Maintenance measurements record events that may or may not affect service, but need to be monitored by service personnel. Because of the speed of the network and the quickness at which SS7 entities respond to problems, traffic measurements are the best way for maintenance personnel to keep track of what is happening in the network and preventing network failures from happening.

Usage measurements are always peg counts and record the number of messages by message type that enter and leave the network. These peg counts are aggregated by a collection process and stored on magnetic tape. The tape is then sent to a Regional Accounting Office (RAO) (Bell Operating Company networks) for processing. The RAO then creates an invoice for its customers, which are long distance service providers and independent telephone companies, charging them for access into the network.

This billing feature can be used to bill all users of the network and helps offset the expense of deploying the SS7 network. The cellular providers and PCS service providers have also been seeking database access from this network. Their access is also through a gateway Signal Transfer Point (STP), with usage measurements recording the access.

There has been some discussion about creating subscriber billing records from the SS7 network. By monitoring the SS7 messages sent through thge network, computers can determine when a call is being connected to a specific subscriber number, and determine the duration of that call. By using the SS7 network rather than the end office as the source for billing records, billing activities can be centralized within the network, providing better efficiency and cost savings to the telephone companies.

In the local SS7 network, the STP receives messages in packet form from the Service Switching Point (SSP). These packets are either related to call connections or database queries. If the packet is a request from an SSP to connect a call, the message must be forwarded to the destination end office where the call will be terminated. The destination is determined by the dialed digits.

If the message is a database query seeking additional information regarding a subscriber or an 800 number translation, the destination will be a database. Database access is provided through another SS7 entity, the Service Control Point (SCP—Fig. 2.4). The SCP does not store the information, but acts as an interface to the mainframe or minicomputer system that houses the requested information.

If the Service Switching Point (SSP) does not know the address of the destination Service Control Point (SCP), the Signal Transfer Point (STP) must provide the address. In this case, the SSP will send a database query to the local STP, with the destination address of the STP. The STP will look at the dialed digits (or *global title digits,* as they are called) and determine, through its own translation tables, the address of the database. This is referred to as *global title translation.* The global title translation provides the subsystem number (address) of the database and the point code of the SCP that interfaces to the database.

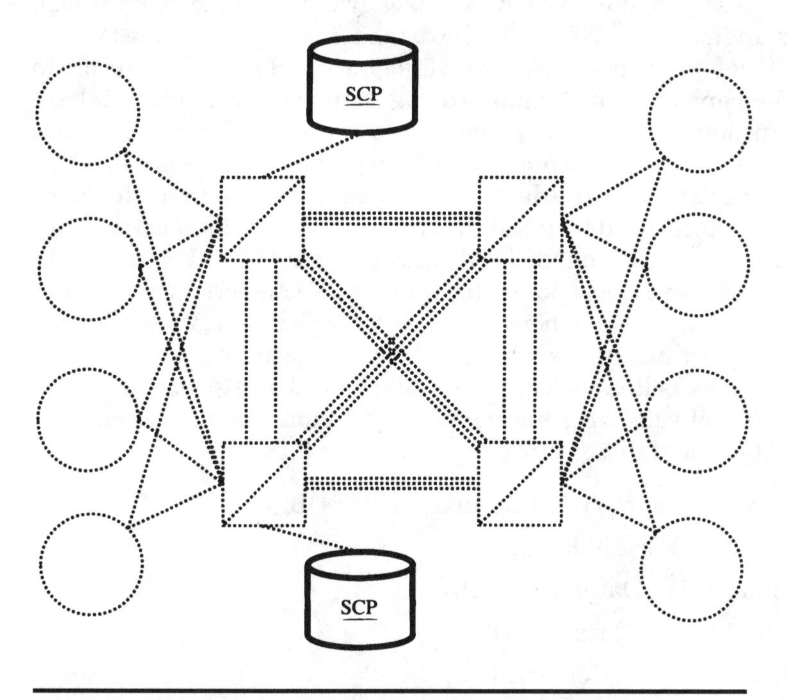

Figure 2.4 This figure shows the relationship of the SCP to the SS7 network.

The STP is the most versatile of all the SS7 entities, providing a wide array of services to the users of the network. Whether it be gateway services or routing functions, the STP is a major component in the network.

For specific information regarding Bellcore requirements for an STP, refer to the Bellcore publication GR-082-CORE, *Signal Transfer Point Generic Requirements.*

Service Control Point (SCP)

The Service Control Point (SCP) serves as an interface to telephone company databases. These databases are used to store information about subscribers' services, routing of special service numbers (such as 800 and 900 numbers), calling card validation and fraud protection, and even Advanced Intelligent Network (AIN) services used when creating a service for a subscriber.

The SCP is usually a computer used as a front end to the database system. Some new SCP applications are being implemented in STPs, providing an integrated solution. In all cases, the address of the SCP is a point code, while the address of the database is a subsystem number.

The SCP function does not necessarily store all the data, but is the interface to the mainframe or minicomputer system that is used for the actual database. These computer systems are usually linked to the SCP through X.25 links. In integrated STP/SCP, the database is resident in the SCP.

The SCP can perform protocol conversion from SS7 to X.25, or it can provide the capability of communicating with the computer database directly, through the use of *primitives.* A primitive is an interface which provides access from one level of the protocol to another level. In the case of the database, the database is considered an application entity, and the protocol used to access and interface to this application entity is the Transaction Capabilities Application Part (TCAP).

The type of database depends on the network. Each service provider has different requirements, and their databases will differ. Bellcore has defined some basic models of databases for achieving the needs of their networks. In addition to the Bellcore networks, cellular providers also use databases for the storage of subscriber information. The databases most commonly used within either of these networks are listed as follows:

- Call Management Services Database (CMSDB)
- Local Number Portability (LNP)
- Line Information Database (LIDB)
- Business Services Database (BSDB)
- Home Location Register (HLR)
- Visitor Location Register (VLR)

Each database contains information for a specific application. Each database is also given an address, called a *subsystem number,* for use in routing queries from Service Switching Points (SSPs) through the SS7 network to the actual database entity. The subsystem numbers are defined by the service provider and are fixed. The following databases are not absolutes; in other words, not every network must have these specific databases. These are the databases used within the Bell Operating Company's networks, and they are mentioned here as a model of database types.

Call Management Services Database (CMSDB)

The CMSDB provides information relating to call processing, network management, and call sampling (for traffic studies). The call-processing portion is what defines the routing instructions for special service numbers such as 800, 976, or 900 numbers. In addition to routing instructions, this database also provides billing information, such as billing address or third-party billing procedures.

CMSDB also provides certain network management functions used to prevent congestion in the network. When congestion occurs in the SS7 network, this database can provide important routing instructions for rerouting messages around the congested node.

Call sampling is used to create reports that indicate the types of telephone calls being made in the telephone network. These reports are then used in traffic studies to determine if additional facilities are needed to handle the voice traffic. The *Service Management System* (SMS) schedules reports for automatic printing and allows administration personnel to update the database records through a terminal interface.

Local Number Portability (LNP)

Local Number Portability is a new application mandated by the Communications Act of 1996. The purpose of LNP is to allow subscribers to change telephone companies without having to change their telephone numbers. This of course changes a lot of methods that have been used for years within the telephone network. The office code (NNX) portion of a telephone number can no longer be used to identify the destination exchange. The number may have been reassigned to a different exchange, with a different office code.

For this reason, when a call is initiated, the originating exchange must first search its routing table to determine if the called number has been flagged as "ported." If the NNX has been flagged as ported, then a query must be sent to determine if the actual called number has been ported, and if so, how the call is to be routed. Even if only one telephone number within an NNX has been ported, the entire NNX is considered ported, and a query must be generated for each and every call with that ported NNX.

The query is sent to an LNP database, where the called number is looked up. If the called number has been ported, then the database will identify the new terminating exchange by giving its Local Routing Number (LRN). The LRN works the same as the NNX code did, providing a unique identity for each and every exchange in the network. This information is then returned to the originating exchange so the call can be routed. If the telephone number has not been ported, then the call is routed as it normally would be.

The implementation of LNP is very new. There are still a lot of questions being asked about this new application. Many of you reading this may already be involved with LNP in some form or fashion. LNP introduced the biggest and costliest challenge to the telephone industry since equal access.

Line Information Database (LIDB)

The LIDB provides information regarding subscribers, such as calling card service, third-party billing instructions, and originating line number screening. Billing is the most important feature of this database. Third-party billing instructions, collect call service, and calling card service all determine how subscribers will be billed for their telephone calls in real time.

In addition to billing instructions, the LIDB also provides calling card validation, preventing fraudulent use of calling cards. The user's personal identification number (PIN) is stored in this database for comparison when a user places a call.

Originating line number screening provides information regarding custom calling features such as call forwarding and speed dialing. These features carry from network to network, as each service provider provides its own distinct calling services.

Business Services Database (BSDB)

This database is mentioned only as a model, since the last publication of Bellcore recommendations still has not defined the applications for this database. The purpose of this database is to allow subscribers to store call-processing instructions, network management procedures, and other data relevant only to their own private network. The telephone company would offer this database as an extra service, allowing large corporate customers to create their own private networks linking PBX equipment across the country. With their own proprietary databases, corporations can alter traffic routing by time of day or congestion modes without altering software in the PBX equipment.

This type of database would be a valuable asset to companies using the flexibility of the Intelligent Network to define their own services through *Service Creation Environments* (SCEs). The network management functionality can provide special routing instructions for calls destined to congested PBXs, a feature popular with inbound call centers using automatic call distributors (ACDs).

Home Location Register (HLR)

The Home Location Register (HLR) is found in cellular networks and is used to store information regarding a cellular subscriber. The HLR stores information regarding billing, as well as services allowed. In addition to these, the current location of the cellular phone is stored in the HLR for retrieval by Mobile Switching Centers.

When a cellular telephone is activated, a cell site receives a signal containing the *Mobile Identification Number* (MIN) and other identification. This information is received by an MSC, which must determine which Home Location Register (HLR) the MIN belongs to. This is accomplished in the same way that normal POTS telephone calls are connected. The MIN is equivalent to a POTS number.

Every few minutes, the cellular phone resends this signal. As the car moves from one cell site to another, the Mobile Switching Center (MSC) updates the location of the mobile in the HLR database. When a call is received into the network for a mobile telephone, the home MSC must determine which HLR to access to obtain the location of the cellular telephone. The HLR informs the MSC of the location, and the call is connected using voice circuits through the appropriate MSC to the cell site servicing the cellular subscriber at that time.

Visitor Location Register (VLR)

The Visitor Location Register (VLR) is used when a cellular telephone is not recognized by the local Mobile Switching Center (MSC). When subscribers roam outside of their "home" areas, the servicing MSC must keep track of their locations and be able to verify the validity of the Mobile Identification Number (MINs). This is done through accessing the Home Location Register (HLR), using the Transaction Capabilities Application Part (TCAP). The VLR is used to store the current locations for the visiting subscribers. The VLR communicates this information to the home HLR as well, allowing the HLR to keep track of the subscribers' locations.

The VLR may or may not be colocated with every MSC. In some networks, there is one VLR and one HLR. The only requirement is that all MSCs be able to access all HLR and VLR databases using the SS7 protocol.

For specific Bellcore requirements relating to Service Control Points (SCPs), refer to the Bellcore publication TR-NWT-001244, *Supplemental Service Control Point (SCP).*

Operations Support Systems (OSS)

In the United States, the Bell Operating Companies have established remote maintenance centers for the monitoring and management of their SS7 networks and voice networks (Fig. 2.5). With the capability of accessing all equipment today through digital interfaces, on-site personnel are no longer required.

The OSS serves many functions. Only those related to SS7 are discussed here. All signaling points within the network can be monitored for the remote maintenance center. The remote maintenance center uses large projection screens and terminals to display information (both textual and graphic) regarding the status of all signaling points and their data links.

Access to any of these signaling points and their components is accomplished through terminals. Bellcore has defined a standard set of commands for use in all of their network equipment, SS7 and otherwise, allowing their maintenance personnel to learn one set of commands for accessing all devices in the network. This eliminates the need for training on specific equipment.

The maintenance personnel in these offices may have to interact with many types of equipment. The equipment may include multiplex-

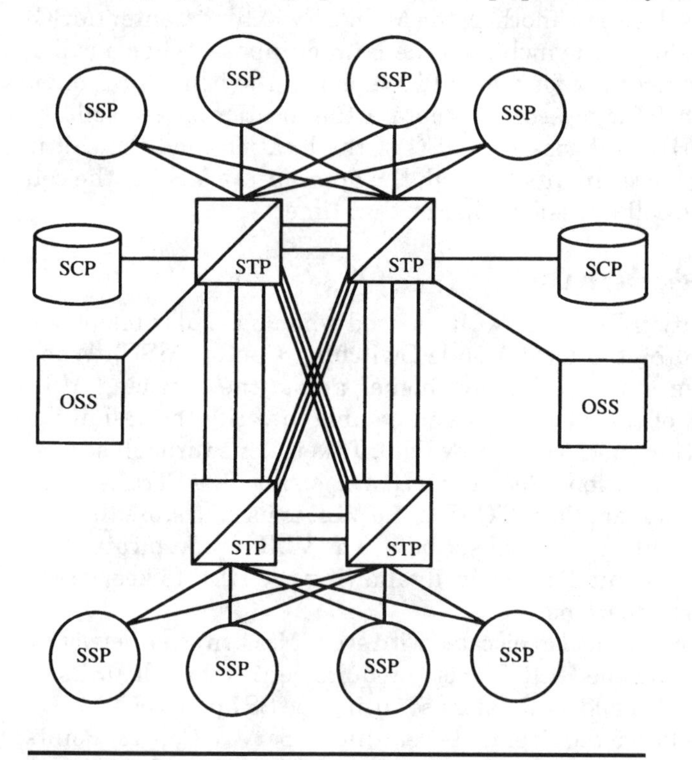

Figure 2.5 The OSS is typically located adjacent to an STP and accesses the network via a signaling link to the STP. Using TCAP and SCCP, the OSS is then capable of sending SS7 messages to any entity within its own network.

ers, digital cross-connects, and switches. Use of a common command set is almost paramount in this type of application.

To update the SCP databases and monitor the performance of the databases, a Service Management System (SMS) is used. The SMS is a standard interface consisting of a command set and graphical interface that can be used to administrate the database, monitor the status of the database, and retrieve measurements pertaining to performance from the database. The SMS also provides a central point for making updates to multiple databases. The database changes are made within the SMS, and then propagated to all the databases in the network. This eliminates the need to visit each and every database site to incorporate new changes and ensures consistent database updates.

Signaling Data Links

All SS7 signaling points are interconnected via signaling data links. These links are 56-kbps, 1.536-Mbps, and ATM data facilities in North America and 64-kbps data facilities in nearly every other portion of the world (the exception to this rule is Japan, where 4.8-kbps links are used). Links are bidirectional, using both a transmit and receive pair for simultaneous transmission in both directions.

There are three modes of signaling which can be used. These three modes depend on the relationship between the link and the entity it services. The simplest mode is referred to as *associated signaling*. This is the mode used when ATM links are implemented. In associated signaling, the link is directly parallel with the voice facility for which it is providing signaling. This, of course, is not the ideal, because it would require a signaling link from the end office to every other end office in the network. There do exist, however, some associated modes of signaling.

Nonassociated signaling uses a separate logical path from the actual voice, as seen in Fig. 2.6. There are usually multiple nodes involved to reach the final destination, while the voice may be a direct path to the destination. Nonassociated signaling is a common occurrence in many SS7 networks.

Quasi-associated signaling (Fig. 2.7) uses a minimal number of nodes to reach the final destination. This is the most favorable method of signaling, because each node introduces additional delay in signaling delivery. For this reason, SS7 networks favor quasi-associated signaling.

Signaling data links (Fig. 2.8) are labeled according to their function. There is no difference between the various links, only in the way the links are utilized during message transfer and how network management interacts with the links (Fig. 2.9).

Links are placed into groups, called *linksets*. All the links in a linkset must have the same adjacent node. The switching equipment will alternate transmission across all the links in a linkset to ensure equal usage of all facilities. Up to 16 links can be assigned to one linkset.

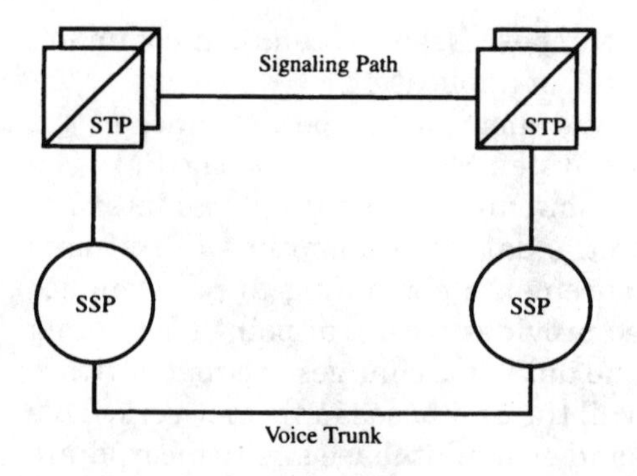

Figure 2.6 Nonassociated signaling involves the use of STPs to reach the remote exchange. As depicted in this figure, to establish a trunk connection between the two exchanges, signaling messages would be sent via SS7 and STPs to the adjacent exchange.

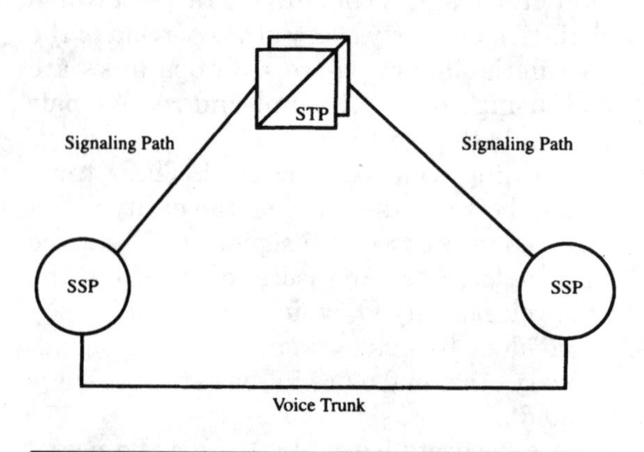

Figure 2.7 In quasi-associated signaling, both SSPs connect to the same STP. The signaling path is still through the STP to the adjacent SSP.

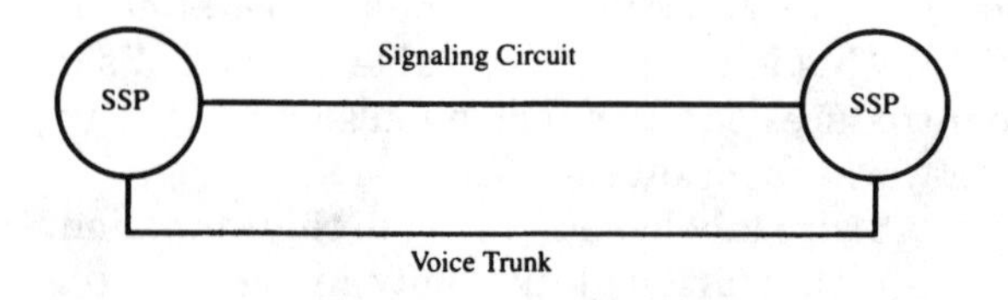

Figure 2.8 In some cases, it may be better to directly connect two SSPs via a signaling link. All SS7 messages related to circuits connecting the two exchanges are sent through this link. A connection is still provided to the home STP using other links to support all other SS7 traffic.

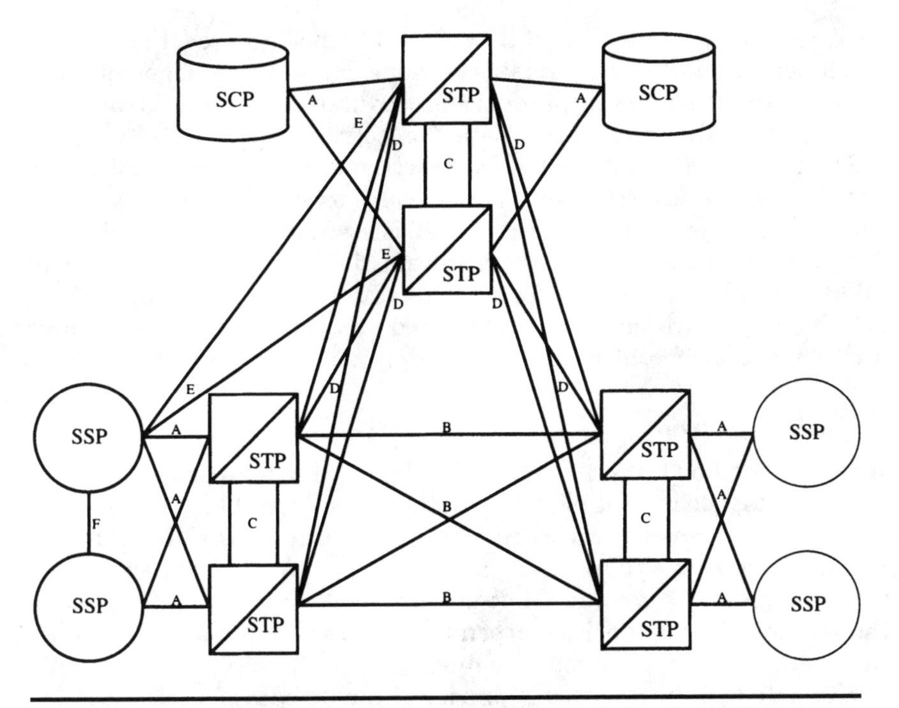

Figure 2.9 All signaling links are labeled according to their location in the network. There is no real physical difference between different links other than network management treatment.

Routes

In addition to linksets, a signaling point must define *routes.* A route is a collection of linksets used to reach a particular destination. A linkset can belong to more than one route. A collection of routes is known as a *routeset.*

A routeset is assigned to a destination. Routesets are necessary because, if only a single route existed and that route were to become unavailable, an alternate route would not be defined and no signaling could be sent to that destination. A routeset provides alternate routes to the same destination in the event that any one route becomes unavailable.

A *destination* is an address entered into the routing table of a remote signaling point. The destinations do not have to be directly adjacent to the signaling point, but they must be a point code which can be reached by the signaling point. A signaling point does not have to know all point codes in between itself and its destinations, only which link or linkset to use to reach its destination. A signaling point can have multiple addresses, if it is necessary to partition a signaling point into multiple functions. For example, a gateway STP used to enter the international network may have multiple point codes: one for the gateway function and another for global title translation services.

Links should always be terrestrial, although satellite links are supported in the standards. Satellite links are unfavorable because of the delay introduced. In the event that satellite links are used, the labeling and functions of the links remain the same. Network management procedures are the same except for the procedures used at level two of the protocol (link alignment, error detection/correction).

Satellite links use a different method of error detection/correction than terrestrial links. *Basic error detection/correction* is used for all terrestrial links, and *preventive cyclic retransmission* (PCR) is used for satellite links. The difference between the two lies in the retransmission mechanism. In basic error detection/correction, an indicator bit is used to indicate retransmission. In PCR, if an acknowledgment is not received for transmitted signaling units within a specific time, all unacknowledged signaling units are retranslated.

Link implementation

When a node has links to a mated STP pair, the links are assigned to two linksets, one linkset per node. Both linksets can then be configured as a combined linkset. A combined linkset contains links to both STP pairs, which means their adjacent signaling point address (point code) will be different. Combined linksets are used for load sharing, where the sending signaling point can send messages to both pairs, spreading the traffic load evenly across the links (Fig. 2.10).

Alternate linksets are used to provide alternate paths for messages. An alternate linkset or link is defined in the signaling points' routing tables and used when congestion conditions occur over the primary links. Figure 2.11 illustrates a typical configuration using alternate links to other nodes in the network, providing diversity in the event of node congestion.

Link performance

Links must remain available for SS7 traffic at all times, with minimal downtime. When a link fails, the other links within its linkset must take the traffic. Likewise, if an SS7 entity (such as an STP) should fail,

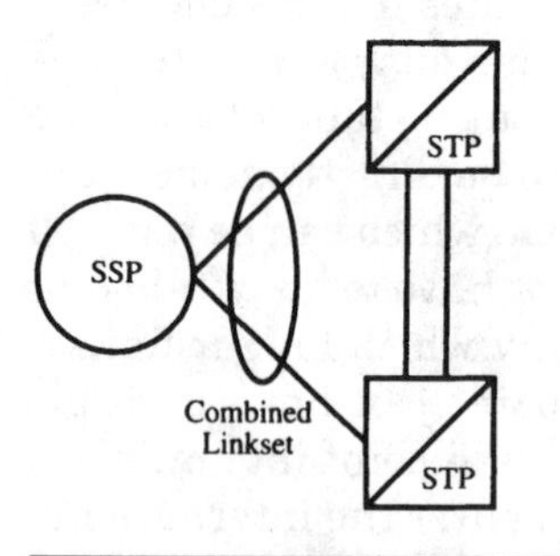

Figure 2.10 A combined linkset connects an adjacent pair. Each link in this figure connects to a different signaling point, but has the same destination.

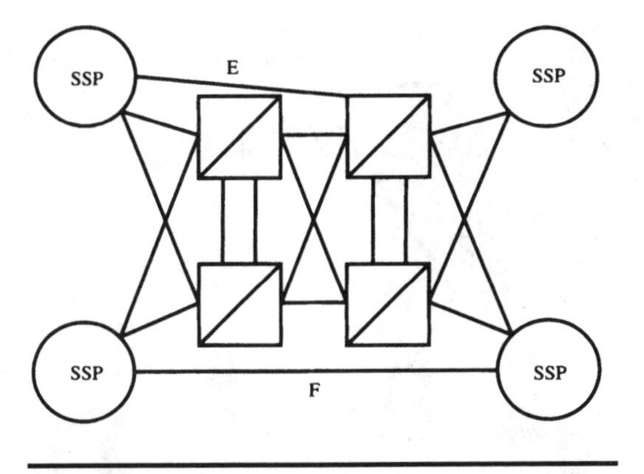

Figure 2.11 In this figure, the E and F links are alternate links and would be used when the primary links become unavailable or congested.

its mate must now assume the load. This means links can suddenly be burdened with more traffic than they can handle. For this reason, SS7 entities are designed to send less than 40 percent traffic on any link.

In the event of a failure, any one link may suddenly be responsible for the failed link's traffic. At 40 percent capacity, there is plenty of room for this traffic. Even at 80 percent capacity, the links still have enough capacity to carry SS7 network management messages in addition to the extra traffic.

You can calculate the number of messages a link will support by first determining the average length (in bytes) of each message. For example, if a DSO link is being used, the transmission rate of that link is 56 kbps. If you divide 56,000 bits into 1 byte (8 bits) you will see that a link can support 7000 bytes per second. Given that constant, you must now consider that the link is engineered to carry 40 percent traffic, which comes to 2800 bytes per second.

If the average message length is 40 bytes (as is the case with most ISUP), your link can carry 70 ISUP messages per second (2800/40=70). You can use this simple formula to calculate capacity of any link, which becomes important when you are sizing your network.

A maximum of 10 minutes downtime per year is allowed for any one linkset. This downtime relates to the ability to send SS7 messages to the destination using levels two and three of the protocol stack. These are stringent rules and are specified in the Bellcore publication GR-246-CORE.

There are six different types of links used in SS7:

- Access links (A)
- Bridge links (B)
- Cross links (C)
- Diagonal links (D)
- Extended links (E)
- Fully associated links (F)

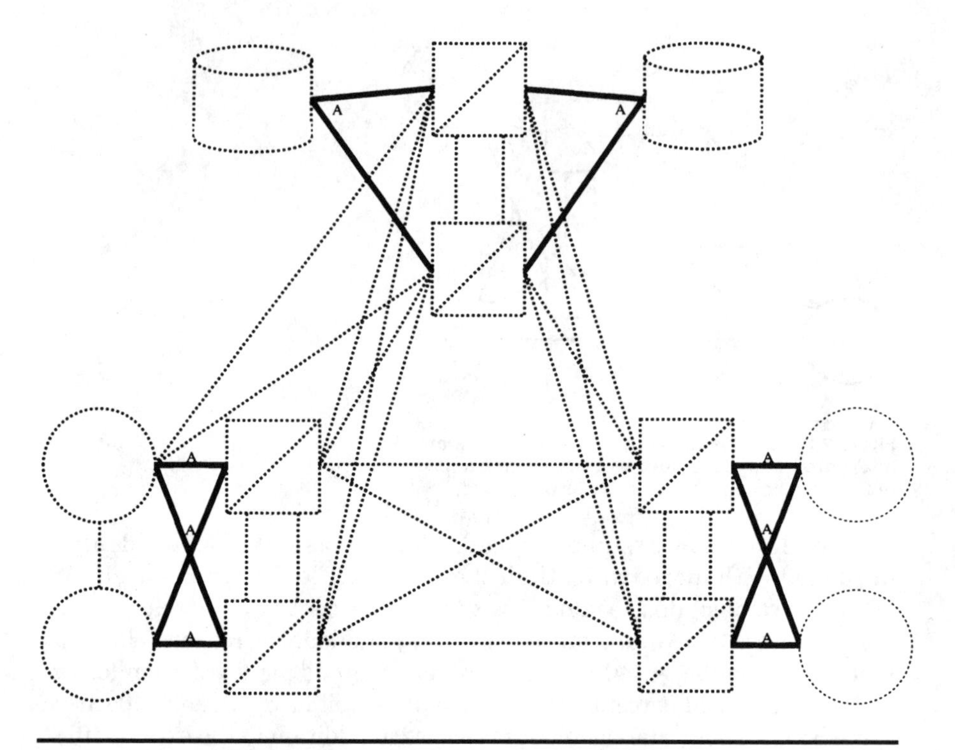

Figure 2.12 Access links connect end signaling points to the SS7 network.

Access links (A)

Access links (Fig. 2.12) are used between the SSP and the STP, or SCP and STP. These links provide access into the network and to databases through the STP. There are always at least two A links, one to each of the home STP pairs. In the event that STPs are not deployed in pairs, there can be one A link; however, this is highly unusual.

The maximum number of A links to any one STP is 16. A links can be configured in a combined linkset, with 16 links to each STP, providing 32 links to the mated pair.

Bridge links (B)

Bridge links are used to connect mated STPs to other mated STPs at the same hierarchical level. Bridge links are deployed in a quad fashion, as seen in Fig. 2.13. A maximum of eight B links can be deployed between mated STPs.

Cross links (C)

Cross links (Fig. 2.14) connect an STP to its mate STP. They are always deployed in pairs, to maintain redundancy in the network. Normal SS7

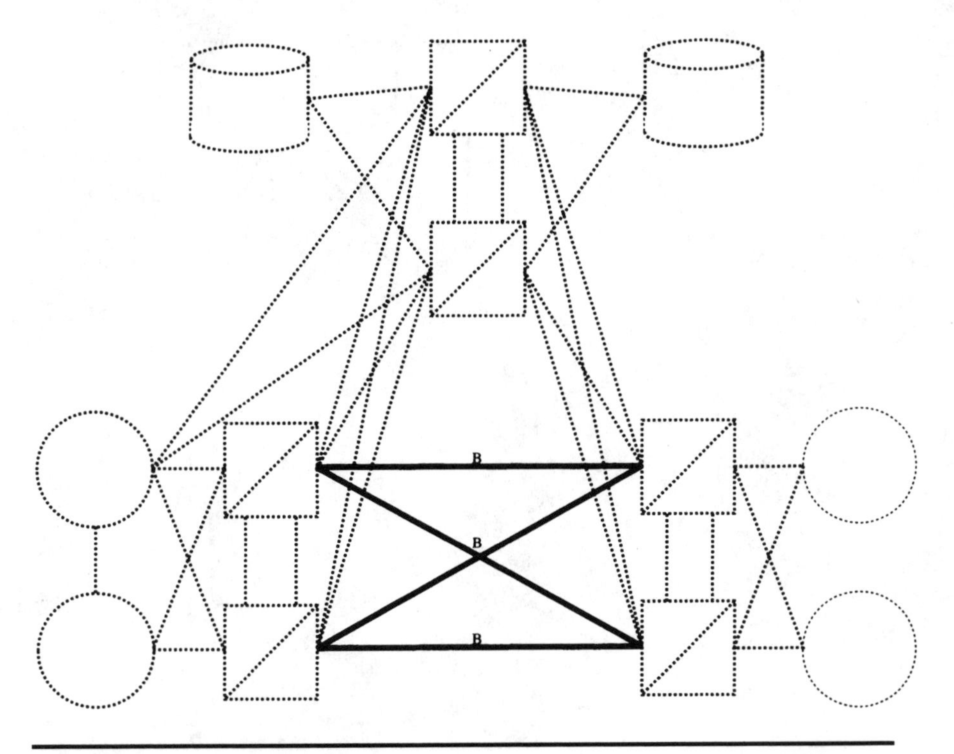

Figure 2.13 Bridge links connect a mated pair of STPs to another mated pair of STPs.

traffic is not routed over these links, except in congestion conditions. The only messages to travel between mated STPs during normal conditions are network management messages. If a node becomes isolated and the only available path is over the C links, then normal SS7 messages can be routed over these links. A maximum of eight C links can be deployed between STP pairs.

Diagonal links (D)

Diagonal links (Fig. 2.15) are used to connect mated STP pairs at a primary hierarchical level to another STP mated pair at a secondary hierarchical level. Not all networks deploy D links, since not all networks use a hierarchical network architecture. D links are deployed in a quad arrangement like B links. A maximum of eight D links may be used between two mated STP pairs.

Extended links (E)

Extended links (Fig. 2.16) are used to connect to remote STP pairs from an SSP. The SSP connects to its home STP pair but, for diversity, may be connected to a remote STP pair as well, using E links. Extended links then become the alternate route for SS7 messages in the event

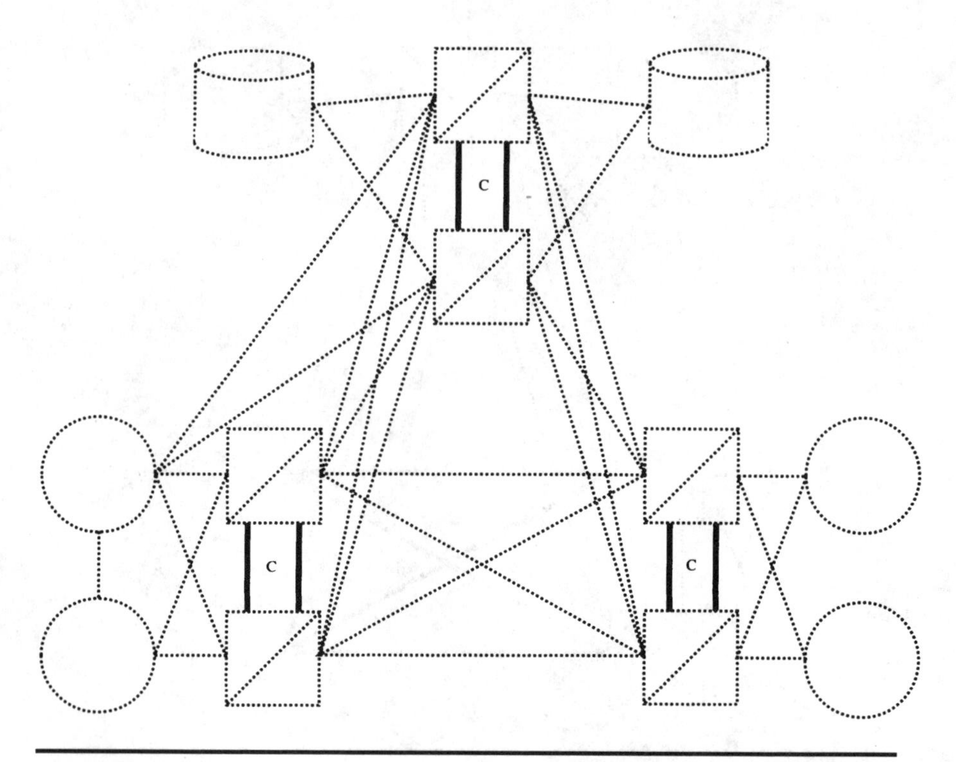

Figure 2.14 Cross links connect an STP to its mate STP, creating a mated pair. Mated pairs are identical in function and configuration, and have the ability of assuming the traffic of their mate in the event the mate fails. The cross links are used to share network management messages, and when no other route is available, to signal traffic.

that congestion should occur within the home STP pairs. A maximum of 16 E links may be used between any remote STP pairs.

Fully associated links (F)

Fully associated links (Fig. 2.17) are used when a large amount of traffic may exist between two SSPs, or when an SSP cannot be connected directly to an STP. F links allow SSPs to use the SS7 protocol and access SS7 databases even when it is not economical to provide a direct connection to an STP pair.

When traffic is particularly heavy between two end offices, the STP may be bypassed altogether, providing that both SSPs are local to each other. Only call setup and teardown procedures would be sent over this linkset.

Physical Link Interfaces

The signaling data links are connected to network equipment using electrical interfaces. These interfaces are industry-standard interfaces defined by standards bodies such as ITU-TS and EIA. The interface

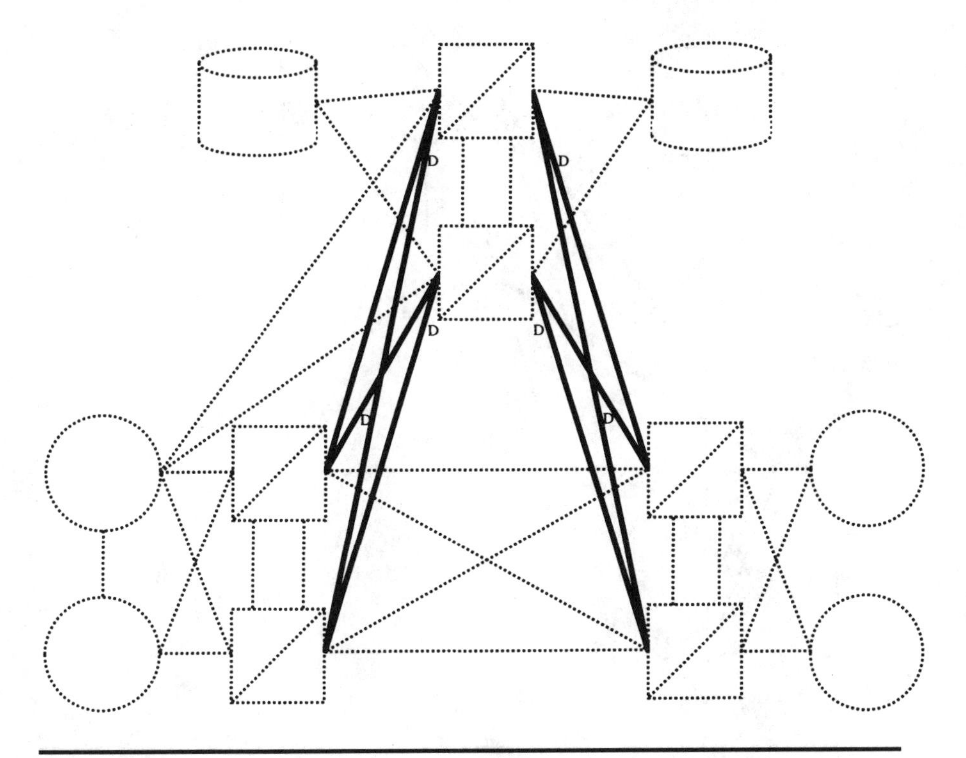

Figure 2.15 Diagonal links connect mated STPs to another mated pair of STPs that are deployed on a higher level in the network hierarchy. In this figure, the D links are used to connect to a pair of regional STPs, which provide access to a regionally located database.

type will depend on the type of equipment used with the links. For example, if a *data service unit* (DSU) is used, a V.35 interface will be needed to connect from the DSU to the signaling point.

Interfaces operate at level one of the SS7 protocol stack, and provide the electrical/optical medium for transmission of data packets within the SS7 network. Following are descriptions of the most commonly used interfaces in the SS7 network.

V.35

This interface is commonly used from a data service unit to the SS7 signaling point. The V.35 interface can also be used from a digital x-connect (DSX) panel. V.35 provides data rates up to 56 or 64 kbps. Slower data rates are supported as well.

The V.35 interface was originally intended for use with high-speed modems. Interfacing analog modems to a digital line at 48 kbps was the first implementation of this interface. Later, the ITU-TS adopted this interface for use in all digital lines, with data rates of 48, 56, 64, and 72 kbps.

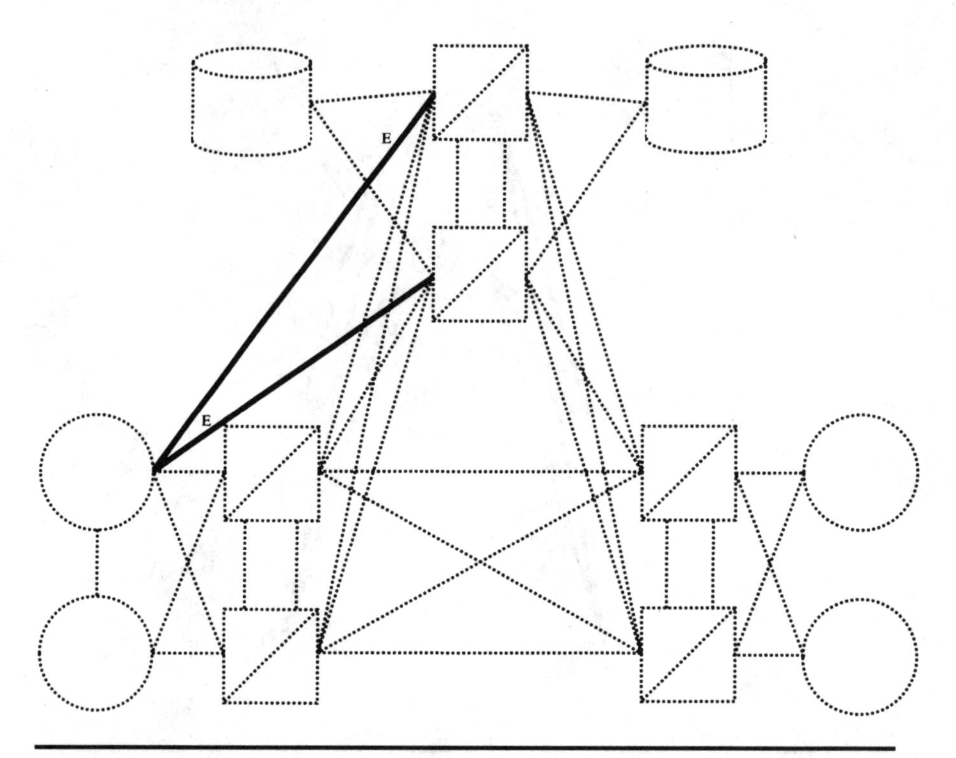

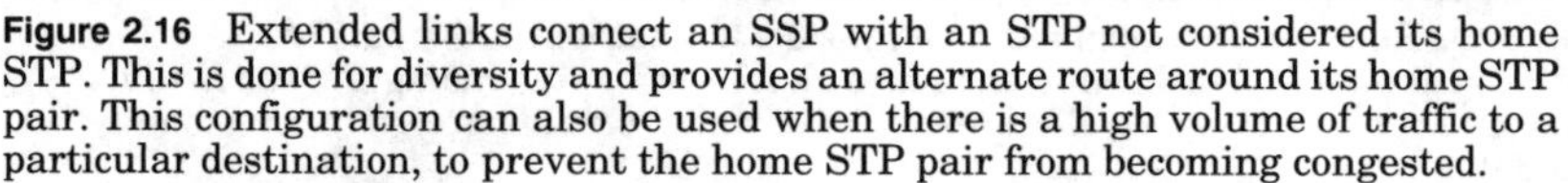

Figure 2.16 Extended links connect an SSP with an STP not considered its home STP. This is done for diversity and provides an alternate route around its home STP pair. This configuration can also be used when there is a high volume of traffic to a particular destination, to prevent the home STP pair from becoming congested.

The ITU-TS Blue Book considers this interface obsolete and no longer recommends its usage. Instead, the ITU-TS Blue Book recommends use of V.36 or other similar standards. The V.35 interface is still very common, however, and is found in many equipment types (Fig. 2.18).

When using a V.35 interface, a clock source must be provided. This clock source is typically provided by the switch itself, but can be provided from an external source. One side of the V.35 connection must be defined as *master* (clock source) and the other as *slave* (uses master clock). In the event the master side should fail, a mechanism should be provided to allow the slave to become master and provide its own clocking.

DS0A (digital signal 0)

This is a 56/64-kbps channel located in a DS1 (or higher) facility. The link can be carried with an existing DS1/DS3 circuit between offices, as long as one of the DS0As is dedicated to SS7 signaling. A DSU/CSU is required to terminate the DS1/DS3 and separate the various DS0As from the circuit.

The DSU/CSU is usually located close to the entrance point of the digital facility into the telephone company building. From there, the

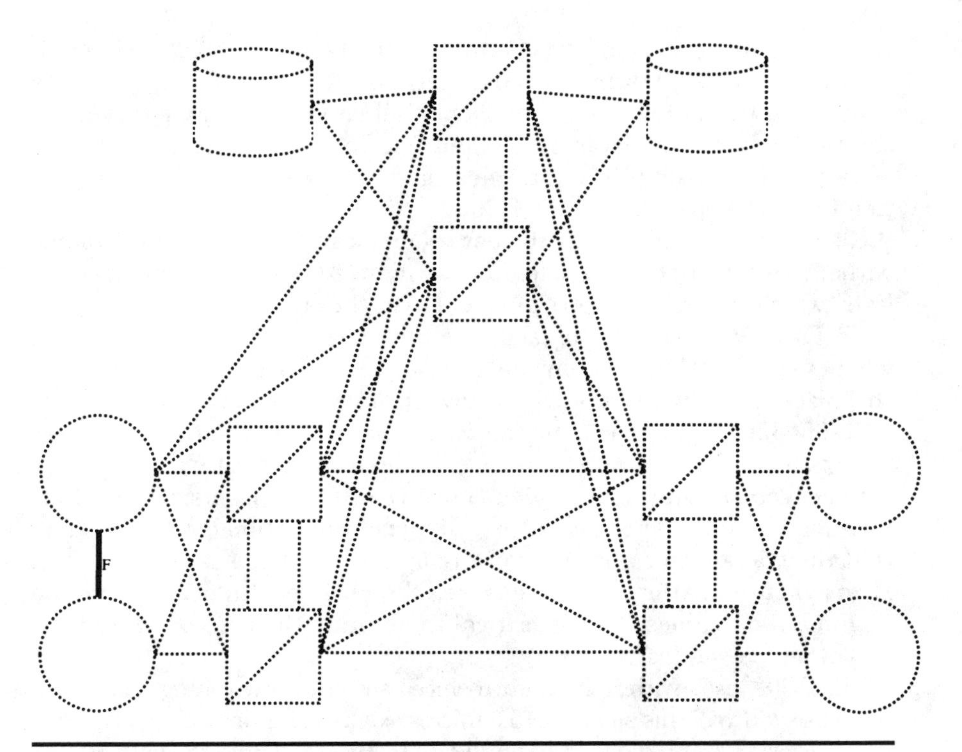

Figure 2.17 Fully associated links connect two SSPs directly, allowing signaling traffic to follow the same path (in parallel) as the voice circuits. This is commonly found when there is a large volume of signaling traffic (ISUP) between two exchanges. This configuration can also be used when there is no direct access to the SS7 network.

Pin 1 - Protective Ground
Pin 2 - Transmitted Data
Pin 3 - Received Data
Pin 4 - Request to Send
Pin 5 - Ready for Sending
Pin 6 - Data Set Ready
Pin 8 - Receive Line Signal Detect
Pin 15 - Tx Signal Element Timing
Pin 17 - Rx Signal Element Timing

Figure 2.18 The ITU V.35 interface may use a 37-pin or a 15-pin connector.

various DS0As are cross-connected through a digital cross-connect to their final destinations.

This is the most commonly used interface in U.S. SS7 networks. Signaling points in the network will usually have the ability to terminate the DS0A circuit without interface adapters. Maximum cable length for a DS0A is 1500 ft. This limitation is not a transmission limitation, but a consideration due to propagation delay. The Bellcore requirements also specify the nominal impedance as 135 Ω with a balanced transmission path in each direction.

In the future, DS0A with clear channel capability may be used. At the time of this publishing, this had not yet been written. However, work is being accomplished in the area of 64-kbps DS0s. The designation for a 64-kbps DS0 will be DS0C.

Clear channel means that the data source can transmit a full 64 kbps without any restrictions on ones density or all zeroes. Currently, the DS1 level, which carries the DS0 signal, enforces the ones-density rules.

Because DS0A is a digital facility, clocking is critical. For this reason, whenever a DS0A is used in the SS7 network, the DS0A must be synchronized according to the Digital Synchronization Network Plan (TA-NPL-000436, *Digital Synchronization Network Plan,* Issue 1: November 1986). Network synchronization is explained in the first chapter.

Synchronization is accomplished at two different levels: bit synchronization and byte synchronization. Bit synchronization is what ensures that the transmitter and receiver are operating at the same data rate. Byte synchronization is what ensures the receiver can properly define alignment of frames. This is critical in defining the beginning and the end of a received frame.

The DS0A used within the central office uses bipolar encoding. The nominal pulse width of this signal is 15.6 microseconds, with rise and fall times of 0.5 μs. Clocking is provided by the Building Integrated Timing Supply (BITS).

The DS0A interface provides reliable digital transfer of data. There are literally no configurable options to worry about in DS0A circuits, except for encoding schemes and data rates. The encoding method used will vary from network to network, depending on the type of multiplexer used.

There has been some discussion regarding a full DS1 interface to replace the DS0A. This would provide 1.544 Mbps on one link. This is not necessary today, but as the traffic mix changes and becomes more complex, this may become a requirement.

The major obstacle in DS0A links is timing. When a DS3 is used between two exchanges, four multiplexers must be used (end to end) before the DS0A signal can get to the signaling point. Whenever there are this many multiplexers, timing synchronization can become a problem.

When the timing on a link is not synchronized between any two multiplexers, the links cannot carry data properly, because the signaling point will not be able to read the data. Remember that timing is used to delineate between bits. If the synchronization between any two devices is not correct, it causes the receiving device to see bits that do not exist.

When using DS0A links, the most common problem encountered is related to losing the clock synchronization, which causes the links to be taken out of service. The fastest correction is to reset all the multiplexers and allow them to resynchronize with one another.

High-Speed Links

There are three interfaces defined to support ATM in the SS7 networks: SONET, DS1, and DS3. This provides a migratory path for telephone companies looking to deploy ATM in their signaling networks. The transmission rates vary.

DS1 supports transmission rates of 1.544 Mbps, while DS3 supports transmission rates of 44.736 Mbps. SONET supports much higher transmission rates. For SS7 signaling links, transmission rates of 51.840 Mbps, 155.520 Mbps, 622.080 Mbps. amd 2.48832 Gbps have been defined using the ATM SAAL protocol in place of the MTP levels 1 and 2. A subset of MTP level 3 is used to deliver BISUP messages over these links. This interface is still being defined. Refer to Bellcore document GR-1417-CORE for more details on the ATM interface.

There are two ways in which telephone companies can deploy ATM in their SS7 networks. Most companies will likely choose to use DS1 interfaces initially, because they already have these facilities available. However, as they migrate to SONET, the DS1 and even DS3 facilities will be replaced. SONET facilities will then be used to carry all traffic through the network.

SS7 messages will be included in this migration. The concept of ATM is to use one facility for all traffic. The SS7 traffic will then be carried with the voice and data traffic. Due to the inherent features of ATM routing, signaling traffic will be routed the same way voice and data are routed in ATM networks, using the Signaling ATM Adaptation Layer (SAAL) instead of MTP. Only a portion of MTP level 3 will be used for routing SS7 messages on ATM links.

Miscellaneous interfaces

The V.35 and DS0A are the most commonly used interfaces for connecting links to network nodes. There are other interfaces used for interconnecting adjunct equipment such as terminals and communications equipment. They are mentioned here so that the reader might understand their usage.

RS-232/V.24. The RS-232 interface is a serial interface designed originally for connecting modems to computer equipment. Over the years, the RS-232 has found many other uses as a serial interface. Printers, terminals, and any other device requiring a serial interface can use the RS-232.

The RS-232 provides a separate transmit and receive path, as well as flow control. However, there are limitations. The maximum cable distance for the RS-232 is 50 ft. In many cases, this is not a problem. But in the central office, this may be unacceptable, since terminal equipment and switches are often placed in different areas. In addition to the distance limitation, RS-232 has a maximum data rate of 19.2 kbps.

The signals provided by the RS-232 interface are divided into four categories: data signals, control signals, timing signals, and grounds. Data signals consist of the transmit and receive paths for the user data.

There are eight control signals provided by the RS-232. Not all of these control signals are needed, as some were developed for use with modems specifically. *Request to send, clear to send,* and *data set ready* are typically used with printers and terminals for flow control. *Data terminal ready, ring indicator, data carrier detector, data modulation detector,* and *speed selector* are all optional control signals used specifically for modems.

The mechanical requirements of the RS-232C call for a 25-pin connector (typically the DB-25 type). Most any connector fitting the application and the equipment can be used. The DB-25 is the most common connector used; however, many smaller connectors are being sought since equipment is getting smaller.

The electrical requirements of the RS-232C are specified as:

- Binary 1 = voltage more negative than −3 V
- Binary 0 = voltage more positive than +3 V
- Signal rate = <20 kbps
- Distance = <15 m

The functional requirements call for an unbalanced transmission path. One ground is provided as a return for both data leads. The other ground is a protective isolation ground. Synchronous transmission is accomplished by sending timing signals over the leads designated as *receiver signal element timing.*

An unbalanced line means that one lead is used to transmit the data, while the common ground is used as the return path. Interference can cause signals to be altered and, because the signal path is only over one lead, the voltage difference can be damaging to the data. In a balanced circuit, the data is sent over one lead with another lead used as the return for the same circuit. This means that current is carried in one direction (data flow) and returned on another lead, creating a complete circuit. Interference may occur, but it will not affect both leads. Thus, balanced circuits are better for data transmission over long distances. RS-232 is limited to short distances because of its use of unbalanced transmission circuits.

When a device is ready to send data, the *request to send* (RTS) lead goes high. The receiving device acknowledges RTS and raises the *clear to send* (CTS) lead high. These are the minimal signaling requirements of the RS-232C. There are many other signaling and control leads which can be incorporated, but most manufacturers find that RTS and CTS are all that are needed. Modems require most of the leads indicated for reliable transmission.

In a modem configuration, connections are established in a different fashion. Modems are most often used by maintenance personnel wishing to connect to a remote SS7 signaling point from their computer workstation. They can then perform maintenance and administration tasks from their location. To understand the sequences that take place when a modem is concerned, let's look in more detail at the steps involved.

When the craftsperson is ready to connect to the modem, he or she will choose a communications software application on the workstation. The application will perform the steps necessary to connect to the modem. The interface will go through various stages before transmission actually begins. When the computer is ready to transmit, the *data terminal ready* (DTR) lead from the workstation will go high.

The modem has now been alerted that the workstation wants to place a call. The workstation will need to send the telephone number of the signaling point to the modem. This can be accomplished over the *transmitted data* pins of the interface. The modem then dials the number over the analog telephone line. Most modems will have a speaker incorporated into the modem, so the user can actually hear the dialtone as the modem goes off-hook and begins dialing the number. This can be extremely helpful when trouble is encountered, because you can hear if the call actually went through or not.

When the line begins ringing, the distant modem should detect *ring generator* on the line. When it detects ring generator, the distant modem will raise the *ring indicator* lead to high to alert the signaling point that there is an incoming call.

The signaling point should then raise the Data Terminal Ready (DTR) lead high to indicate it is ready to receive data. This indicates to the modem that it should answer (go off-hook) and establish a connection with the distant modem.

The distant modem then answers the line and places a carrier signal to the calling modem. At the same time, the called modem raises the Data Set Ready (DSR) lead high to indicate to the signaling point that it has answered the incoming call and it is ready to communicate.

The calling modem then sets Data Set Ready (DSR) high to indicate to the workstation that a connection has been established and it is ready to transmit data. The calling modem will return a carrier signal to the called modem so that full-duplex transmission can be established. The called modem will set the *carrier detect* (CD) lead high to indicate receipt of a carrier signal.

The workstation then sets Request to Send (RTS) high, indicating it is ready to transmit data. The modem responds and sets Clear to Send (CTS) high. This is an acknowledgment to the workstation. The workstation then begins transmitting the data in serial fashion (1 bit at a time) over the *transmitted data* (TD) lead, using the carrier signal to send the data. The carrier signal is modulated (using any means of modulation) to represent the bit stream in an audible tone.

The called modem receives the modulated data, demodulates it, and sends it to the signaling point over the *received data* (RD) lead. When the transmission is complete, either modem can drop the connection. The Request to

Send (RTS) lead is set low (off), which causes the called modem to set Clear to Send (CTS) low. The carrier is dropped and the connection is released.

This entire procedure can be monitored at any end of the circuit by using a breakout box or RS-232 monitoring device (a simple device with a series of LEDs for each circuit). Whenever a modem is connected to a signaling point or any other communications equipment in the network, it is strongly suggested that such a tool be kept in the toolkit for troubleshooting the connection. Most modem troubles can be detected and isolated through the use of such a tool.

V.24 is the ITU version, after which the RS-232 was designed. The V.24 interface provides the same functions as the RS-232 interface, plus additional signals for automatic calling. In addition to the RS-232 signals, the V.24 provides many more signals for timing, control, and data transmission.

RS-449. The RS-449 interface was intended to replace the RS-232, providing increased distance and higher data rates. However, manufacturers have been using the RS-232 for so long, and there is already a large number of these interfaces in use today, that the RS-449 has not shared widespread acceptance.

In the PC market, there really is no incentive for the RS-449, because distance is not a problem. Devices using a serial interface are typically found right next to the computer. But where distance and speed are issues, the RS-449 is a better interface.

RS-449 supports distances up to 200 ft with a data rate of 2 Mbps. In addition to the enhanced performance, the RS-449 interface provides 37 basic circuits and 10 additional circuits to support loopback testing and other maintenance functions.

The mechanical requirements specified for the RS-449 call for a 37-pin connector for the basic interface, and a separate 9-pin connector if the secondary channel is used. The electrical requirements show a significant improvement over RS-232 interfaces, which are limited to an unbalanced line.

The RS-423-A standard specifies an unbalanced mode for the RS-449 interface, while the RS-422-A standard specifies the balanced mode. In a balanced mode, the electrical characteristics are:

- 100 kbps at 1200 m
- 10 Mbps at 12 m

In an unbalanced mode, the performance is not as good, but is still an improvement over the RS-232 standard. In unbalanced mode, the performance rating is:

- 3 kbps at 1000 m
- 300 kbps at 10 m

Chapter

3

Overview of a Protocol

Before looking at the OSI model, let us examine the functions of a protocol. A protocol is a set of rules governing the way data will be transmitted and received over data communications networks. Protocols must provide reliable, error-free transmission of user data, as well as network management functions. Protocols *packetize* the user data into data *envelopes,* some being of a fixed length while others can be variable lengths, depending on the protocol used.

Protocols are used whenever a serial bit stream is used. The protocol defines the order in which the bits will be sent and also appends information for use by the network in routing and management of the network. This appended information is used only by the protocol and is transparent to the user.

Some protocols, such as SS7, actually send predefined messages to the other nodes in the network. Messages can be used at any layer above layer one and are commonly found at layers two and three. A typical example of a protocol message is the *initial address message* (IAM) sent by the SS7 protocol to establish a connection on a voice circuit between two end offices. Other messages exist for SS7 and will be discussed in greater detail in the other chapters. Predefined messages are an excellent way to send network management functions and handle data error procedures.

Other functions of a protocol include the segmentation of blocks of data for easier transmission over the network and reassembly at the receiving node. When sending multiple blocks of associated data, procedures must be provided that allow the blocks to be identified in the order they were sent and reassembled as such. In large networks, these data blocks can be sent in order but received out of order.

There are three basic modes of operation for a protocol, depending on the type of network. A circuit-switched network protocol establishes a connection on a specific circuit, and then sends the data on that circuit. The circuit used depends on the destination of the data. A good example of a typical circuit-switched network is the Public Switched Telephone Network (PSTN), which uses various circuits for the transmission of voice from one exchange to another.

Once the transmission has been completed, the circuit is released and is ready to carry another transmission. The protocol must manage the connection and release when transmission has been completed, and must also maintain the connection during the data transmission.

Another type of network is a local area network (LAN). LANs use different types of protocols, but the method of transmission is usually very similar. The topology of a LAN is usually a bus topology or a ring topology. In both topologies, the data is transmitted out on the LAN, with an address attached in a protocol header.

When a data terminal recognizes its address, it reads the data. Some mechanism must be used within the protocol to remove the data from the LAN once it has been read. This differs from one protocol to the next. These types of networks only allow one message at a time to be transmitted across the LAN.

Packet-switching networks provide multiple paths to the same destination. Each message has both an originating address and a destination address. The addresses are used to route the message through the network. Unlike LANs, a packet-switched network allows many messages to be transmitted simultaneously across the network.

The circuits used for this type of network are always connected, and transmission is taking place continuously. The direction the message takes from one node to the next depends on the packet address. Each packet provides enough information regarding the data to allow the packet to reach its destination without establishing a connection between the two devices. The X.25 and SS7 networks are both packet-switched networks.

In any protocol stack, there are several layers of addressing used. Typically, at least three layers of addressing can be found. Each device on the network must have its own unique physical address. The node address identifies the particular device within its own network. The layer two protocols are users of this address, since they are responsible for the routing from one device to the next adjacent device.

The next address layer is that of the network itself. This address is used when sending messages between two networks. This address can usually be found in layer three of most protocols. The network address is used by those devices which interconnect two or more networks (such as a router).

Once a message reaches its final destination, the logical address within the destination node must be provided to identify which operation or application entity within the node is to receive the data. An application entity is a function within a network node, such as file transfer or electronic mail. Application does not imply something like word processing (in the network sense).

In the SS7 network, application entities are objects such as IS-41, which allows Mobile Switching Centers in the cellular network to exchange data from one to the other, using the services of SS7 protocols.

As the information is handed from one layer to the next, the protocol appends control information. This control information is used to ensure that the data is received in the same order it was sent, and allows the protocol to monitor the status of every connection and automatically correct problems that may occur.

Control information includes sequence numbering and flow control. This function is usually found at layers two through four, but can also be found in the higher layers. In the SS7 protocols, levels two through four provide varying levels of control.

As mentioned earlier, segmentation and reassembly are also tasks of the protocol. This is necessary when large blocks of data must be transmitted across the network. Large blocks of data can be time consuming and, if an error occurs during transmission, can cause congestion on the network while retransmitting.

For this reason, blocks are broken down into smaller chunks, which make it faster and easier to control and transmit through the network. When a retransmission becomes necessary, only a small portion of the original data must be retransmitted, saving valuable network resources.

Encapsulation is the process of appending the original data with additional control information and protocol headers. This information is stripped off the message by the receiving node at the same layer it was appended. This information is transparent to the user.

Connection control is one of the most important tasks of a protocol. Connections must usually be established not only between two devices, but between two application entities as well. These logical connections must be maintained throughout the data transmission. The establishment of a logical connection ensures reliable data transfer, with the use of positive and negative acknowledgments to advise the adjacent node of transmission status.

Sequence numbering is also used in these types of services to ensure that the data is received in the same order it was transmitted. This type of protocol service is referred to as *connection-oriented*. Each node may have multiple logical connections established at one time.

When the data transmission is complete, the logical connection must be released to allow another application entity to establish a connection and transmit data. Protocol messages (such as connect requests and disconnects) are used to manage these logical connections.

Connectionless services are supported in many protocols. Connectionless services allow data to be transmitted without establishing a logical connection between two application entities. The data is simply transmitted with enough information to allow the receiver to know how to process the data.

Sequence numbering and retransmission are not used with connectionless services. This type of service is not reliable and is typically found in applications such as electronic mail.

The SS7 network provides support for both types of services, but uses mostly connectionless for data transfer. However, despite its use of connectionless services, the protocol in SS7 provides mechanisms that allow emulation of connection-oriented services.

Flow control is used in most protocols to control the flow of messages to a particular node. This function is particularly important in SS7 networks, because it is used to prevent congestion in any one signaling point. With flow control, protocol messages can be used to alert adjacent nodes of the congestion situation and to invoke rerouting functions.

Stopping the flow of messages to any one node is also necessary in some cases when a node becomes unavailable and is unable to process messages. The protocols in SS7 are able to perform this task without human intervention. Often, congestion or outages can occur and routing can be changed without anyone even knowing what occurred until after the events have taken place and the problem has been resolved.

Error detection and correction are ways for protocols to determine if the data it is carrying has been corrupted. The methods for error detection vary, but they almost always rely on some technique such as a *cyclic redundancy check* (CRC). This runs an equation on the bit stream before it is transmitted and places the sum into a check sum field.

When the data is received by the distant node, the same equation is run on the data again. The receiving node then checks the sum and compares it against the sum in the check sum field. If they match, no error occurred. If they do not match, then an error occurred and the packet or data is discarded.

Overview of the OSI Model

The OSI model was developed and published in 1982 by the International Standards Organization (ISO) for use in mainframe environments. This protocol provides the procedures and mechanisms necessary for mainframe computers to communicate with other devices, including terminals and modems. Since the OSI model was developed after SS7, there will obviously be some discrepancies between the two protocols. Yet the functions and processes outlined in the OSI model were already in practice when SS7 was developed (such as layering protocol functions).

The OSI model divides data transmission into three distinct functions. There is the application itself, which is not included in any of these three functions. The application, or process, may be something like file transfer or electronic mail. The process is the user of the protocol, and will be the entity transmitting data over the network.

The process will depend on a service or function within the protocol that will allow it to pass its data to the network for transmission. Before this can happen, information must be appended, and certain tasks must be performed first. These tasks are the responsibility of the process layers.

The process layers use protocols which are unique to the application which uses them. The application has specific requirements of the protocol, yet the protocols used at the lower layers need not be concerned with any of these functions, so they remain independent and transparent to the process layers.

The process layers interface to the transport layers, which provide the mechanisms necessary to reliably transfer data over the network. The transport layers include error detection and correction, as well as other tasks such as sequencing of the individual segments.

Process layers are not dependent upon any particular network protocol. In fact, a successful protocol should be able to use the services of any network protocol. This is the main objective behind the OSI model. The various layers should be independent of one another and should be able to use any protocol over the network.

The network protocol provides the mechanisms for actually routing the data over the network and getting it to its destination node. The network protocol has no knowledge of the process addresses and does not work with any of the transport information. Its only concern is moving the packet from one node to another node within the network.

Routing is accomplished by reading the device address and the network address. This is the only information needed by the network protocol. Some sequencing may also be used, but this is not to be confused with the sequencing used by the process layers. The sequencing at this layer is simply used to ensure that all the packets that were transmitted were indeed received. No order is necessarily implied by the numbering.

The network protocols are usually divided into two parts: node-to-node transfer and network-to-network routing. Node-to-node transfer is only concerned with the transmission of a data packet between two physical entities. This takes place over a physical connection between the two entities, which ensures that the sequenced data is received in the proper order.

The network-to-network part is concerned with the routing of information between two networks. This layer typically uses the network addressing and is not concerned with the device address in many cases. In fact, with many LAN protocols, the network layer does not know the device address, since it is located in a different layer of the protocol header.

This layered approach provides specific functions and is used for specific applications. By using a layered approach, changes to the protocol do not affect all the layers. This is important to network users. Network equipment works at specific layers, rather than at all layers. If a change is made to the protocol, the equipment needs to be changed only if the change affects the layer at which the equipment operates.

As we look at the OSI model, you will begin to see how the tasks assigned at each layer can easily be independent from the other layers. Begin thinking of simple devices used in networks (such as routers and bridges) and what their functions are in the network, and then match these devices to the layer at which they operate.

The OSI model addresses all the functions previously mentioned and divides the functions into seven different layers. Each layer provides a service to the layer above and below it. For example, the physical layer provides a service to the data link layer. The data link layer provides services to the network layer. Yet each layer is independent, and should the function change at any one layer, it should not impact the other layers.

The OSI model defines the following seven layers, as shown in Fig. 3.1:

- Application (layer 7)
- Presentation (layer 6)
- Session (layer 5)
- Transport (layer 4)
- Network (layer 3)
- Data link (layer 2)
- Physical (layer 1)

We will look first at the bottom layer, the physical layer.

Physical layer

The physical layer is the layer responsible for converting the digital data into a bit stream for transmission over the network. The physical layer must provide the electrical characteristics needed to transmit over the interface being used. Conversion of the digital signal from elec-

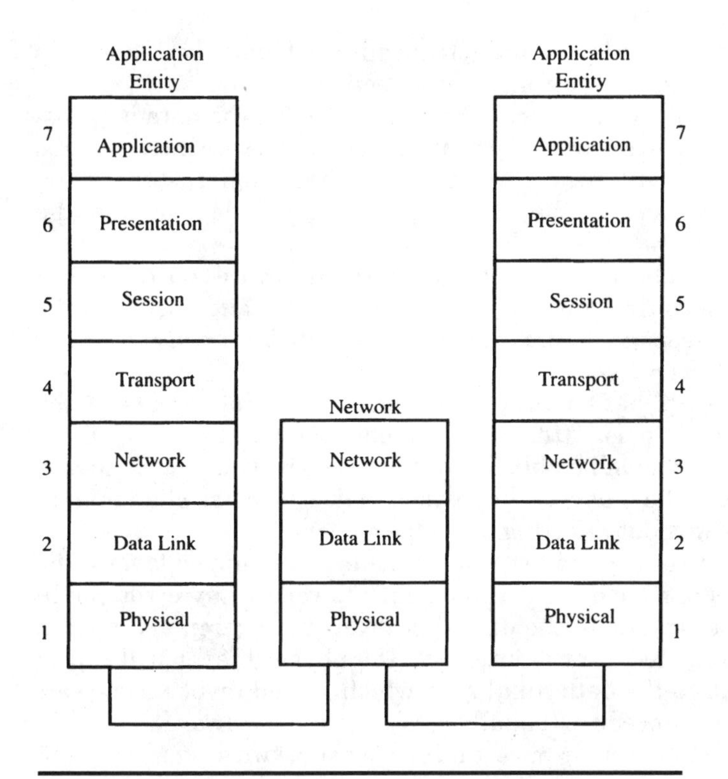

Figure 3.1 The OSI model defines seven layers of functions. The first three layers are primarily network functions, used to transport information from source to destination.

trical to audible (as in the case of a modem) and even light (as in fiber optics) is the responsibility of the physical layer.

The responsibilities of the physical layer can be divided into several tasks. The most basic of these tasks is mechanical. The interface itself is a mechanical connection from the device to the physical medium which will be used to actually transmit the digital bit stream.

The mechanical specifications are dependent on the standard used. An RS-232 interface may use a DB-25 type of connection, while a V.35 will use an AMP connector. The mechanical specifications do not specify the electrical characteristics of the interface. The electrical characteristics are called out separately and are not dependent on the connector.

The electrical characteristics of an interface will depend on the medium being used and the type of interface. Other factors, such as distance and the type of signals being transmitted, also play an important part when choosing a standard. The electrical properties of an interface include the signals used to actually transmit the bit stream, as well as control signals used to maintain the connection.

In the case of interfaces such as RS-232 and V.35, the data is transmitted on separate wires from the control signals. These control signals have nothing to do with the control information found in the upper layers of the protocol. In fact, they, too, are completely independent of the upper layers.

The control signals at the physical layer are hardware controlled and are used for flow control and for maintaining a connection between the two devices. These signals may include *data terminal ready* (DTR) or *request to send* (RTS).

The upper layers have no knowledge of the control signals at the physical layer and do not attempt to influence their status at any time. The physical layer must be able to work on its own. This independence allows changes to be made to the physical layer (such as changing an interface type) without affecting the upper layers.

While the control signals are independent of the upper layers, the physical layer does have the responsibility to report any error conditions or line loss events to the data link layer. For example, if a clock signal is suddenly lost on the interface, the physical layer will report the loss of clock to the data link layer, which should invoke some sort of error recovery procedure (usually a reset of the hardware).

Electrical signals in their purest form consist of two states: on or off. To be more specific, an interface uses two voltage levels to represent binary digits. These levels remain constant until the binary digit changes to an opposite value. This is known as *nonreturn to zero* (NRZ). As long as there is a binary 1 being represented, the voltage level remains high. When there is a change in the bit stream and a binary zero occurs, the voltage level changes to low.

There are two terms used to describe these transitions. The bit rate is the number of actual bits that can be transmitted over a line in one second. This is significantly different from the baud rate. The baud rate has nothing to do with the digital connection. The baud is a measurement of analog transitions that occur when using a device such as a modem. When using a modem, the digital signals are converted into analog audible tones which are then transmitted at varying frequencies over the telephone line. Each time there is a frequency change, this is known as a baud. Understanding this fundamental difference between these two terms is important.

Many professionals mistake the two terms as being the same. Actually, it is quite possible to have a baud rate higher than a bit rate. When speaking of all digital facilities, where modems are not used, the proper term is bit rate. The baud rate is measured after the digital interface, on the analog side of the line.

Line encoding is the process of altering the bit stream to force more transitions or fewer transitions, depending on the method of line encoding used. There are two reasons line encoding becomes necessary.

Digital signals cannot travel long distances and maintain their voltage levels. After some distance, the voltage level begins to drop. This can be critical when there are long series of consecutive 1s. To correct this problem, additional power can be applied to increase the wattage, which provides more push to the signal. This, too, is unfavorable, because it requires larger power supplies, which give off more heat.

The optimum solution is to be able to use low power and maintain communications over long distances. To accomplish this there are many types of line encoding. The simplest method is called *alternate mark inversion* (AMI). With AMI, every occurrence of a binary 1 causes the voltage level to change. For example, if the voltage level is at a positive level and there is another binary 1 in the bit stream, then the signal changes to a negative value. The use of negative voltages allows lower power requirements to be met.

There are many other methods used at the physical layer to overcome problems that occur when transmitting digital signals over long and short distances. None of these techniques is defined in the OSI model. The OSI model simply defines the processes that must take place, the various functions of the physical layer, the mechanical and electrical descriptions, and the services provided by the interface.

Data link layer

The OSI model defines the data link layer as the means to provide reliable communications between two devices. It is important to understand that the data link layer is only concerned with the data transmission between two devices, and not the entire network. Communications through a network are handled at a higher layer.

The data link layer provides the services and functions necessary to transmit a bit stream between two devices using some method of sequencing and error detection and correction. Any management functions provided are only from the perspective of the physical interface used to interconnect the two devices. There is no knowledge of any other connections to the two devices from the perspective of the data link layer.

In addition to providing the services for reliable data transfer between two devices, the data link layer must also interface with the network layer above it and the physical layer below it. This is accomplished through the use of primitives. A primitive is a protocol between two layers. Since this interface is primarily software controlled, these primitives remain transparent to the end user.

The primitive to the network layer is used to pass received data from the physical layer to the network layer. Before this data is passed along, any information appended by the data link layer at the distant device must be removed. This includes sequence numbers and check sum fields.

When data is to be transmitted, it is passed to the data link layer from the network layer using this same interface. The data link layer must then append information to the original data. This information may include a device address, sequence number, and check bit sum. The address does not have to be that of the adjacent device. In fact, it would not make much sense if the address was always that of the adjacent device, since that is the only destination known at this layer.

The destination address is usually the final destination for this data transmission. When received by another device, the device must search a routing table to determine how to route the data to the destination address provided at the data link layer.

Not all protocols use addressing at the data link layer. In SS7, the addressing is somewhat different from that of other protocols. This will be discussed with the network layer, because the routing function is typically found at layer three, the network layer.

Sequencing in SS7 is provided at layer two to ensure that data is received in the same order it was transmitted. If data transmission was always reliable and never errored, sequencing would not be necessary. But this is never the case, and data transmission can get lost. When this occurs, the distant device has no indication that data was transmitted and lost.

Sequence numbering in SS7 provides a mechanism by which the distant device can tell if data was transmitted and then lost, because the next data packet received will contain a sequence number that is not sequential with the previously received packet. Sequence numbers can be of any range, although they usually fall within two ranges. Modulo 8 provides sequencing in the range 0 to 7. Modulo 128 provides sequencing in the range 0 to 127.

However, even though the protocol allows 127 packets to be transmitted without acknowledgment, very few networks will allow this. A "window" size is configured in all network equipment to prevent retransmission of too many packets. The idea is to send a burst of packets (say 15, for example) and if acknowledgment is not received after *n* seconds, retransmit the 15 packets. This is better than retransmitting 115 packets.

When acknowledging receipt of a message, the sequence number of the received packet is provided in the acknowledgment. Not every sequence number is acknowledged individually. One acknowledgment can be sent for a range of sequence numbers. For example, if an acknowledgment is sent with the sequence number of 6, and the last acknowledgment had a value of 3, the acknowledgment is for sequence numbers 4, 5, and 6.

Errors can be detected by use of a check sum field. When an error is detected, the recovery procedure requests a retransmission from the originator of the errored packet. The errored packet is then discarded. There are many methods for requesting a retransmission, depending on the protocol used.

When a message is received, the data link layer must determine where the packet begins and what type of packet it is. Each packet is preceded by some sort of delimiter, or *flag*. A flag is a specific bit pattern used before every packet. This bit pattern may never be duplicated in the packet itself, because the data link layer will consider that octet as the beginning of a new packet.

For this reason, whenever a pattern is used for a flag, there must be some technique for ensuring the bit pattern is never duplicated. The most common method is the use of bit stuffing. Bit stuffing inserts a bit in a fixed location (such as after every fifth consecutive binary 1).

The packet type will vary depending on the protocol. In some protocols, there are many different types of packets. A packet may be a supervisory packet, information packet, or unnumbered packet. SS7 uses three types of packets (called *signal units* in SS7).

The most important function of the data link layer is link management. The data link layer must be responsible for the integrity of the data link. When an error is discovered by the physical layer (such as loss of timing), the data link layer is notified. The data link layer then invokes some method of error recovery to restore the link. In the case of a loss of timing, the link may be taken out of service and then reset. This allows the link to realign itself with the source clock.

Flow control is an important part of link management. In the data link layer, flow control can be performed through the use of protocol messages. In SS7 networks, a special signal unit is used with protocol messages which indicates congestion conditions at an adjacent node or that an adjacent node is out of service and unable to process any messages.

Flow control initiates rerouting of messages by the upper layers of the protocol stack, so that messages will not be lost. This is not a function of the data link layer, but the data link layer must report congestion and out-of-service events to the network layer so that routing procedures can be invoked.

So far we have discussed only the procedures of a point-to-point configuration. Not all protocols and networks use point-to-point configurations. Many network topologies may use multipoint configurations as well, with duplex or half-duplex transmission. The data link layer is impacted by the configuration, and its services and functions will differ depending on the configuration.

For the purposes of this book, we will only discuss point-to-point, since all SS7 networks use point-to-point configurations between signaling points. Full-duplex allows transmission in both directions simultaneously, as is the case in SS7 networks. This requires two separate paths per link. A half-duplex link uses only one path, but simultaneous bidirectional transmission is not possible.

SS7 networks use a simple data link layer protocol. Because of the point-to-point configuration and the nature of the transmissions, this layer does not require much complexity. In other networks, it may be necessary for the data link layer to inquire before sending a data packet. Without a positive acknowledgment, transmission cannot take place.

In SS7, data is transmitted continuously, from a variety of sources. This transmission is always asynchronous in nature and does not require a session to be established with the receiving device. In fact, SS7 protocols are all connectionless-type protocols. Connection-oriented services are not used in today's SS7 networks.

Network layer

The network layer provides routing services for data packets received from another node. In the case of a packet-switched network, packets may come in to a node from a variety of locations. It is up to the network layer to examine the destination address and determine the link to be used to reach that destination.

The network layer is responsible for data transmission across networks. The transport layer provides connection to an entity within a device, while the network layer provides a transparent transfer of the data for the transport layer. The network layer allows the transport layer to free itself from the worries of internetwork data transfer.

There are two methods of reaching a destination. In some protocols, there is the requirement of establishing a virtual connection with another node. Other protocols use *datagrams,* packets of information which contain all the control information necessary to advise the destination how to process the received packet.

A virtual connection is established by sending a call request to another node. The purpose of the *virtual circuit* is to establish a consistent path through the network for all associated messages to follow. This method is used to overcome the inherent problem with packet switches, routing associated messages in multiple directions and resulting in packets being received out of sequence.

When a virtual circuit has been established, the packets that follow use the same path through the network, ensuring that all messages are received in the same sequence they were sent. This method is not favorable, because it reduces the reliability factor in the network. If a node in the path becomes congested, messages are delayed. If a circuit fails, there are no alternate circuits, and the message is lost.

Many packet-switching networks use datagram services to route packets throughout the network. This enhances the performance of the network, as messages can be routed dynamically based on the status of the circuits and the nodes in the network. When congestion occurs at any one node, messages are quickly rerouted in another direction, avoiding the congested node.

This is much like the routing used in SS7. Although SS7 uses a datagram-type service, it also utilizes certain procedures for specific types of messages that emulate a virtual circuit. The difference is in the network management of SS7. Even though a message is routed over a virtual circuit, if a circuit should fail in that path or a node become congested, it can be rerouted. SS7 enjoys the best of both worlds.

The addressing at this layer typically incorporates a multitier addressing scheme. The station or nodal address is found in the data link layer, while the network layer provides a higher level of addressing. Above the network layer is yet another layer of addressing: the logical connection which is the final destination for all protocol messages. The logical address resides within a network entity.

SS7 addressing differs from this, in that all addressing is located within the network layer. Addressing of the node, the network, and even a group of signaling points within a regional area is accomplished with what is called a *point code*. The point code uniquely identifies all entities in the ANSI SS7 network.

Quality of Service (QoS) is a parameter that is used by the routing function to identify the quality of transmission that must be provided. For example, if a particular message requires sequencing and special handling, the network layer must identify the level of processing required to route the message throughout the network. This parameter is used by the network management function when congestion occurs or when messages get lost.

The SS7 protocol provides several mechanisms for QoS, including a priority parameter for prioritizing message types. The priority of a message determines when a message can be discarded and when it must be routed no matter what.

There are network management functions at the network layer as well. When we discussed the data link layer, we discussed management procedures at the link level. Remember that the data link layer has no knowledge of the rest of the network. It is only concerned with the adjacent node to which it is connected.

Link management is the sole responsibility of the data link layer. As we discussed before, the status of the link is not broadcast throughout the network. This is of local significance only. However, if the status of the node itself should be impacted (perhaps by causing congestion), then the rest of the network must be notified.

This is the responsibility of the network layer. The network layer sends network management messages throughout the network, or, at least, to all of its adjacent nodes to advise them of degrading service at that node. This allows other nodes to make decisions about routing messages in different directions, around the troubled node.

In many cases, the affected node sends a network management message to all of its adjacent nodes. They, in turn, must decide whether or not another network management message needs to be sent to all of their adjacent nodes, hence, broadcasting out to the rest of the network. This usually depends on the type of network management message that is received.

As is the case with all levels, an interface to the layer above and below it is necessary. The OSI model talks about the use of *service data units* (SDUs). These are messages sent between layers of the protocol stack that contain the actual user data, as well as information appended by the protocol (such as control information). This is passed in either direction, depending on the flow of the message. If a message has been received, it is always passed in the upward direction. If a message is being prepared for transmission, it is always passed in the downward direction.

In networks that use point-to-point architecture, there is little use for a network protocol. This is certainly the case in local area networks (LANs). For this reason, protocols used in LANs do not use this layer, unless other networks are bridged to the LAN. When other networks must be accessed by the LAN (internetworking), then the network layer becomes a necessity. In the SS7 network, the network layer is also important, since this network consists of many individual networks all bridged together.

The OSI model also talks about the difference between *data terminal equipment* (DTE) and *data communications equipment* (DCE). In OSI terms, the DTE is an entity that originates a data message and uses the services of the network to send this data to its destination, another DTE.

The DCE is the network device responsible for the actual handling and relaying of the message through the network. A DCE device can be a modem, router, packet switch, or any other intermediate node in the network for which the message is not the destination. The purpose of the DCE is to route the message to its destination, nothing else.

A DTE device is further defined to work at all seven layers of the OSI model, while a DCE device works at only the first three layers of the OSI model. These first three layers are the only layers necessary for actually transmitting data over the network.

In these simple terms, we can easily identify the Service Switching Point (SSP) in the SS7 network as a DTE device. The Signaling Transfer Point (STP) could be considered a DCE (although there are some functions of the STP which might qualify it as a DTE as well). The Service Control Point (SCP) could be considered a DTE.

The easiest way to remember this is to identify the end points of the network. The end points are where messages originate and terminate. The intermediate devices in the network work only at the first three layers and are considered DCEs.

In the world of networking, one of the most difficult achievements is the ability to interwork with other networks, despite the differences in the protocols. This means that network layer and data link layer procedures must be converted. Conversion is not as simple as it may seem. Many times, one protocol may have procedures and functions not found in another.

When a message is received into a network from another, unlike network, the interface between the two networks (the gateway) must provide direct one-to-one mapping of the message and all of its parameters to the equivalent in the other protocol. This can be difficult if such procedures and parameters do not exist and have no equivalents. The rule is to try and provide some sort of alternative when possible.

The conversion must always be transparent to the upper layers, which are not typically affected. Remember that the network layer operates independently of the upper layers, providing a service to the upper layers. When this service changes, the upper layers should not be affected.

In the SS7 network, interworking sometimes occurs at all levels of the protocol stack. Not only does the network layer require conversion, but the application layers must be converted as well, in order for the upper layers to be compatible between networks. This is done through the use of gateway STPs or protocol converters.

Understanding the network layer can help you understand the routing and network management that must take place within any network. Let us now take a look at the next layer, the transport layer.

Transport layer

The transport layer is used to ensure reliable communications over the network. This means that data must be received without error, in sequence, and without loss of segments. The transport layer can be sophisticated or simple. However, if layer three is not capable of providing reliable transfer of data, then layer four must possess the ability to fulfill the role.

In essence, the transport layer relies on the reliability of the network layer so that it does not have to concern itself with this role. When a reliable network layer is provided, the transport layer is very simple. But when the network layer is not reliable, the reliability factor must be built into the transport layer. Such is the case with protocols such as frame relay, which does not use any of the control parameters found in other protocols.

Addressing at this layer consists of the *service access point* (SAP). The SAP is a logical address within a node. The logical address is the interface from the network segments of the protocol to the upper layers.

Because the connection is taking place between two different devices, the transport layer must have some knowledge about the addresses in the other device. This is accomplished in a couple of different ways. The easiest method is to use predefined addresses for common entities. By using predefined addresses, all systems can address logical entities at the transport layer without having to query the distant device about addressing.

Another method is to broadcast the address any time a new function is added. This is commonly used in some local area network (LAN) protocols today, and allows functions not commonly used or too specialized to be predefined to notify other nodes of their function and address. The transport layer is the only layer that needs this information, because it is responsible for the connection and termination.

The OSI model also talks about a naming convention, in which the particular task or logical function is called by name. This means that another device must provide the lookup capability of finding the physical address for the task name. This is commonly used in SS7 networks, where the signaling points may not know the actual address, but know the task with which they wish to interface.

This is a very favorable method in large networks, because it allows nodes to route to a function without having to know every address in the network. If another entity can provide the physical address, it saves memory space at each of the end nodes.

In X.25 networks, the transport layer also provides a multiplexing service. Virtual circuits may be used by many users, but only one transport service is used by all. The transport service must be able to multiplex its services among the many different users, even if they all come in on the same link. The transport service then splits the users to their various service access points (SAPs).

This function is not used in SS7 networks. In fact, the transport layer function is not even defined in SS7 today. As we will discuss a little later, the transport layer is not used in SS7 networks because SS7 does not presently support connection-oriented services.

Connection-oriented services, even with reliable network protocols, require the services of the transport layer to ensure the connection establishment and to maintain the connection. Flow control is included in this layer to manage the data flow through the connection. The data flow is controlled to the layer below, which is the network layer.

It is clear that the OSI intended the transport layer to be used as a backup to the network layer, providing additional mechanisms for reliable data transfer. In today's networks, this is not an issue. Today's networks use reliable mediums and do not suffer from the maladies of networks five years ago. This is certainly evident to those who use modems for network access.

Not too many years ago, modem transmission was very unreliable at high speeds. Today, modem speeds of 14.4 kbps are possible, because the telephone circuits have been improved. This is also the case with network mediums.

With protocols such as frame relay, where there are no control parameters, the transport layer becomes important. The philosophy in many of these networks is to let the upper layers worry about flow control and error detection/correction. This allows the lower layers to be simple and, thus, faster and cleaner. With dependable facilities, error detection does not become much of an issue.

Session layer

The session layer is responsible for establishing a dialog, or session, with another entity. The session layer must also define the type of dialog to be established. This, in itself, implies a connection-oriented service.

The session layer also provides flow control procedures. Flow control at this layer is imposed on the interface to the transport layer. The peer entity at the remote destination does not interact with this flow control, as it is of local significance only.

The session layer also manages what is called *synchronization points.* These are *dialog units.* An example of a dialog unit may be multiple file transfers, each file being one dialog unit. For example, if an entity needs to send several files to another remote entity, the session layer can establish each file as one synchronization unit. The entity can require that an acknowledgment be received for each synchronization unit before another can be sent. This is to ensure that each file is received properly before sending more data.

If a large data transfer is to take place and the transmission must be interrupted (for maintenance purposes or another task of a higher priority), the session layer must remember where the file transfer was interrupted so that it may start up in the place it left off. The session layer is not responsible for saving any data received, only for marking the place of interruption and continuing on from that point.

The OSI model also specifies the use of a token at the session layer. The token is passed by the session layer to grant permission to transmit data. There are several types of tokens defined. One token grants permission to transmit data, another sets the synchronization points, and a third is used to release a connection.

Tokens are passed from one session layer user to another. Only the holders of a token may transmit data (if they are holding the data token). The holder of a token may pass the token to the adjacent user as well.

As with the transport layer, the session layer is needed only when using connection-oriented protocols. If only connectionless services are provided, there is no reason to use this layer. In SS7 networks, the session layer is not necessary, because SS7 does not support connection-oriented services.

Presentation layer

While the application layer is concerned about the user's perspective, or view, of data, the presentation layer concerns itself with the view taken by the lower layer protocols. It is at this layer that data encryption and compression are found.

Perhaps the best description of the presentation layer is to consider the function of compression. If data must be compressed before it is transmitted over the network, the presentation layer must perform the compression and provide a format (or syntax) that the session layer is going to be able to use.

The syntax of the data at the presentation layer does not necessarily match that of the layer above. The only requirement at this layer is to provide the data in a syntax that can be sent over the network and received at the distant node. The peer presentation layer at the distant node must be capable of decompressing the data for the upper layers.

Another function at this layer is encryption. Encryption involves scrambling the data in some format that can be descrambled at the distant end. The purpose of encryption is to provide security over the network.

The encryption technique used must be transparent to the session layer and to all layers below it. The presentation layer at the distant end is responsible for descrambling the data. In today's networks, encryption and compression are about the only applications really suited for this layer. In previous networks, where mainframes had to communicate with terminals, the presentation layer was used to present the data on the terminal.

Syntax is used by programmers who must write the procedures in software code for the various network devices. A standard notation for data is used in most programming languages. This layer uses an abstract syntax (such as Abstract Syntax Notation One, or ASN-1) to represent data types.

ASN-1 is the syntax used in SS7 applications. This syntax is commonly found in many network protocols and is widely used throughout the industry.

Application layer

The application layer in the OSI model is the interface between the application entity and the OSI model. This interface is the first stage in processing the received data for transmission over the network.

The services listed in the OSI model relating to the application layer include information transfer, identification of the intended receiver, availability of the receiver, and any other functions not already defined in the lower layers.

Some examples of applications provided by the application layer include file transfer, job transfer, message exchange, and remote login. It is this layer that also ensures that an addressed entity, once it has committed to another entity, cannot be interfered with by another entity. A database could be left in an unknown state if this were to be allowed.

Another principle to remember about this layer is that the application layer views data in the same perspective as the user. In other words, while the rest of the layers will view the data from a network transmission perspective, this layer must view the data the way the user will see it. Thus, the data must be reconstructed as it was originally, before it can be passed on to the application.

Overview of the SS7 Protocol Stack

The SS7 protocol differs somewhat from the OSI model (Fig. 3.2). While the OSI model consists of seven different layers, the SS7 standard uses only four levels. The term "level" is used in the same context as "layers."

The functions carried out by these four levels correspond with the OSI model's seven layers. Some of the functions called for in the OSI model have no purpose in the SS7 network and are, therefore, undefined.

It should also be noted that the functions in the SS7 protocol have been refined over the years and tailored for the specific requirements of the SS7 network. For this reason, there are many discrepancies between the two protocols and their corresponding functions.

Regardless of the differences, the SS7 protocol has proven to be a highly reliable packet-switching protocol, providing all of the services and functions required by the telephone service providers. This protocol continues to evolve as the network grows and the services provided by the telephone companies change.

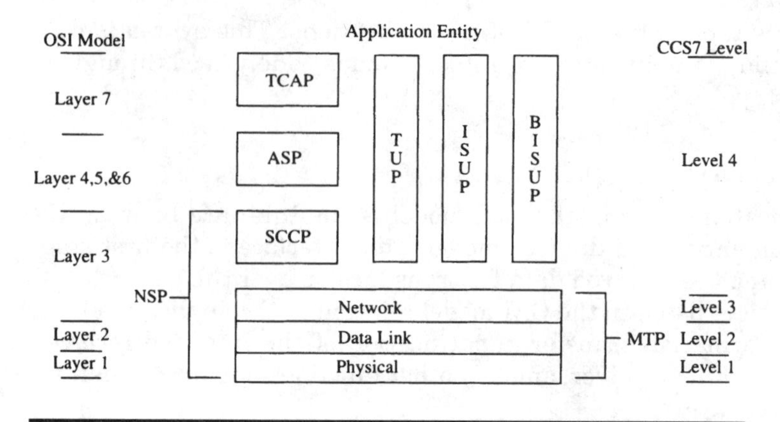

Figure 3.2 The SS7 protocol stack consists of only four levels and does not perfectly align with the OSI model. This is due in part to the fact that SS7 was developed before the OSI model. Many of the principles were in place, however, which explains the similarities.

Level one—physical level

The physical level in SS7 is virtually the same as that of the OSI model. The OSI model does not specify any specific interface to be used, as this will always differ from network to network. In SS7, we can specify which interfaces will be used, since the Bellcore standard and the ANSI standards all call for one of two types of interface—the DS0A or V.35.

The DS0A interface is the most favored for this application, with the V.35 acceptable as the second choice. There is no inherent value for using DS0A in SS7 networks, other than the fact that DS0A is already available. Because central offices are already using DS3 and DS1 facilities to link to one another, the DS0A interface is readily available in all central offices.

As telephone companies migrate to broadband networks, ATM will become the transport for SS7 messages. Bellcore is presently studying the possibility of using a full DS1 facility for a link interface in conjunction with broadband services.

Standards are already being prepared for broadband links using the ATM protocol to carry SS7 messages over DS1 or SONET facilities.

The SS7 standard does not specify any one interface for use. In fact, the standards allow the protocol to use any interface at any rate. Performance requirements impact Bellcore requirements for switching entities and, in some cases, even the ITU-TS performance standards will determine the type of interface to be used.

The theory, however, is that the protocol should be able to use any type of interface and any type of medium, maintaining true transparency throughout the layers. The other factors, of course, are distance and transmission rates needed to support the traffic mixes in each unique network.

Level two—data link level

The data link level of the SS7 protocol stack provides the SS7 network with error detection/correction and sequenced delivery of all SS7 message packets. As with the OSI model, this level is concerned only with the transmission of data from one node to the next node in the network. It does not concern itself with the final destination of the message. As the message travels from node to node, each node examines the dialed digits (contained in level four) and uses that information to determine the next route for the message. Level two is provided the information by level three, which determines message routing. Level two then provides the functions necessary to transmit the packet to the next node.

Level two does not provide the routing for SS7. This is a level-three function. Level two provides only the mechanisms needed to ensure reliable transfer of the data over the network. This is accomplished in several ways. First, level two provides the sequencing of messages between nodes. The sequence numbering is only of significance on one link. Each link will use its own sequencing series and will be independent of the other links.

The sequencing numbering is used by this layer to determine if any messages have been lost during transmission. A lost message indicates an error, which is counted by an error counter maintained by level three. After significant errors, the link is taken out of service and the network begins diagnostics and recovery procedures.

Another error-checking function maintained at level two is the *frame check sequence*. SS7 uses CRC-16 for error checking of the user data. The purpose of this mechanism is to maintain data integrity. The bit stream is subjected to the CRC-16 equation, and the remainder is placed into the FCS field. When the message is received by the distant node, the same equation is used again, only this time the value is compared with the value in the FCS field of the received message.

If there is an error in a message, or a message is lost, level two is responsible for requesting a retransmission. The retransmission may be accompanied with a message containing user data (user data in this context refers to level-four information). Unlike most protocols, where retransmissions are unique messages which do not carry any bearer information, the SS7 method maintains this function at the lower level, allowing the upper layers to function independently.

This allows retransmission requests to be sent to the distant node while also sending a layer-four message. This also allows higher throughput of SS7 traffic rather than network management messages.

A length indicator is provided to allow level two to determine what type of packet (signal unit) it is receiving. Level two must know the type of signal unit being received so it knows how to process the message. If it determines that there is information intended for a higher layer, then this level will pass the contents of the message up to the network level, or level three.

Level three—network level

The network level provides three functions: routing, message discrimination, and distribution. All three functions depend on the services of level two. When a message is received, it is passed by level two to level three for message discrimination.

Message discrimination determines who the message is addressed to. If the message contains the local address (of the receiving node), then the message is passed to message distribution. If the message is not addressed to the local node, then it is passed to the message-routing function. The message-routing function reads the called and calling party addresses in the message to determine which physical address to route to. The called and calling party addresses can be considered logical addresses, and the physical address the node address.

The physical address in SS7 networks is referred to as a point code. Every node in the network must have a unique point code. The routing function determines which point code to route the message to based on information stored in its administrable routing tables. These routing tables are maintained by the service providers themselves and are network dependent.

The point code in many cases is not the final destination for a message, but the adjacent point code for this node. This allows messages to be routed through the network and rerouted in the event of a network failure to another node. The routing scheme is determined by the network providers and can vary depending on philosophy.

Message distribution is used when message discrimination determines that the address is a local address. Message distribution is responsible for identifying which user part the message is addressed to (based on the service information octet field of the message) and routes the message to its internal user.

There are three network management functions at this level. Link management, route management, and traffic management are all level-three functions. Each type of network management uses different mechanisms to achieve results.

The link management function uses the *Link Status Signal Unit* (LSSU) to notify adjacent nodes of link problems. A link problem does not necessarily mean that the link cannot transmit messages. Software errors or processor problems on link interface cards can cause a link to become unusable.

When this occurs, it is quite possible for a link to remain operational at level two and even level three, but nonoperational at level four.

When this occurs, the adjacent node must be notified that the indicated link cannot be used for traffic, because there is a problem at the affected signaling point.

Level three sends LSSUs via level two to the adjacent node, indicating the problems with the link and advising of its status. The link can be removed from service (which means that no MSUs are transmitted over the affected link) and diagnostics can begin. Diagnostics consist of realigning the link or resynchronizing the link.

Realignment occurs when traffic is removed, all counters are reset to zero, all timers are reset to zero, and Fill-In Signal Units (FISUs) are transmitted for a prescribed duration of time, called the *proving period.* The duration of the proving period is dependent on the type of link being used. Bellcore has specified that the proving period for a DSO at 56 kbps shall be 2.3 seconds for normal proving and 0.6 seconds for emergency proving periods. At 64 kbps, the normal proving period duration is defined at 2.0 seconds and emergency proving period at 0.5 seconds. When a 1.536-Mbps link is used, the normal proving period is defined at 30 seconds and emergency proving defined at 5 seconds. During the proving period, any errors that may occur with the FISUs' transmission are counted.

When link management has determined that too many errors have occurred on the link, the entire process begins over again, with timers and counters being reset to zero and FISUs being transmitted for a prescribed duration of time.

Another form of link management entails the use of changeover and changeback messages. These are sent using *Message Signal Units* (MSUs) and advise the adjacent node to begin sending traffic over another link. The alternate link must be within the same linkset. During the time that all MSUs are being rerouted over different links, the affected link is being realigned by level three.

A changeback message is sent to advise the adjacent node that traffic may be sent over the affected link once again, since it has been restored to service. The changeback message is typically followed by a changeback acknowledgment message.

Route management provides the mechanisms for rerouting traffic around nodes which have failed or have become congested. This is a function of level three and works with the link management function.

Usually, when a link management message has been received, if the route of the node is affected, it may trigger the generation of a routing message, depending on the impact on other nodes. Route management is used to inform other nodes in the network of the status of a particular node which has become unavailable or congested. This differs from link management, which only notifies an adjacent node about link status.

Route management messages use the MSU and are generated by nodes adjacent to affected nodes and not usually by the affected nodes themselves. These messages are the *transfer-prohibited, transfer-restricted* messages and are discussed in Chap. 6, "Message Transfer Part Level Three."

Traffic management is used as a flow control mechanism. Flow control is used in the event that a node has become congested, but only at a single level. For example, if a particular user part is not available (such as the ISDN User Part), a traffic management message can be directed at adjacent nodes informing them that ISUP at a particular node is not available, without having any impact on Transaction Capabilities Application Part (TCAP) messages to the same node.

Traffic management, then, is different from the previous two functions in that it deals with a specific user part within an affected node, rather than with the entire entity. This mechanism allows the network to control the flow of certain messages based on protocol, without impeding other traffic that should not be affected.

Level four—user parts

Level four in the SS7 network consists of several different protocols, all called *user parts* and *application parts.* For basic telephone call connection and disconnect, the Telephone User Part (TUP) or ISDN User Part (ISUP) protocols are used. TUP is used in Europe and other countries following ITU-TS standards, while ISUP is used primarily in North America.

To access network databases, the Transaction Capabilities Application Part (TCAP) protocol is used. TCAP supports the functions required to connect to an external database, perform a query of the database, and retrieve information. The information or data retrieved is then sent back in the form of a TCAP message to the signaling point that requested it.

TCAP also supports remote control of other entities in the network. A network switch can invoke a feature or a function in another network switch by sending a TCAP message from one entity to another.

TCAP is being used more and more as the network evolves into a more intelligent network, capable of many self-invoked functions. With the inclusion of cellular networks into the SS7 networks, TCAP will increase in usage for roaming and other cellular functions.

The Operations, Maintenance, and Administration Part (OMAP) is really an application entity that uses the services of the Transaction Capabilities Application Part. The standard describes the syntax used for OMAP, relying on the Abstract Syntax Notation Number One (ASN-1) standard. This is used to provide communications and control functions throughout the network via a remote operations center terminal. This terminal is typically located in a remote maintenance center, where control over all network elements is possible. Administration of system databases, maintenance access, and performance monitoring are all parts of these centers.

The Mobile Application Part (MAP) is a relatively new level-four protocol used in cellular networks. The purpose of this protocol is to provide a mechanism by which cellular subscriber information may be passed from one cellular network to another. The MAP parameters include information such as the mobile identification number (MIN) and the serial number of the radio unit itself. This information is most often used by the IS-41 protocol during roaming procedures.

There are other level-four functions, which are discussed in much greater detail in later chapters. For now, an understanding of the differences between the OSI model and the SS7 protocol stack is all that is necessary. While all the functions called for in the OSI model are addressed in the SS7 protocols, the SS7 protocol stack is condensed and does not address connection-oriented services used to establish a "session" with another user.

In addition to providing connection requests in the voice network, SS7 also provides for database access from any entity in the network. This is the most important feature of the SS7 network, and the main reason SS7 has been deployed in the Public Switched Telephone Network all over the world, so that all telephone companies can share subscriber information and call-handling procedures on a call-by-call basis.

SS7 Protocols

Now that we have discussed the various layers, or levels, of the SS7 protocol, let us examine the protocols used within these levels to accomplish the specific functions called for at each level. The protocols used within SS7 each have a specific application and are used according to the services they provide the network.

Levels one, two, and three are combined into one *part,* the Message Transfer Part (MTP). MTP provides the rest of the levels with node-to-node transmission, providing basic error detection/correction schemes and message sequencing. In addition, MTP also provides routing, message discrimination, and distribution functions within a node.

When a database transaction is requested, MTP is accompanied by another higher level protocol, the Signaling Connection Control Part (SCCP). SCCP provides the addressing necessary to route a message to the correct database. Database addresses are called *subsystem numbers,* and are the logical addresses used by the protocols to route to the appropriate database entity.

In the event that the subsystem number is not known by an originating node, the dialed digits or other similar information is provided in a *called address field.* This information is then used for routing the message through the network. At some point, before the database is reached, the called party address must be translated into a point code and a subsystem number.

The point code is of the Service Control Point connecting to the database, while the subsystem number is the logical address of the database itself. Once the SCP is reached, the subsystem number may be sent over another type of network, such as an X.25 network.

The SCCP message is then returned with the proper routing instructions to the end office requesting the *global title*. SCCP is also used as the level-three protocol supporting the Transaction Capabilities Application Part (TCAP). TCAP is the protocol used for all database transactions. SCCP is required for routing TCAP messages to their proper database.

Another function of the SCCP protocol is to provide end-to-end routing, which is not possible with MTP. SCCP provides the means for routing a message transparently through the network using intermediate nodes as routers without the need to know the individual addresses of each of the intermediate nodes.

The addressing provided in the SCCP field allows each of the intermediate nodes to route based on the address in the SCCP protocol. The signaling points then base their routing on the SCCP address and generate the routing label for use by level-three routing.

Although the standards often show a correlation between SCCP and the ISDN User Part (ISUP), there is no current definition supporting such services. SCCP at this time is used only in conjunction with TCAP protocol messages.

ISUP is the protocol used to set up and tear down telephone connections between end offices. This protocol was derived from the Telephone User Part (TUP), which is the ITU-TS equivalent to ISUP, but offers the added benefit of supporting Intelligent Networking functions and ISDN services. ISUP is used throughout the United States today, and provides not only call connection services within the Public Switched Telephone Network, but also links the cellular network and the PCS network to the public telephone network.

Broadband ISUP (BISUP) is used for setting up and tearing down connections on ATM facilities. BISUP is still being refined, but most of the message structure and functions of BISUP have been documented. BISUP will gradually replace ISUP as ATM is rolled out into the network.

Through the use of these protocols, SS7 is able to provide a variety of services not obtainable with the previous signaling methods. SS7 is a message-based packet-switching network, capable of growing with the technology it must support. Because of SS7, the telephone companies have had to change their philosophies regarding service and are now finding themselves in a new industry—data communications.

Chapter

4

Overview of Signal Units

Overview of the Signal Units

SS7 is a packet-switching network and uses data packets just like X.25 and other packet-switching technologies. A packet contains all the information necessary to route data through a network, without establishing a connection to the destination.

There are three basic methods of switching in a network: circuit switching, message switching, and packet switching. Circuit switching uses a physical connection between two entities for transmitting a data stream. The circuit remains connected until both entities have completed transmission. A good example of circuit-switching networks is the Public Switched Telephone Network (PSTN).

Message switching came about in the 1970s and 1980s and uses a message structure to route data through a network. The data is accompanied by an address and a message (which serves as an instruction to the receiver). The data is sent in its entirety and does not include any error-checking schemes or flow control.

Packet switching arranges the data into a packet or a group of packets and transmits it in the form of a complete packet, providing all the information needed to route and process the received data. Included in packet-switching networks are network management and error detection/correction.

Packet switching is a more efficient way of networking, and it makes better use of facilities. In the event of network failures or any other problems in the network, the packet protocol can dynamically change the routing for a particular destination and can provide higher reliability through error checking and correction.

Usually, the packet-switching networks use different types of packets depending on the function. For example, if sending a data packet in an X.25 network, the packet type is called an *information frame.* The information frame has a distinct format and provides parameters specific to the transfer of data through the network.

If a packet is received in error and the packet must be retransmitted, a supervisory frame is used to inform the originator of the data packet that the data was in error and it needs to retransmit the errored packet.

The *supervisory frame* contains the parameters needed to inform another node of an error, but it does not support the transmission of any data. This packet then serves a very unique purpose and cannot be used for anything else.

SS7 uses three different structures of packets, called *signal units.* These signal units provide three different levels of service in the SS7 network. The SS7 protocol uses all three signal units for transmission of network management information, depending on the level of management. Information is sent using only one type of signal unit.

Another unique factor about SS7 networks is the source of information. In most networks, we are sending data from a user to another user. In SS7 networks, the user is the telephone network. The information is control and signaling information from telephone company switches and computers, which must be shared from one device to another. This makes the SS7 network a machine-to-machine network, rather than a user-to-user network. There is very little human intervention in this network, because most of the procedures and processes are automated and do not require any operator control.

A signal unit is nothing more than a packet, but in SS7, there are many applications, requiring different packet structures and capabilities. The applications found in the SS7 network vary from standard networks. There are circuit-related applications and non-circuit-related applications. These are the two basic foundations used to identify the functions within the network.

Circuit-related applications are directly related to the connection and disconnection of telephone circuits used to connect telephone subscribers. These circuits can be analog voice trunks or digital data circuits. They are located in a separate network outside of the signaling network and are used for the sole purpose of connecting telephone subscribers to other telephone subscribers.

The SS7 network does not have anything to do with the voice and data in these circuits other than identifying the type of data and voice transmission to take place (for example, data rates and encoding methods used at the voice interfaces).

Non-circuit-related applications consist of all other traffic in the SS7 network. To support these circuit-related functions of the Public Switched Telephone Network (PSTN), telephone switches must be able to communicate with one another. Whether they are requesting information from a database stored in a central computer system or invoking a feature in a remote telephone switch, there needs to be a protocol for all other aspects of the telephone network not related to a specific circuit.

Network management information is another type of communication which must be supported within the network. This is the automated part of the network, which allows signaling points within the network to automatically recover from failures and signaling point outages. Network management is completely autonomous in SS7 networks.

All signal units rely on the services of the Message Transfer Part (MTP) for routing, network management and link management, and basic error detection and correction. Without MTP, the signal units are worthless. MTP is a lower level protocol and is found in all signal units.

Anytime information is to be transferred through the network, from one signaling point to another, the Message Signal Unit (MSU) is used. It is called the "message" signal unit because SS7, like many protocols, uses data messages to convey information to another entity in the network. Information is considered control information or network management information.

The MSU provides the fields of the MTP protocol, as well as an additional two fields: the *service indicator octet* (SIO) and the *service information field* (SIF).

The SIO is used by level three to identify the type of protocol used at level four (i.e., ISUP or TCAP) and the type of standard. A standard can be a national standard or an international standard. If the protocol at level four is based on the ITU-TS standard, then the SIO field would indicate this as an international standard. If the protocol is any other type of protocol (such as ANSI), the SIO field would indicate this as a national protocol.

This information is used by level three and the message discrimination function to determine the type of signal unit, the protocol, and how it should be decoded. There are also spare bits in the SIO field which can be used for priorities in the ANSI standard, or for other functions in private networks.

The SIF is used to transfer control information, as well as the routing label used by level three. This field can contain up to 272 octets and is used by network management, ISUP, TCAP, MAP, and any other protocols which may be developed over time.

Not all information in this field is considered level-four information. For example, in the case of network management information, it is level three. The same is true if SCCP is used to transport TCAP. The SCCP portion of the message is found in the SIF field of the MSU. SCCP is considered to be level-three information.

Link status information is carried using another signal unit type called the Link Status Signal Unit (LSSU). The LSSU is actually used by level three at one node to transmit status regarding the link on which it is being carried to its adjacent node. The LSSU is never used to carry link status messages through the network. It is only used to communicate this status between two adjacent signaling points.

When no traffic is being sent and the network is idle, the Fill-In Signal Unit (FISU) is sent to provide constant error checking on the link. This allows the SS7 network to maintain its high reliability, because even though no information is being sent, the signaling points can still perform error detection on the FISU to determine if the link is beginning to deteriorate.

In addition to the FISU transmission, the MTP protocol is constantly monitoring the status of the link. The MTP is used in all three signal units. These signal units are explained in more detail in the following section.

Fill-In Signal Unit (FISU)

The lowest level signal unit—that is, one that provides the lowest level of service—is the Fill-In Signal Unit (FISU). The FISU acts as a flag in the SS7 network. When there is no payload to be delivered and the network is idle, FISUs are sent (Fig. 4.1).

This is different than in any other network, where flags are transmitted. A flag is usually a one-byte pattern, consisting of a 0, six 1s, and followed by a 0 (01111110). This one-byte pattern is used to maintain clock synchronization in many asynchronous networks. There is no "intelligence" in this type of pattern, however, so it serves no other purpose.

In the event that a data link begins to degrade, there is no indication that the link can carry traffic any longer until a transmission is attempted. By this time, it is too late. The transmission will fail and the link will have to go through diagnostics.

In the SS7 network, in order to maintain a high level of reliability, the FISU is used. This signal unit does not provide any information,

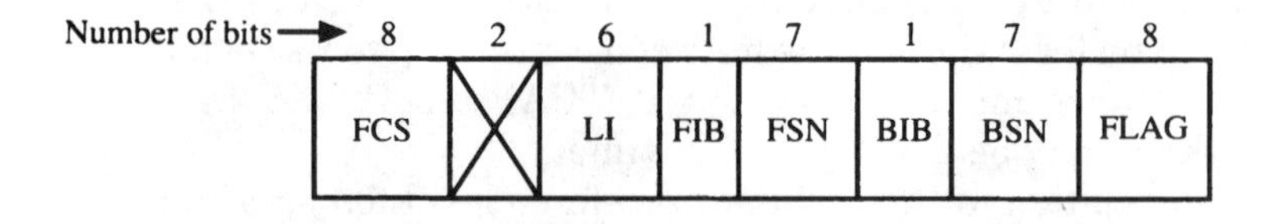

Figure 4.1 The Fill-In Signal Unit (FISU) consists of the components necessary for routing (per level-three MTP) and is sent during idle periods, instead of flags. By sending the FISU, a signaling point can verify the integrity of a link by checking the FCS field for errors.

but it does contain a minimal amount of information. The sequence numbers, for instance, can be used to acknowledge a previous signal unit.

The most significant field in the FISU is the Frame Check Sequence (FCS) field. This field is used by level three to determine if there are any errors in the FISU. This is based on the bits in all the fields of the FISU. The FCS is found in all signal units and is used to transport the remainder of the CRC-16 equation performed by the transmitting signaling point.

The CRC-16 is the error-checking mechanism implemented by all transmitting signaling points when transmitting a signal unit. The purpose is to provide the remainder of the CRC equation to the receiver of a signal unit. The receiver then uses the same CRC equation and compares it to the value received.

By using this field for error checking in the FISU, the Message Transfer Part (MTP) can constantly evaluate the status of any link, even during periods of idle traffic. In the event that a link has degraded to a point where it is causing too many errors, the link can be taken out of service by the MTP link management function before it is needed for actual traffic.

The FISU can also be used to acknowledge a previously received signal unit. This is done by sending an FISU with a backward sequence number equal to the forward sequence number of the signal unit being acknowledged. In other words, the backward sequence number identifies the sequence number of the last good signal unit received.

In the event that a signal unit is received and rejected by the MTP at level two, the FISU can be used to send back a negative acknowledgment. A negative acknowledgment requires the use of a *backward indicator bit* (BIB). Usually, the BIB and the *forward indicator bit* (FIB) are of the same value. However, when there is a negative acknowledgment, the BIB is toggled, assuming an opposite value from the FIB. This signifies a retransmission request.

The receiver of a Fill-In Signal Unit (FISU) or any other signal unit with opposite values in the indicator bits examines the backward sequence number (BSN) to determine which signal units need to be retransmitted. This is more efficient than using specialized supervisory frames, as other protocols use.

When there are no errors, the indicator bits maintain the same value; that is, both the forward indicator bit (FIB) and the backward indicator bit (BIB) are exactly the same. When a retransmission occurs, the retransmitted signal unit is sent with the indicator bits set to the same value.

When the retransmission is acknowledged, the originator of the retransmission request toggles the FIB to match the BIB, and maintains this value until another retransmission is required. This procedure is explained in full detail in Chap. 5.

The Fill-In Signal Unit (FISU) is never retransmitted if it is found in error. In fact, when FISUs are sent through the network, the sequence numbers do not increment (*forward sequence numbers,* or FSNs). There is no reason to retransmit these signal units because they do not provide any information. They are used only to monitor the integrity of a signaling link.

The forward sequence number (FSN) assumes the value of the last Message Signal Unit (MSU) sent by the transmitting signaling point, and stays at the same value until another Message Signal Unit (MSU) or Link Status Signal Unit (LSSU) is transmitted. The backward sequence number (BSN) follows the same procedure, unless used to acknowledge a previously received MSU.

The length indicator field in the Fill-In Signal Unit (FISU) is always set to zero. The length indicator identifies the type of signal unit being received. The length is that of the information field, which does not exist in the FISU. The total length of an FISU is static, at 48 bits long.

Although many drawings and publications will show both an opening and closing flag, there is only one opening flag and no closing flag. The opening flag of one signal unit is the closing flag of the previous signal unit. This is defined in Bellcore TR-NWT-000246.

Link Status Signal Unit (LSSU)

The Link Status Signal Unit (LSSU) is sent between two signaling points to indicate the status of the signaling link on which it is carried. Therefore, the LSSU is only of significance between two signaling points and does not get broadcast through the network (Fig. 4.2).

When a link is determined to have failed, the signaling point that detects the error condition is responsible for alerting its adjacent signaling point that the link is no longer available. The types of error conditions that warrant this procedure are alignment problems.

Alignment of a link means that all signal units received are of the correct length, and there are no ones-density violations. A ones-density violation occurs when the bit stream has more than five consecutive

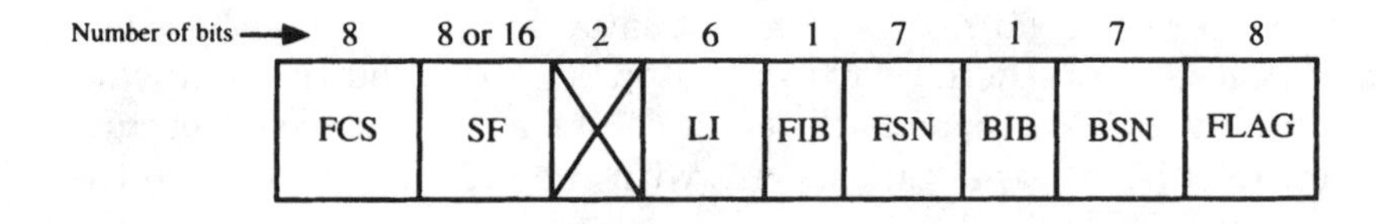

Figure 4.2 The Link Status Signal Unit (LSSU) is used by level-three MTP to send link status information to an adjacent signaling point. The LSSU is not broadcast to any other signaling points.

1s, considered by the protocol as a flag. With link management, and level-two functionality, this should never occur. Yet when it does, the link must be taken out of service and realigned.

Realignment is the procedure used by level two and level three to correct a link problem. The real problem is usually within a processor at either end of the link. Therefore, the processor that is at fault must be corrected. The first step is to remove all traffic from this link. The LSSU is sent to the adjacent signaling point to inform the adjacent node that all traffic should be removed from the link and the link is being realigned. No acknowledgment is required; this is simply an information signal unit.

There are two reasons that this procedure is necessary. To begin with, the two signaling points run independently of one another. Each end of a signaling link has its own processor. This processor and its accompanying software are what provide the functionality of levels two and three. In the event the processor should fail or be unable to process any more Message Signal Units (MSUs), only that processor would be aware of the trouble. The adjacent processor thinks that the link is working fine and is capable of carrying traffic. It is for this reason that the LSSU is necessary.

When level two determines there is a problem (as notified by level three), it will transmit an LSSU identifying the problem. The status indicators in the status field of the signal unit identify the specific status of the link; i.e., the link is in alignment or there is a processor outage at the originating node.

The fact that the link is capable of sending this LSSU indicates the ability to send lower level traffic, despite the inability to process upper level traffic. Depending on the status of the link, the receiving signaling point may send a network management message in the backward direction (to its adjacent signaling points), indicating the inability to reach a particular signaling point. This would occur only if the route to a certain destination became inaccessible because of the link failures.

This means that the LSSU works in conjunction with other network management functions. The link management function uses the LSSU to notify adjacent signaling points of link status, while the route management function uses the MSU to notify adjacent signaling points of problems with a route to a destination.

The Link Status Signal Unit (LSSU) consists of the same components as the Fill-In Signal Unit (FISU), with the addition of the *status field* (SF). The status field carries the link status information for the link on which it is carried. The LSSU is not transmitted on parallel links and does not carry information about other links. The status field indicates the status of the link on which the LSSU is carried.

Again, this implies that the link did not have a hard failure. A hard failure is one in which no traffic can be carried by the link, such as in the case of a backhoe digging up a facility with SS7 links. The LSSU relies on level two and level three still being functional on the link.

As with the Fill-In Signal Unit (FISU), when an error occurs within the signal unit, the LSSU is not retransmitted. An errored LSSU is discarded, and the error is counted as an error on the link.

The value of the length indicator in an LSSU is either a 1 or a 2. Currently, the LSSU status field is always one octet in length. Until further definitions are made for additional status indications, this rule will probably not change.

Message Signal Unit (MSU)

The Message Signal Unit (MSU—Fig. 4.3) provides the structure for transmitting all other protocol types. This includes the ISDN User Part (ISUP), the Transaction Capabilities Application Part (TCAP), and the Mobile Application Part (MAP). The difference between the MSU and the previous two signal units is the addition of the Service Indicator Octet (SIO) and the service information field (SIF).

The service indicator octet is used by level-three message discrimination to determine the type of protocol being presented in the Message Signal Unit. This allows message discrimination to identify who the user will be at level four. For example, if the Service Indicator Octet (SIO) indicates the protocol to be ISDN User Part (ISUP), then ISUP will be the user at level four.

The SIO also identifies the version of protocol being presented: international or national. International applies only to the ITU-TS standard compliant protocols. This is used when connecting to the international plane of the SS7 network. The national protocol applies to all other standards, including the ANSI standard used in the United States. It should be noted, however, that national does not imply ANSI. There are many national standards used throughout the world. For example, in Germany, the national standard would be 1TR7, in Hong Kong it would be the Hong Kong standard, and in New Zealand it would be the New Zealand standard.

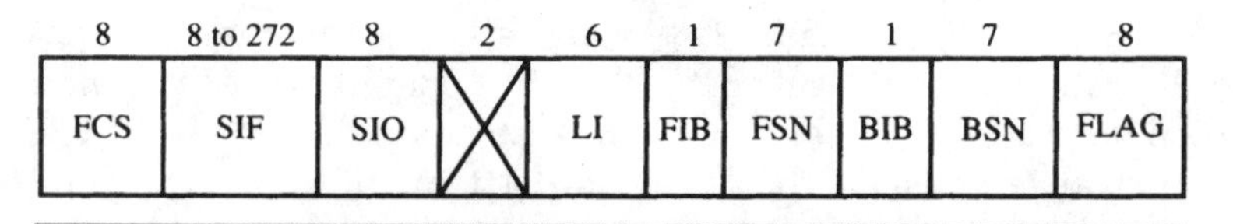

Figure 4.3 The Message Signal Unit (MSU) is used to deliver level-four information to its destination. Level-four information is found in the variable Signaling Information Field (SIF).

The use of two planes in the network allows all nations to interwork on the international level, using a gateway signaling point (usually a Signal Transfer Point) to gain access into the international network from the national side, and vice versa.

Individual countries can then use their own individual flavors of the SS7 protocols, depending on the requirements of their own unique networks, without impacting the entire SS7 network worldwide.

All nations must comply with the ITU-TS standards at the international plane and use some method of interworking between the two planes. Interworking will almost always require protocol conversion. The SIO can be used to determine when this will be necessary.

Another reason this parameter is important is because of the difference between the point codes used between international and national. International point codes are formatted as a three-bit zone identification, an eight-bit area or network identification, and a three-bit signaling point identification. National point codes can be any variation, as long as the total field length remains the same. The ANSI standard is an exception to this rule, where the point code is a 24-bit point code, eight bits for a network identification, eight bits for a cluster identification, and eight bits for a member identification.

The MSU provides a signaling information field with a capacity of up to 272 octets of user data. In the case of SS7, user data consists of any data from an upper layer (such as ISUP or TCAP). The signaling information field (SIF) does not necessarily have to be used for level-four information. Network management is also a user of the signaling information field. Network management is a function of level three.

The length indicator of the Message Signal Unit (MSU) can be any value over 2 and up to 64. The length indicator is a six-bit field, which limits it to the range it can represent. Yet the signaling information field (SIF) can be up to 272 octets long.

When the service information field exceeds 64 octets in length, the length indicator of the MSU remains at 63. This is not an issue with the protocol, because the only purpose of this field is to allow level-three message discrimination to be able to determine the type of signal unit being received. There is no other use for this field.

Based on this fact, it is safe to say that any value in the length indicator over 2 is always an MSU, and any value over 2 is really insignificant. There is no reason to expand this field, because the exact length is not important to level three.

Primitives

In order for the various levels to interface with one another, some method of standard interfacing must be implemented. The use of prim-

X	Generic Name	Specific Name	Parameter

X = MTP or N (SCCP)

Figure 4.4 Primitives are used to communicate with the various levels of the protocol stack within a network entity. Primitives are not seen in the network, but reside in software at each signaling point. This figure depicts the structure of a primitive.

itives is not unique to SS7, although the particular primitive types used in this protocol are unique.

Communications between levels two and three and between levels three and four are all software controlled. We do not see any communications over the network, although we will see the results over the network. A primitive is the method used by software to pass information to the next level, in either direction.

The important thing to understand is that a primitive is pure software. There is nothing for us to see or examine, unless we are looking at the source code of a signaling point itself. Primitives are discussed here for those who are actively writing software for SS7 network products and need to understand the full picture of what is taking place.

As seen in Fig. 4.4, the primitive provides four fields. The first field, marked by the "X," indicates the originator of the primitive. If the Message Transfer Part (MTP) is passing information up to the ISDN User Part (ISUP), then the first field would indicate "MTP."

The next field is the *generic name*. The generic name identifies the type of information being provided. For example, if information regarding the address of the originator (such as the calling party address) is being sent from ISUP to MTP, the generic name would be "unitdata." The generic name will differ, depending on the level. For example, the Signaling Connection Control Part will have different generic names than, say, the ISUP. The functionality remains the same.

The field after the generic name is the *specific name*. The specific name describes the action that is to take place. The specific name can be any one of the following:

- Request
- Indication
- Response
- Confirmation

A *request* is used to invoke some type of service from another level. For example, in the case of network management, there may be the need to start a procedure. The request would be used to invoke that procedure at a higher level.

An *indication* is used to inform the requesting level that the requested service has been invoked. This is like an acknowledgment between levels. Using the preceding case, when a user part invokes a management procedure, it will send an indication to MTP to inform MTP of the invocation.

A *response* is sent to complete a particular transaction between a service element and a user. A user, in this sense, is the protocol, while the service element is something such as SCCP or the upper user parts. The response is used only when a service has been previously invoked and an indication has been sent.

The *confirmation* is sent to inform the user part that a connection has been established or a requested service has been invoked. Confirmation, in some aspects, is similar to an acknowledgment.

There are many procedures that surround the various primitives, depending on the level they are communicating with and which user part they are interfacing to. The purpose of the primitive, once again, is to provide a means of communication between the various protocol levels within a signaling point.

Overview of SS7 Protocols

As we have discussed in the previous chapters, the SS7 network uses many different protocols. Each protocol is used for a specific purpose and provides the necessary functionality to accomplish specific tasks.

In this section, we will look at the various protocols used in SS7 networks, and discuss their usage and applications. This section will only provide an overview of the various protocols. For a more specific explanation of these protocols, refer to their respective individual chapters.

The SS7 network provides some basic services to the Public Switched Telephone Network. The impetus behind deploying this network was to remove all signaling information away from the voice network. In the early days of common channel signaling, this certainly seemed enough to justify the usage of another network. However, the SS7 network slowly evolved into much more than just a signaling network. It has also evolved into a control network.

The word "control" implies a lot of different things. From the voice network perspective, control refers directly to the ability to control features and tasks in a remote telephone switch or centralized computer. The user of this remote control capability is usually another telephone switch or another computer system.

This network obviously forms the basis for the Intelligent Network. Without the mechanisms for supporting remote control or other network entities, the Intelligent Network would not be possible.

However, the initial purpose of signaling cannot be ignored either, especially since today this remains as the principal function of this network. As the Intelligent Network evolves, this will gradually change.

As we discussed in the first chapter, the SS7 network evolved from the earlier CCS6 network, which was more limited, yet of similar technology. The primary difference between the two technologies is in the protocols used and their structures. CCS6 used a very stringent structure, with fixed-length signal units. This did not allow for variable-length signal units, and limited the protocol as far as the type of information that could be provided.

It was for this reason that SS7 was structured the way it is today. By providing a basic structure, which various protocols can depend on for transport, it allows the upper level protocols to be more dynamic. This means they can grow and evolve with the network without affecting the transport mechanism. This, of course, was the main limitation in CCS6 networks. The structure was too constrained and did not allow for sufficient growth of any kind in the upper layers, due largely to the absence of an independent transport function.

Message Transfer Part (MTP)

The Message Transfer Part (MTP) is the transport protocol used by all other SS7 protocols in the SS7 network. This protocol is actually divided into three different levels. In comparison with the OSI model, MTP provides the same functionality as layers one, two, and three.

The physical level of MTP allows for the use of any digital-type interface supporting the data rate required by most networks. Common interfaces in most SS7 networks today include DS0A and V.35.

The physical level, or level one, works independently of all other levels. This allows the upper levels to evolve to meet the ever-changing demands of the network without affecting the interface.

There is one exception to this rule, and that applies to the new broadband networks. There is a question as to whether or not existing interfaces will be sufficient for supporting the signaling in broadband networks. Bellcore has released preliminary standards on the usage of a full Digital Signal 1 (DS1) facility, at 1.544 Mbps, as a signaling link.

The DS1 is commonly found in ISDN networks and is used by T-1 trunks as the basic carrier. Usage as a signaling link will reduce the number of multiplexers used in the network, since the use of DS0A requires a DS3 between two exchanges to be demultiplexed at two levels before the DS0A can be derived. Use of a DS1 will eliminate one multiplexer in these cases.

The level-two function of MTP provides the functions necessary to provide basic error detection and correction for all signal units. This protocol is concerned only with the delivery of signal units between two exchanges or signaling points. There is no consideration outside of the signaling link.

This implies that level two has no knowledge of the final destination. This is a fair assumption. Level two does not need to concern itself with this information. In the true spirit of the OSI model, this is left up to upper levels. Level two provides reliable transfer of information over a signaling link to the adjacent signaling point. Once the information reaches the adjacent signaling point, it is up to level three to determine how to route the message any further.

Level two is maintained at the signaling link level. Each circuit card in an SS7 device must be able to provide and support this functionality independently of the rest of the system. For example, if there are several links connecting to the same signaling point, each link runs independently and does not concern itself with the activities of the other links.

Sequence numbering is a function found at this level. Now that we understand that this level works independently of all other levels and all other links, we can assume that the sequence numbering is significant only on each particular link. In other words, if one link is transmitting messages using sequence numbers one, two, and three, there is no synchronization of sequence numbers on the other links. They may be using a completely different range of numbers, as long as they are all sequential. Each link maintains its own sequence numbering.

This is also true for the adjacent signaling point on the same link. One signaling point can be sending sequence number 10, while the adjacent signaling point is sending sequence number 121. This is due to the fact that these links have independent processing that is not synchronized, allowing links to be much more efficient.

Another function of level two is error checking. There are two methods of error checking: *basic* and *preventive cyclic redundancy* (PCR). Basic error detection/correction is used with all terrestrial signaling links. This is by far the most favorable, because it is much more efficient than PCR.

PCR is used only with satellite signaling links, and it uses constant retransmission rather than error checking. With basic error detection and correction, when an error is detected, a retransmission is requested. The sequence number is provided for the last received signal unit that was good, allowing the originator of the bad signal unit to be able to determine which signal units to retransmit.

In PCR, all transmitted signal units are retransmitted automatically during idle periods, until they are acknowledged. Once they have been acknowledged, they are dropped from the transmission buffer. They are continually retransmitted until the distant end acknowledges them. This, of course, is not efficient use of the network and creates a lot of overhead.

The reason for this method lies in the propagation delay introduced when using satellite signaling links. If a signal unit is sent, a retransmission may cross an acknowledgment because of the delay encountered. This would lead to some confusing situations in which a signal unit is retransmitted (due to a time-out, for example) and, at the same time, an acknowledgment is received. The receiving end would also find itself somewhat confused if it sent an acknowledgment, only to find the same signal unit being retransmitted.

Procedures to alleviate this are implemented whenever satellite is used. The general rule is not to use satellite for signaling links whenever possible, but when there are no other alternatives, the protocol will support the use of satellite and microwave as well.

Level two also detects the presence of an opening flag for the delineation of an incoming signal unit. The flag is always a fixed pattern of 01111110, and is located in the first octet of the signal unit. As mentioned earlier, the opening flag is also the closing flag of the previous signal unit.

Message Transfer Part (MTP) level three provides four functions: message routing, message discrimination, message distribution, and network management. Network management is probably the most important. Network management maintains the integrity of individual signaling links by continuously monitoring them and counting the number of errors which occur on any single link.

When excessive errors have been counted, the link is removed from service (messages are blocked from the link) and the link is reinitialized. Since most errors are the result of clock signal degeneration and other related factors, resynchronization of the link usually resolves any problems that may occur.

When a link is said to be functioning properly and messages are of the correct length, the link is in alignment. When messages are received that are not the correct length or if there is a ones-density violation, the link is said to be out of alignment.

Network management can rectify this problem. There are several functions within network management. Each function looks after a specific area of the network. They are:

- Link management
- Traffic management
- Route management

Link management is concerned with the integrity of an individual link. While this is a level-three function, it relies on the service of level two to indicate when there is a problem on a link. The types of problems are typically errors, such as signal unit length and synchronization.

Link management is capable of blocking messages from a particular link and notifying the adjacent signaling point to do the same. Once again, level two is utilized to alert the adjacent signaling point of a problem. The Link Status Signal Unit (LSSU) is used to inform an adjacent signaling point of the status of a link.

Link management does not inform other signaling points in the network of its problems. Only adjacent signaling points need to be concerned about link troubles. Therefore, link management is a local function, and does not directly affect the performance of the overall network.

There is a subtle impact, however, on the rest of the network when links fail. Link failures can cause traffic to reroute to another link, causing that link to become congested. If too much traffic is directed to another link and the processor cannot keep up with it, the signaling point can be considered congested.

When a signaling point becomes congested, the adjacent signaling points are notified to reroute all traffic around the congested signaling point. This can result in delays in the network, and can even create congestion in other signaling points.

Link management is also responsible for activating and deactivating links and, in some cases, even automatic allocation. *Automatic allocation* is a feature offered in some SSPs that provide both voice circuits and SS7 links. Automatic allocation removes voice circuits from service and automatically places them in service as SS7 signaling links. The circuits must be preconfigured for this capability. Not all systems offer this capability, but when it is offered, it can be valuable in handling sudden bursts in link demand.

Traffic management provides the mechanisms for routing traffic around failed links within a linkset. Traffic management uses the Message Signal Unit (MSU) to send changeover and changeback messages to an adjacent node, informing the adjacent node of the failed links.

This is not to be confused with link management, which is responsible for turning links up and taking links out of service. Link management is what controls the status of a link and informs the adjacent signaling point of the link status. The difference lies in the mechanism used to inform the adjacent signaling point.

Link management uses the LSSU, which is carried on the link that is affected. Traffic management uses another link within the same linkset, and is used when a link fails for any reason to advise the adjacent signaling point to use another link within the same linkset. This mechanism is necessary when a link is unable to carry any level of traffic, such as when a backhoe digs up a link facility.

Route management is used to advise other signaling points in the network about the inability of one signaling point to reach another signaling point. For example, if a signaling point becomes inaccessible by an adjacent signaling point, the adjacent signaling point will send a route management message to its adjacent signaling points to advise them that it can no longer reach the specified point code.

Transfer-restricted and transfer-prohibited messages are two of the most commonly used route management messages. In the event that a link becomes unavailable and link management sends a link management message to an adjacent signaling point, it is possible that, in time, if congestion occurs, route management will be implemented to alert other signaling points in the network (only those adjacent to the originating signaling point) that the destination (or affected signaling point) can no longer be reached.

Besides the network management procedures just described, there are three other major functions within level three:

- Message discrimination
- Message distribution
- Message routing

Message discrimination uses the routing label of the Message Signal Unit (MSU) to determine, first, whom a message is addressed to. If the routing label contains the address of the local signaling point, then the message is handed off to message distribution. If the address is of another signaling point, the message is handed off to message routing.

Message distribution uses the service indicator octet (SIO) to determine who the user of a message is. If the SIO indicates that the user part is ISUP, the message is handed off to the ISDN User Part (ISUP). If the service indicator octet indicates the user part is the Transaction Capabilities Application Part (TCAP), the message is handed off to TCAP.

Message routing attaches a new routing label to an outgoing message and determines which signaling link should be used to route the call. The signaling points routing table works with this function in determining the destination point code and the linkset that should be used to reach the destination.

Signaling Connection Control Part (SCCP)

The Signaling Connection Control Part (SCCP) is used only with the Transaction Capabilities Application Part (TCAP), although the standards indicate its use with the ISDN User Part (ISUP).

The purpose of SCCP is to provide the means for end-to-end routing. The Message Transfer Part (MTP) is only capable of point-to-point routing. This means that a message can be routed based only on the physical links available from a signaling point.

SCCP provides the addressing to route a message through the entire network. This information is used at each signaling point by MTP level-three routing to determine which linkset to use.

The difference between MTP and SCCP is the way the information is used, and the nature of the addresses. The MTP provides both the *origination point code* (OPC) and the *destination point code* (DPC). In both cases, the point code is from a node-to-node perspective.

In the case of SCCP, the address consists of three parts: called/calling party, point code, and subsystem number. The routing can be based on any of the three, although when routing by point code, the address is a combination of the point code and the subsystem number.

When routing a TCAP message, the signaling point must be able to identify the destination, which is almost always a computer database or a specific signaling point. In many cases, there may not be any dialed digits associated with the transaction (although in today's applications this is not the case). SCCP provides the addressing needed by MTP to route a TCAP message through the network.

The address information in SCCP remains fairly static, unless the point code and subsystem number are unknown to the originator. In this case, a Signal Transfer Point (STP) will have to provide translation. This is usually the case when a number is dialed that cannot be routed by the network.

The digits provided in the called party address are called *global title digits.* When the signaling point originating the message does not know the point code or the subsystem number of the database that will be providing a routing number for the requesting exchange, the global title digits have to be used by MTP level three for routing. At some point, the point code and subsystem number have to be provided so the message can reach its final destination. This function is known as *global title translation* and is usually provided by the Signal Transfer Point adjacent to the destination database.

When a number is dialed, such as an 800 or 900 number, the network cannot route the call based on conventional routing methods. This is because the numbering plan uses the area code of a number to determine which area in the nation's network the call should be routed to (the area being handled by a specific toll office). Likewise, the prefix usually denotes a specific central office that can route this call to the subscriber. In the case of 800 and 900 numbers, these do not have area codes that denote a toll office.

The SS7 network will provide a routing number by which the end exchange can route the call. This requires the services of the Transaction Capabilities Application Part (TCAP) and the Signaling Connection Control Part (SCCP). The called party address of SCCP will provide the dialed digits, although not all the digits are necessary. Only the area code (800) and the prefix are necessary.

The number is compared in a database, which provides the routing number to TCAP. The routing number is then returned to the requesting exchange via TCAP and SCCP, so that a connection can be established for the call.

This is just one example of how SCCP can be used. The called party address does not have to be dialed digits. In the cellular network, the mobile identification number (MIN) is placed in the called party address for roaming information.

When used by the ISDN User Part (ISUP), SCCP will allow ISUP messages associated with an already established connection to be routed using end-to-end routing, the same as TCAP messages. This functionality has not yet been implemented in SS7 networks; however, with new services and the evolution of the Intelligent Network, this may become necessary.

ISDN User Part (ISUP)

The ISDN User Part (ISUP) is a circuit-related protocol, used for establishing circuit connections and maintaining the connections throughout a call. ISUP is associated only with voice and data calls and does not presently support broadband technologies such as Frame Relay and Asynchronous Transfer Mode (ATM). These new technologies will be addressed by a new version of ISUP called *Broadband ISUP* (BISUP).

Broadband ISUP (BISUP) is presently under development by the ITU-TS. BISUP will provide the mechanisms and parameters necessary to support the bandwidth and Quality of Service requirements of these services.

ISUP supports both analog and digital voice circuits, and was adopted by ANSI to replace the Telephone User Part (TUP). The Telephone User Part (TUP) does not support data transmission or digital circuits. ISUP added the parameters necessary to support digital circuits and data transmission.

ISUP is compatible with the ISDN protocol, which was developed as an extension of SS7 to the subscriber. There is direct mapping of ISDN message types to ISUP message types, even though the message types are not the same. The purpose of the ISDN compatibility is to allow subscriber switches to send signaling information to remote subscriber switches during the call connection phase. After the connection has been established, ISUP supports communications between the two end-point subscriber switches. This feature may be necessary to support caller-invoked features, such as conference calling or automatic callback. The ability to invoke features and share information between two subscriber switches and/or networks is the unique capability of ISUP and the purpose for its development.

Broadband ISDN User Part (BISUP)

To support the new broadband ISDN and Asynchronous Transfer Mode (ATM) architectures, the ISUP protocol was modified. This new version of ISUP provides additional message types and parameters, which provide the support necessary for ATM and broadband networks.

The most significant difference between the ISUP and the BISUP protocols is in the circuit assignment procedures and the type of circuits supported. ATM and broadband ISDN circuits are virtual circuits rather then physical circuits. This places new demands on the SS7 network, because it must be capable of assigning these virtual circuits and maintaining them. Because of the number of circuits available in broadband networks, a new circuit-numbering convention was adopted.

In addition to the new circuit requirements, broadband networks also support the dynamic allocation of bandwidth, on a per-call basis. Now, when a call connection is established, the available bandwidth for that call is negotiated between the originating exchange and the destination exchange.

There are a few other, more subtle, changes that appear in this newer version of the ISUP protocol. They have been described in more detail in the ISUP chapter (Chap. 9). In addition to the procedure descriptions, there is also a section which explains the various message types and their parameters.

Telephone User Part (TUP)

The Telephone User Part (TUP) is used in international networks. This protocol is compatible with the ISDN User Part (ISUP), with differences between the two mainly in the message type and parameters. Regardless of these differences, the two protocols can be mapped to one another successfully, even if a one-to-one mapping relationship does not exist. TUP is being replaced by ISUP at the international level as well.

The United States and ANSI decided early on to replace this protocol with the ISUP protocol, in order to support the evolving services provided in many U.S. networks. The international market is just now learning of the possibilities that ISUP provides and is slowly evolving to this protocol.

Before ISUP support for data services and digital facilities was provided through a now-obsolete protocol called the Data User Part (DUP). DUP is no longer used in U.S. networks and has been omitted from this book. Because of the migration towards ISUP at the international level, TUP has been omitted from this book as well.

Probably the most noticeable difference between the structures of ISUP and TUP is in the header field. The ISUP protocol uses message types, whereas the TUP protocol uses an H0/H1 header. The H0/H1 header was originated from the CCS6 protocols. Messages are grouped into classes, which are represented by the H0 field. The H1 field denotes the specific message within that class. ISUP uses message types without any classification. This provides more flexibility in the protocol, and more room for growth.

Transaction Capabilities Application Part (TCAP)

The Transaction Capabilities Application Part (TCAP) is probably the most versatile of all the SS7 protocols. TCAP is used for two purposes: accessing remote databases and invoking features in remote network entities.

A network entity does not have to be a switch. Any network device, provided it is equipped with the proper interfaces and can provide all four levels of SS7 support, can be accessed by this protocol.

In today's networks, TCAP is limited to database access, although more and more networks are providing new advanced services which require the use of TCAP to invoke those services. Custom calling features provided by the Intelligent Network will most certainly require the support of TCAP to invoke features and services in remote switches.

The TCAP protocol is not being used to its fullest potential today. The mechanisms currently provided in this protocol reach far beyond database access. As the Intelligent Network evolves, the traffic mix in all SS7 networks will rapidly change to consist of mostly TCAP traffic.

The TCAP protocol has been designed to provide for remote control of other network entities, which, in itself, holds many possibilities. For example, a subscriber wishes to change his or her telephone service. Normally, this would require a telephone call to the telephone company, which could remotely access the subscriber database and add the new service to the customer record. A service order would be generated and the new services programmed into the switch serving the subscriber. With TCAP capability, subscribers could enter the order entry system themselves. With an interface between the order entry system and the switch, subscribers could then change the program within the switch serving their telephone numbers.

Of course, no one would expect a subscriber to understand the intricacies of a telephone office switch, so front-end interfaces using icons would be required to facilitate the subscriber. Sound far-fetched? Not really, since this is what the Intelligent Network will do when it is complete.

TCAP is an integral part of the Intelligent Network, the cellular network, and, soon, the broadband services network. It is probably somewhat of a cliché, but it is accurate to say that TCAP is ahead of its time.

Chapter

5

Message Transfer Part (MTP)

The Message Transfer Part (MTP) acts as the carrier for all SS7 messages, providing reliable transfer of messages from one signaling point to another. This function includes levels one, two, and three. In addition to providing signaling point to signaling point communications, the MTP also provides error detection and correction.

The methodologies of MTP are very similar to those used in other bit-oriented protocols (BOPs), such as X.25. Sequence-numbering and error-checking mechanisms are very similar.

The signal unit structure used in all SS7 messages provides all the information required by MTP level two and level three. Flow control is provided through the use of a special signal unit called the Link Status Signal Unit (LSSU), described in this chapter.

MTP is defined in the ITU-TS documents Q.701 through Q.704, Q.706, and Q.707. They can be found in Bellcore document TR-NWT-000246, Volume One, Chapter 1.111.1 through 1.111.8. ANSI publications referring to the MTP protocol are numbered T1.111-1992, "Functional Description of the Signaling Message Transfer Part (MTP)."

The Bellcore recommendations add reliability and versatility to the network. The Bellcore publications are almost identical to the ANSI and ITU-TS publications, other than the additions made by Bellcore.

Description of MTP

The Message Transfer Part (MTP) provides all functions of layers one, two, and three in the OSI model. We have already discussed the types of interfaces used at level one of the SS7 network. These are industry-standard interfaces and do not necessarily require in-depth discussion here. Level two of MTP provides error detection/correction, as well as error checking through the check bit field. In addition to these two fundamental functions, the following is provided:

- Signal unit delimitation
- Signal unit alignment
- Signal unit error detection
- Signal unit error correction
- Signaling link initial alignment
- Signaling link error monitoring
- Flow control

Signal unit delimitation

Every signal unit is preceded by a flag. The flag is an eight-bit pattern, beginning with 0 and followed by six consecutive 1s, ending in 0 (01111110). The flag is used to signify the beginning of a signal unit and the end of the preceding signal unit. While the protocol actually allows both an opening and a closing flag, in the United States, only one flag is used.

Signal unit delimitation is important to the upper layers. In most networks, there is traffic constantly flowing through each signaling point, even though the messages may not contain any information.

Signal unit alignment

A link is considered in alignment when signal units are received in sequence, without ones-density violations, and with the proper number of octets (based on the message type). The signal unit must be a total length of eight-bit multiples. If the signal unit is not in eight-bit multiples, or if the signaling information field (SIF) of a Message Signal Unit (MSU) exceeds the 272-octet capacity, the signal unit received is considered in error.

The link is not taken out of service until there has been an excessive number of errors. This is determined by a counter, the *signal unit error rate monitor* (SUERM). This counter is used to count the total number of errors on a signaling link. Each link keeps its own unique counter.

The purpose of the counter is to determine when an excessive number of errors has occurred (64) and take the link out of service. The type of errors is limited to alignment errors. The typical cause of alignment errors is usually clock signals not being properly synchronized on both ends of a link.

The network management procedure at level three is responsible for realigning the link (by taking it out of service and resynchronizing the link). Level two is responsible for reporting any errors to level-three link management.

When the link is taken out of service, the link must be tested for integrity before it is available for Message Signal Units (MSUs) again. This process is known as the *alignment procedure*. There are two types of alignment procedures used: normal alignment procedure and emergency alignment procedure. Both these methods are discussed in greater detail later in this section.

Signal unit error detection

Errors are detected using the check bit field and the sequence number of the signal unit. If the check bit field is in error, the signal unit is discarded and a negative acknowledgment is sent to the originating signaling point. An error is also counted by the signal unit error rate monitor (SUERM).

The level-two timer T7, "Excessive delay of acknowledgment," prevents a signaling point from waiting too long for a positive or negative acknowledgment. Usually, an acknowledgment is sent when a signaling point becomes idle and does not have any more traffic to transmit. When congestion occurs at a signaling point, or an extreme amount of traffic is present, it is possible that T7 could time out and force retransmission of messages.

The recommended value for T7 is 11.5 seconds. This, of course, depends on the network. While the actual value of T7 is optional, the timer is usually a nonadministrable timer, which means that once it is set, it cannot be changed by system administration.

Signal unit error correction

When an error is detected in a signal unit, the signal unit is discarded. Level-two MTP counts the error (SUERM or errored interval monitor) and requests a retransmission if basic error control is being used. Preventative cyclic retransmission (PCR) treats errors differently and is explained in greater detail later in this chapter.

When excessive errors are detected on any one link, the link is taken out of service. The link is then placed through an alignment procedure to test the link and place it back into service automatically. The link will not be placed back into service until it has passed the "proving period" of the alignment procedure.

Signaling link error monitoring

Three types of error rate monitors are used. Two are used when the link is in service and the other is used when the link is going through an alignment procedure. The SUERM and the errored interval monitor are used while the link is in service, and are often referred to as the "leaky bucket" technique, because of the way they decrement the counter after *n* number of good signal units or intervals.

With the SUERM, each signal unit received with an error increments a counter. Every 256th signal unit received without error decrements the counter (hence the nickname "leaky bucket"). When the counter reaches a value of 64, the link is removed from service and alignment procedures begin. This method is used with 56/54-kbps links.

When 1.536-Mbps links are used, the errored interval monitor method is used. The link is monitored for a determined period of time (defined by Bellcore as 100 milliseconds). If a flag is lost during the interval, or a signal unit received in error, then the interval is considered in error. An errored interval increments a counter, in the same fashion as the SUERM. The counter is decremented when 9308 intervals have passed without error (also defined by Bellcore). When the counter reaches a value of 144,292 intervals, the link is removed from service and alignment procedures are started. The actual values may be different, especially in international networks, but these are the recommended values defined by Bellcore.

The *alignment error rate monitor* (AERM) is used during the alignment procedure and is an incremental counter. Each time an error is encountered during alignment, the counter is incremented by 1. When the AERM determines that there have been excessive errors, it causes the link to be taken out of service, and the alignment procedure begins again.

Flow control

Flow control allows traffic to be throttled when level two becomes congested at a distant signaling point. The Link Status Signal Unit (LSSU) is used to send congestion indications to the transmitting nodes. When the LSSU of busy is received, the receiving signaling point then stops sending Message Signal Units (MSUs) until the congestion is abated.

Flow control also uses a priority for signal unit types to ensure that important signal units such as MSUs are transmitted, even during a congestion condition. Congestion is not the only condition indicated by flow control. Processor failures are also indicated by level two, meaning that level two can no longer communicate with levels three or four.

Flow control at this level should not be confused with traffic management at level four. Level-two flow control is isolated to individual links and does not indicate status of the signaling point. In addition, flow control at level two provides priority consideration for signal units, with no regard to the user part.

This is different from level-three network management, which does give consideration to the user of a signal unit. Level-three network management controls the flow of messages to a particular level-four user, while flow control at level two controls the flow of messages to a link processor.

If the congestion condition should continue, the link will be taken out of service and realigned, using the alignment procedure. This prevents a link from becoming "locked" in the congestion state.

Structure of MTP Level Two

The signal unit components used by level two are shown in Fig. 5.1. These components can be found in all three types of signaling units. The backward sequence number and the forward sequence number are used for sequencing packets and are used by level two to ensure that all transmitted packets are received. They are also used for positive and negative acknowledgments. The indicator bits are used to request a retransmission. The length indicator allows level two to determine the type of signal unit being sent, and the cyclic redundancy check (CRC) field is used to detect data errors in the signal unit.

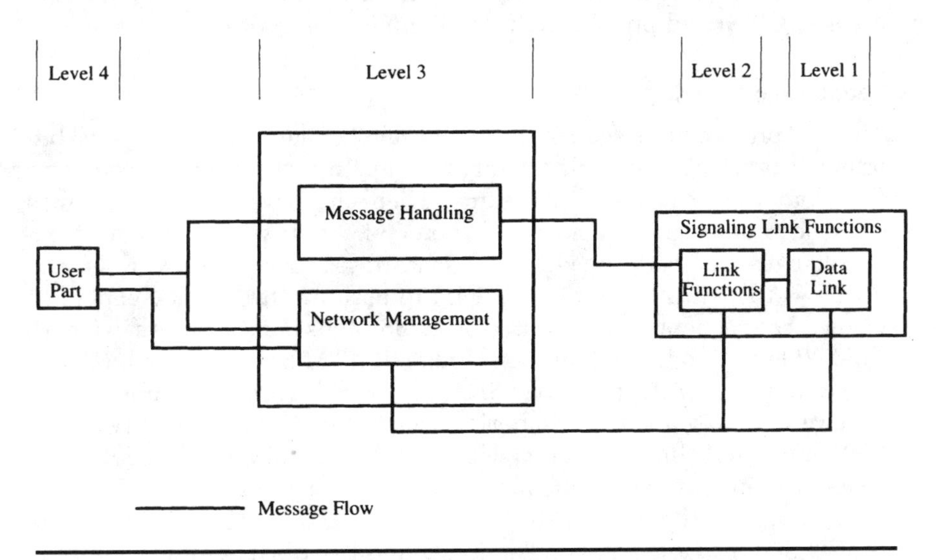

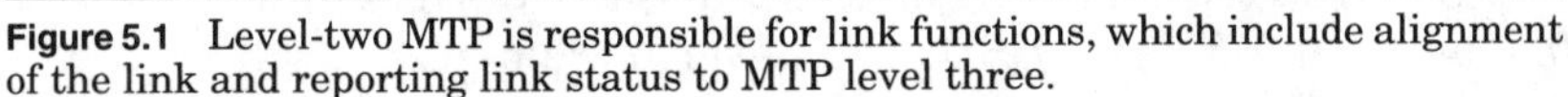

Figure 5.1 Level-two MTP is responsible for link functions, which include alignment of the link and reporting link status to MTP level three.

Flag

The flag is used to indicate the beginning of a signal unit and the end of a signal unit. As mentioned in the previous chapter, the flag in U.S. networks is used to indicate both the beginning of one signal unit and the end of another. In some other networks, there can be both an opening and a closing flag.

The flag bit pattern can be duplicated within the information field of a Message Signal Unit (MSU) (ones-density violation), causing an error to occur. To prevent the data from duplicating the flag pattern, the transmitting signaling point uses *bit stuffing.* Bit stuffing is the process of inserting an extra bit before transmission after every series of five consecutive 1s. The bit value is always 0.

By inserting a 0 after every five consecutive 1s pattern before transmitting a signal unit, the transmitting signaling point can ensure that there is never an occurrence of six consecutive 1s except for the flag, which is inserted just before transmitting the signal unit.

The receiving signaling point, upon receipt of a signal unit with five consecutive 1s, removes the inserted 0s from the signal unit. Because this is an absolute rule, a 0 is always going to be inserted after five consecutive 1s, and always removed after five consecutive 1s.

When a ones-density violation occurs, the signal unit is considered out of alignment and level three is notified of a link failure. The link then enters what is referred to as octet counting mode. During this mode, every octet is counted, and the number of errors per octet is monitored rather than errors per signal unit. The link is then taken out of service (OOS) and put through an alignment procedure.

Sequence numbering

The SS7 protocol uses sequence numbering like many other layer-two bit-oriented protocols. The SS7 technique is just a little different from, say, X.25, but the principle is the same. There are several ways in which sequence numbering is achieved. When 56-kbps links are used, a seven-bit sequence number is used. Both a forward sequence number (FSN) and a backward sequence number (BSN) are used on these links (explained below). The sequence number used on 1.536-Mbps links is 12 bits in length (if MTP level two is used on these links). If ATM links are used, MTP is replaced by the SAAL protocol, which uses a 24-bit sequence number.

Here is how sequence numbering works within MTP level two. The FSN indicates the number of the signal unit now being sent. This sequence number is incremented by 1 after every signal unit transmission, except in the case of the Fill-In Signal Unit (FISU) or the Link Status Signal Unit (LSSU). The FISU and the LSSU assume the FSN of the last sent MSU or LSSU, and never increment.

The BSN is used to acknowledge received signaling units. For example, if sequence numbers 1 through 7 have been sent and received by the distant signaling point, the next signal unit sent by the receiving signaling point could have a BSN of 7, which is acknowledging that all signal units, 1 through 7, were received without error.

The transmitting signaling point maintains all transmitted signal units in its transmit buffer until an acknowledgment is received. When a signal unit is received, the BSN is examined to determine which signal units are being acknowledged. All acknowledged signaling units are then dropped from the transmit buffer.

If there are signal units remaining in the buffer unacknowledged, they will remain until timer T7, "Excessive delay of acknowledgment," times out. When T7 times out, a link failure indication is given to level three. This will cause the link to be taken out of service and placed through the alignment procedure.

Signal units received are checked for integrity (check bit field or CRC), and they are also checked for proper length. A signal unit must be at least six octets in length, or it is discarded and the error rate monitor is incremented. A negative acknowledgment is then sent to request a retransmission of the bad signal unit. A signal unit's length must be in eight-bit multiples, or the signal unit is in error.

The Signal Unit Error Rate Monitor (SUERM) is an incremental counter that is incremented by 1 whenever an error is encountered. An error includes signal units received out of sequence, with a bad CRC or one of improper length. After 256 signal units have been received without error (consecutive signaling units), the SUERM is decremented by 1. This technique has earned it the nickname of the "leaky bucket."

When the SUERM reaches a value of 64 errors, a link failure is reported to level three and the link is taken out of service and placed through an alignment procedure. Level three controls the link management function and directs level two during the alignment procedure. Level two does not initiate the alignment procedure; it simply reports the errors and takes direction from level-three link management.

Indicator bits

The indicator bits are used to request a retransmission. There are two indicator bits: a forward indicator bit and a backward indicator bit. During normal conditions, both indicator bits should be of the same value (0 or 1). When a retransmission is being requested, the signal unit being sent by the signaling point requesting the retransmission will have an inverted backward indicator bit. The forward indicator bit retains its original value. This indicates to the distant signaling point that an error occurred and retransmission must take place.

The backward sequence number indicates the last signal unit received without error. The receiver of the retransmission request will then retransmit everything in its transmit buffer with a sequence number higher than the backward sequence number of the retransmission request. The indicator bits in the retransmitted message are independent of the originating exchange's indicator bits and do not provide any indication. Therefore, they will both be of the same value during the retransmission.

When the retransmitted signal units reach the distant exchange that originated the retransmission request, an acknowledgment will be sent. The forward indicator bit in the acknowledgment will be toggled to match the backward indicator bit and will remain at this value until another retransmission request.

The process we are describing is referred to as *basic error detection/correction.* These procedures are discussed further in the following section.

Length indicator

The length indicator is used by level two to determine which type of signal unit is being sent. The values of the length indicator can be:

- 0 (Fill-In Signal Unit)
- 1 or 2 (Link Status Signal Unit)
- 3 or more (Message Signal Unit)

The length indicator should match the length of the field immediately following it, before the CRC field. This field does not exist in the Fill-In Signal Unit (FISU). In the Link Status Signal Unit (LSSU) this is the Status Field (SF), which can be one or two octets in length. If the signal unit is a Message Signal Unit (MSU), there are two fields, the service indicator octet (SIO) and the signaling information field (SIF). The SIO is an eight-bit field, while the SIF is a variable-length field used by level four.

The maximum length of the SIF is 272 octets. Obviously, the length indicator (LI) cannot accommodate such a large number. Therefore, whenever the SIF length equals 62 octets or higher, the length indicator (LI) is set to the value of 63. Only level two uses this LI, and its only purpose is to indicate what type of a signal unit is being received. Length indicators can be found throughout the level-four packet to indicate the length of the variable fields within level four.

Cyclic redundancy check (CRC)

The CRC field is the last field in the signal unit. This field is calculated using the fields immediately following the flag, up to the check bit field itself. The fundamental process is rather simple. The transmitting signaling point performs the check before bit stuffing and transmission. The remainder is carried in the transmission in the *frame check sequence* field. The receiving signaling point then performs a similar calculation and compares the remainder to the frame check sequence field of the received signal unit. If there is a discrepancy, the signal unit is discarded and an error is counted in the Signal Unit Error Rate Monitor (SUERM). The CRC-16 method of error checking is used in SS7.

Basic Error Control Method

So far, we have discussed how a link is placed into service when it is initially started and a little about negative acknowledgments and retransmission. After a link has been placed in service, error detection/correction is used at level two to maintain proper transmission of SS7 messages. There are two methods of error control: basic and preventive cyclic redundancy (PCR). First, we will discuss the basic error control method.

Basic error detection/correction is used whenever a signaling link uses a terrestrial facility. "Land" links can be of any type of medium—it makes no difference to the Message Transfer Part (MTP) at this level. Basic error detection/correction is the most favored method.

Basic error control uses the indicator bits within the Message Transfer Part (MTP) portion of the signal unit to request retransmission of signal units received with errors (Fig. 5.2). When a Message Signal Unit (MSU) is transmitted, the transmitting signaling point sets the forward indicator bit (FIB) and the backward indicator bit (BIB) to be the same. This is the state the indicator bits should always be in when there are no errors. Both the FIB and the BIB should be of the same value, but it makes no difference if the value is a 1 or a 0.

When an error is detected by a signaling point, the errored signal unit is discarded and a negative acknowledgment is sent to the transmitting signaling point. The negative acknowledgment may use a Fill-In Signal Unit (FISU), a Link Status Signal Unit (LSSU), or a Message Signal Unit (MSU). The backward indicator bit (BIB) is inverted to indicate a retransmission request. Again, the actual value is of no sig-

8	8 to 272	8	2	6	1	7	1	7	8
FCS	SIF	SIO	X	LI	FIB "0"	FSN	BIB "1"	BSN	FLAG

Figure 5.2 This figure illustrates what a retransmission request would look like. The backward indicator bit (BIB) has been changed from its previous value, while the forward indicator bit (FIB) remains the same.

nificance; just the fact that the bit is different from the forward indicator bit (FIB) is of significance here.

The backward sequence number (BSN) should acknowledge receipt of the last good Message Signal Unit (MSU). The BSN is then used by level two at the transmitting signaling point to determine which signal units in the transmit buffer to retransmit. All those signal units which have been acknowledged are removed from the buffer, and the remaining signal units are then retransmitted.

When the retransmission begins, the FIB is inverted to match the value of the BIB of the received retransmission request. Both indicator bits should now match again. They retain this value until a retransmission is requested again, in which case the BIB will invert to a value different from the FIB.

Sequence numbers in the SS7 network can be a value from 0 to 127. This is known as modulo 128, meaning 128 sequence numbers can be sent before recycling again. The number of sequence numbers that can be sent before an acknowledgment must be received is referred to as the *window size.*

The larger the window size, typically, the better the throughput. This is only true, however, when you have reliable transmission. When the link has too many errors, retransmissions can congest the signaling point at the link level. When there is a large number of signal units to retransmit, the problem is compounded.

The window size of any signaling point is something that must be determined based on the type of transmissions, the capacity of the link, and the average number of retransmissions that occur.

Figure 5.3 shows a sequence of good transmissions, with positive acknowledgments. The positive acknowledgment is the sending of a signal unit in the opposite direction with an acknowledgment (the backward sequence number) for previously transmitted signal units. When the acknowledgment is received, the signal units can be dropped from the transmission buffer.

It should be mentioned here that an acknowledgment does not have to be received after every signal unit. A common practice in all asynchronous protocols is to allow several signal units to be acknowledged in one acknowledgment.

Likewise, an acknowledgment could be received for only some of the transmitted signal units. This does not indicate an error. As seen in Fig. 5.4, the acknowledgment could come a little later. Sometimes processors get busy and cannot acknowledge everything at once. As long as the timer T7, "Excessive delay of acknowledgment," does not expire, this does not pose a problem.

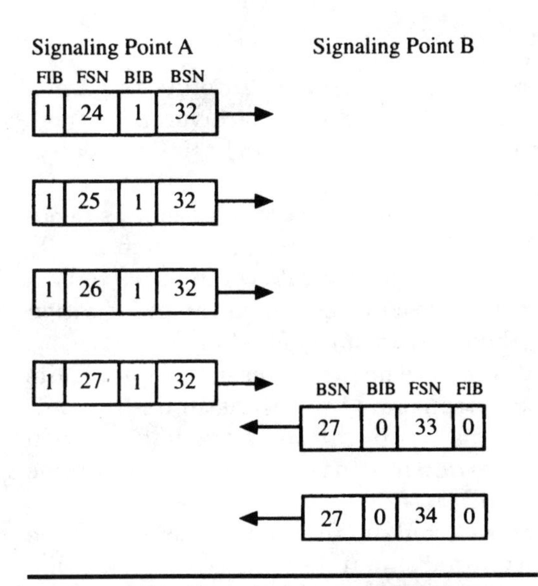

Figure 5.3 This figure depicts a successful transmission sequence between two exchanges. The BSN acknowledges receipt of the previously sent messages. Unlike most protocols that indicate what sequence number they expect next, the BSN value is of the last received sequence number.

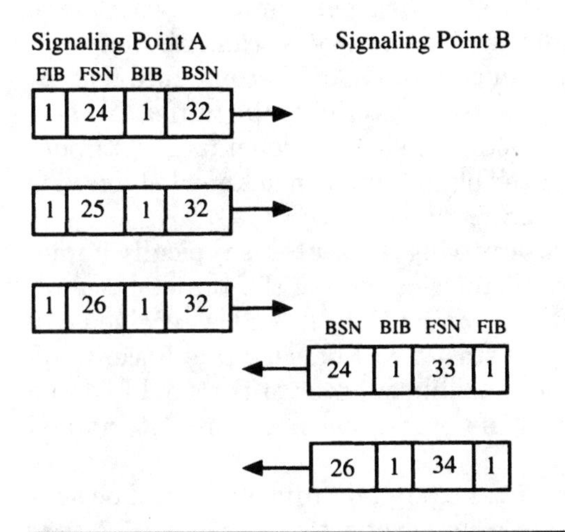

Figure 5.4 This figure appears as if there may have been an error; however, no error occurred. Signaling point B sent an acknowledgment for sequence number 24 and, then sent acknowledgment of 25 and 26. Timers allow this asynchronous acknowledgment of messages without causing retransmissions.

Preventative Cyclic Retransmission (PCR)

Preventive cyclic retransmission is used whenever satellite transmission is required for signaling links. This is not the favored method, since it uses a higher number of retransmissions than basic error detection/correction does.

PCR does not use the indicator bits when a retransmission is needed. Instead, all unacknowledged signal units are automatically retransmitted during an idle period. Indicator bits could prove to be a frustrating tool with satellite transmission, because of the propagation delay introduced with this method of transmission.

The PCR method waits until there are no more Message Signal Units (MSUs) to be sent, and then it retransmits everything in its transmit buffer. Only unacknowledged signal units will be in the buffer, so, in essence, the signaling point is retransmitting what it perceives to be signal units that have not been received.

If a signal unit has been received, and another of the same sequence number is received, the retransmitted signal unit is discarded. If a signal unit is received with a new signal unit, it is processed as normal. Eventually, the distant signaling point will send an acknowledgment for all received signal units.

If a signaling point is sending retransmissions and it receives additional MSUs to transmit, it stops the retransmission and sends the newly received MSUs. The distant signaling point must be capable of determining which are new MSUs and which are retransmissions.

If the transmit buffer of a signaling point should become full, it stops sending any MSUs and sends the entire contents of the buffer. No new MSUs can be sent during this period. If an acknowledgment is still not received, the buffer is retransmitted again, until an acknowledgment is received. This is known as a *forced retransmission*.

To prevent this condition from occurring, a counter is typically implemented to allow the forced retransmission before the buffer becomes full and MSUs become lost. If the buffer is full, MSUs received are going to be discarded, and a busy status will be sent to adjacent signaling points. By using an administrable counter, a threshold can be set to force retransmission when the buffer reaches a predetermined capacity.

The only rule in PCR is the number of signal units which can be sent without acknowledgment and the number of octets which can be sent without receiving an acknowledgment. No more than 127 signal units can be sent without acknowledgment. The number of octets must not exceed the time to send a signal unit and receive an acknowledgment.

Figure 5.5 depicts a sequence of events in which several MSUs have been sent, but not yet acknowledged. The transmitting signaling point

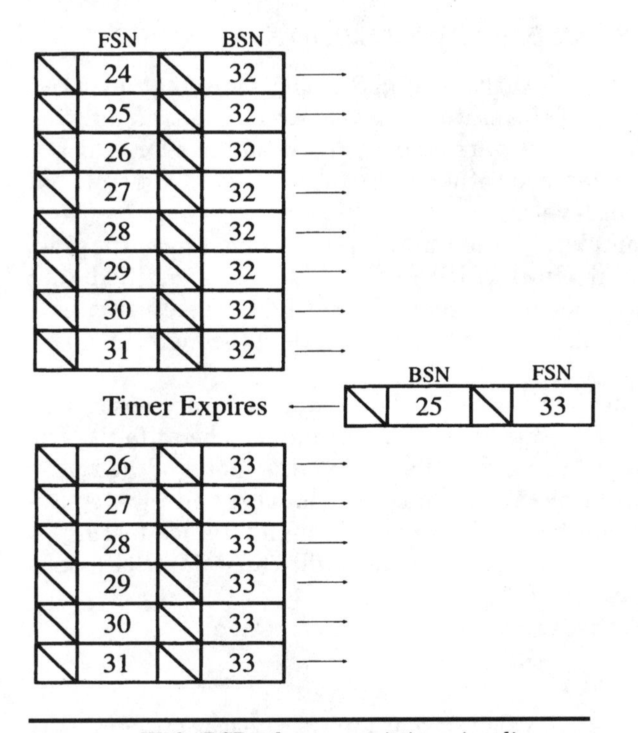

Figure 5.5 With PCR, the transmitting signaling point maintains all messages in its transmit buffer until an acknowledgment is received. The signaling point retransmits all unacknowledged messages after expiration of a timer, or when there is nothing else to transmit.

does not have any more MSUs to transmit, so it automatically begins retransmitting what is in its transmit buffer.

During the retransmission, a group of new MSUs is generated. This results in the retransmission being stopped and the new MSUs being transmitted. Once these have been transmitted, the retransmission begins again.

Finally, the distant signaling point has reached an idle state and begins sending acknowledgments back to the originating signaling point. The theory behind this method is that eventually a signal unit will reach the other signaling point, with or without retransmission. The retransmission may flood the link and push the link beyond its 40 percent capacity, but the messages are guaranteed a higher rate of success if continually transmitted in this fashion.

Structure of the Link Status Signal Unit (LSSU)

The Link Status Signal Unit (LSSU—Fig. 5.6) is also used at level two to notify adjacent nodes of the status of level two at the transmitting signaling point. Link status is carried over the same link for which it applies, and is not carried over other links (assuming, of course, that the link is functional at level two).

If level two fails completely, then nothing is received over the link, including Fill-In Signal Units (FISUs). Level three is notified that acknowledgments have not been received, and level three initiates a link failure recovery. Link recovery is accomplished through the alignment procedure.

The LSSU is transmitted by the hardware and software associated with a particular link. The link itself may or may not be at fault. The trouble most usually can be found at the termination point of the link—for instance, a link interface card. Since this is where level-two software resides, this should be the first point of maintenance testing. A hardware failure at the link interface card will cause level-two software to initiate LSSUs, as well as software.

The LSSU provides three types of status information:

- Level two is congested
- Level two cannot access levels three or four (processor outage)
- The alignment procedure has been implemented

The status field (SF) is used to provide the information to the adjacent signaling point. The information does not include the link number, which means that this signal unit cannot be used to notify nodes about link status over other signaling links. It pertains only to the link on which it was received.

Another important concept with the LSSU is that the status is really level two and three status in the transmitting signaling point, rather than the transmission facility. Throughout the various SS7 publications, when they talk about the signaling link status and signaling link functions, they are really talking about the level-two functionality in the signaling point itself.

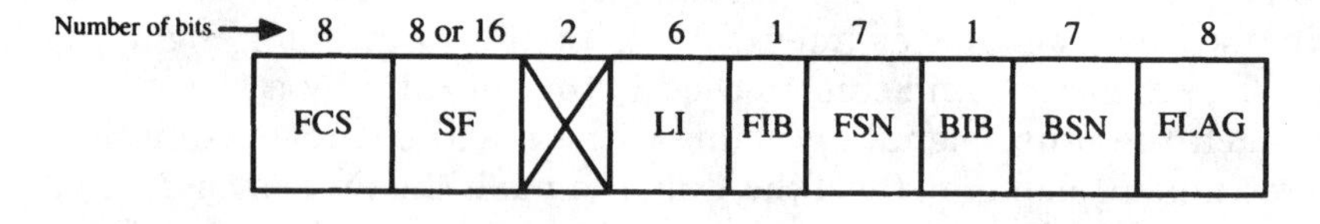

Figure 5.6 Components of the LSSU.

The status field is an eight- or 16-bit field, although only three bits are actually used. The three most significant bits provide the actual status information, while the remaining bits are set to 0.

The *status indicator "busy"* (SIB) indicates that level two is congested at the transmitting signaling point. When a signaling point receives an SIB, it stops the transmission of all Message Signal Units (MSUs) and begins sending Fill-In Signal Units (FISUs). If the condition persists for three to six seconds, level three is informed of a link failure. Level three then initiates the alignment procedure to begin on the affected link.

The congested signaling point will continue sending SIBs at regular intervals (timer T5) until the condition abates. Timer T5 is a value of 80 to 120 milliseconds. Any MSUs already received by the transmitting signaling point will be acknowledged, but no new MSUs should be sent.

At the receiving signaling point, the timer T7 is reset every time an SIB is received. Timer T7 is used for excessive delay of acknowledgment. If T7 should time out, the link is considered at fault, and level three initiates the alignment procedure. To prevent this from occurring, the T7 timer gets reset after every SIB.

To prevent an excessive delay caused by a received SIB, upon receipt of an SIB, a signaling point begins timer T6. This timer is used to time the congestion period, preventing congestion at a remote signaling point from causing the network to bottleneck. If T6 should time out, the link is considered at fault, and level three initiates the alignment procedure on the affected link. The value for T6 is one to six seconds.

To indicate that congestion has subsided and normal processing can begin, the affected signaling point will begin transmitting MSUs again and stop the transmission of SIBs. The receiving signaling point recognizes the absence of SIBs and begins receiving MSUs as normal.

The *status indicator "processor outage"* (SIPO) indicates that the transmitting signaling point cannot communicate with levels three and four. This could be caused by a central processor unit (CPU) failure or a complete nodal failure. If maintenance personnel have manually placed a link out of service, the link will send SIPOs to the adjacent signaling point as well. Level two is usually functional, since it resides within the link interface hardware. The signaling point sends an SIPO to notify the distant signaling point to stop sending MSUs.

Upon receipt of an SIPO, transmission of all MSUs is stopped, and FISUs are sent to the affected signaling point. The transmitting and receiving nodes stop level-two timers T5, T6, and T7. If the condition persists for too long, then the link is failed, and level three initiates the alignment procedure on the affected link.

Level three, even though there has been a processor outage, may still be capable of controlling level-two functions. When this is the case, level three may request level two to empty its buffer (both receive buffer and transmission buffer). When this is the case, all MSUs in the buffer are discarded. When an MSU is received from a remote signaling point, level two of the affected signaling point sends an FISU with the forward sequence number (FSN) and the forward indicator bit (FIB) set to the same value as the backward sequence number (BSN) and the backward indicator bit (BIB) of the last MSU received from the remote signaling point. Normal processing of all messages then resumes.

The Link Status Signal Unit (LSSU) also sends a *status indicator out-of-alignment* (SIO). This condition occurs when a signal unit is received that has a ones-density violation (the data field simulated a flag) or the signaling information field exceeded its maximum capacity of 272 octets. The SIO is sent when a link is failed and the alignment procedure is initiated.

An LSSU of *out of service* (SIOS) indicates that the sending signaling point cannot receive or transmit any Message Signal Units (MSUs) for reasons other than a processor outage. Upon receipt of an SIOS, the receiving signaling point stops the transmission of MSUs and begins transmitting Fill-In Signal Units (FISUs). The SIOS is also sent at the beginning of the alignment procedure.

Link status of normal (SIN) or *emergency* (SIE) indicates that the transmitting signaling point has initiated the alignment procedure. The link is made unavailable for MSUs and only FISUs are transmitted to the affected signaling point until a proving period has been passed (see the section on proving periods further in this chapter). After successful completion of a proving period, MSUs can be transmitted over the link to the affected signaling point.

In the event that a Link Status Signal Unit (LSSU) is received with errors, the receiving signaling point discards the signal unit. Retransmission is not requested of the LSSU. The forward sequence number of an LSSU does not increment, but assumes the value of the last transmitted MSU. The backward sequence number does increment when an acknowledgment is being sent to the distant signaling point.

The LSSU is processed within level two and does not get passed to level three. However, level two may pass control information to level three, depending on the status of the LSSU. For example, if an LSSU with a "busy" status is received, level two notifies level three to stop transmission of MSUs.

It is also important to remember that the LSSU works independently of network management, which is a level-three function. In fact, network management uses the MSU to send management information to other nodes and can use any link or route to reach adjacent nodes.

Level three is used when links have failed altogether and LSSUs cannot be transmitted over the affected link.

Most level-two problems are caused by hardware failures. Therefore, they do not require the same level of sophistication as software problems. Level-three and -four errors are normally software related, and they require more sophisticated reporting mechanisms. The recovery procedures used by maintenance personnel will also be very different for level-three and -four problems from those at level two.

Signal Unit Alignment Procedure

Alignment is the sending of correct signal units with correct length and without ones-density violations. When an error occurs due to either of these events, the affected signaling point will begin the alignment procedure.

The purpose of the alignment procedure is to reestablish the timing and alignment of signal units so that the affected signaling point can determine where signaling units begin and end. As previously mentioned, the out-of-alignment condition occurs when the flag has been simulated within the data (ones-density violation) or the signaling information field is too long (longer than 272 octets), which would indicate that a flag was missed.

The procedure resets both the transmitting and receiving nodes at level two and does not affect other links at either signaling point. The procedure also provides testing for a given period of time to ensure that the link transmission is reliable, preventing further errors from occurring.

There are two alignment procedures used: normal alignment and emergency alignment. Normal alignment is used when there are other links associated with the affected link (such as in a linkset). The other links must be to the same destination. An emergency alignment is used when there are no other links to the adjacent signaling point within the linkset. The emergency alignment goes through the same procedure, but within a shorter time period. Level three is responsible for determining which alignment procedure to use.

There are four states entered during alignment. Timers associated with each state ensure that the signaling point does not get stuck in any one state. When any of the timers times out, the alignment starts over again. The following explanation describes each state and the events that occur during these states.

State 00—idle

This state indicates that the procedure is suspended, and it is the first state entered. State 00 is resumed whenever the alignment procedure is aborted (due to excessive errors). During the time a link is in "proving," level-three network management reroutes signal units to other links. If a link should fail the proving period, level three places the link back into state 00 for a specified time period (level-three timer T17).

To prevent rapid link oscillation between in-service states and out-of-service states, level-three timer T32 is used. When a link is placed back into the alignment procedure, timer T32 is started. If the link fails during T32, the link is placed back into state 00, until T32 expires. Attempts by the remote signaling point to begin alignment of the link are ignored during this time. LSSUs with status indication of out of service (SIOS) are sent by the signaling point until the T32 expires.

State 01—not aligned

This is entered when initiated by level three. The LSSU is used to send a status of out of alignment (SIO). When this state is entered, the level-two timer T2 is started. Timer T2 ("not aligned") for normal alignment is set to 11.5 or 23 seconds. If the signaling nodes at either end use a mechanism for the automatic allocation of signaling links, the timer T2 must be set differently at each end.

Nodes have the option of assigning links for signaling but making them inactive (at which time, they are available for other applications, such as voice transmission). The signaling point has the capability through level three of interrupting the inactive links from their present applications and reassigning them to carry signaling traffic on an as-needed basis (such as in the event of signaling link failures within a linkset). It is with this type of application that T2 must be set differently at either end.

State 01 remains until level three initiates the next state transition. During this state, the signaling point transmits LSSUs with a status indication of SIO.

State 02—aligned

This state indicates that the link is aligned and is capable of detecting flags and signal units without error. Remember that "out of alignment" means that the signaling point can no longer delineate signal units based upon receipt of a flag. What this really implies is that the link has lost its timing, and it no longer recognizes the beginning and end of a signal unit. This state is indicating that the link is now capable of detecting flags and recognizes the boundaries within the signal unit itself.

During the time that the link is in state 02, level-two timer T3 is started. When the link leaves state 02 and enters state 03, T3 is stopped. If T3 should time out, the link is returned to state 00, and the process begins all over again.

State 03—proving

The proving period is used to test the integrity of the link and level two at the signaling point. During the proving period, LSSUs with the value of SIN or SIE are sent, and errors are counted. There are two proving periods: normal and emergency. The proving period of normal lasts for 2.3 seconds, during which time, no more than four errors may occur while in state 03. The alignment error rate monitor keeps count of all errors received during the proving period. The AERM is an incremental counter, which counts all transmission errors, including CRC errors and ones-density violations.

During the proving period, FISUs are sent on the link. The LSSU of normal or emergency (SIN or SIE) is also sent to indicate that the link is in proving.

The emergency proving period lasts for a duration of 0.6 seconds, during which time, no more than one error may occur. This is also monitored by the AERM. When excessive errors have occurred according to the procedure, the link is returned to state 00, and the process begins all over again.

If a link cannot be restored, it is returned to the idle state and the alignment procedure is repeated, until either the link is restored or maintenance personnel detect the repetitive failure and take corrective action. When a link is found to be continually failing or in constant alignment, it usually indicates an equipment failure, which can be resolved by manual intervention (replacement of the link interface card usually fixes the problem). Maintenance personnel should always be monitoring the status of signaling links and watching for links that continuously fail. Traffic measurements provided by most signaling points can be a useful tool in determining the number of failures during a given period of time.

Level-three alignment processes

After a link has successfully passed the alignment procedure, the link is returned to an in-service state, where MSUs are transmitted, and normal processing is allowed. The link is placed into a probationary period by level three, which lasts a period of level-three timer T33. If the link fails during the probation, it is placed back into suspension (state 00), during which time, level two sends LSSUs with a status indication of out of service (SIOS). All attempts to place the link into alignment are ignored until level-three timer T34 expires.

Timer T34, "Suspension timer for link oscillation," is also used to prevent links from rapidly oscillating from in-service to out-of-service state. This set of level-three timers, T32, T33, and T34, are all used to filter link oscillation and are only required in Bell System networks at Signal Transfer Points (STPs).

Upon completion of the alignment procedure at level two, level three begins a signaling link test, to determine if the link is capable of carrying traffic. This is an option in the ANSI recommendations, but it is a requirement within the Bell System networks.

The Bell System networks also use a mechanism for the automatic allocation of signaling links in the event that there is a signaling link failure within the linkset and additional links are required. This procedure uses a predefined set of links which are used for other applications (such as voice transmission) and, when needed, places them into the alignment procedure for preparation as signaling links within the SS7 network.

This process ensures that there are always links available, even when the designated active links have failed and cannot be restored. Level three is responsible for the activation and restoration of these links, and for assigning them to the proper linkset. Link management at level three is the function that looks after this.

In summary, MTP is a vital part of the SS7 network. This protocol provides the transport services for messages between signaling points. All messages traveling through the network rely on MTP for reaching the adjacent signaling point without error.

The level-three network management functions of MTP ensure the integrity of the network, managing the various functions of routing and link alignment. This vital part of the network allows the SS7 entities to work around failures and congestion conditions without human intervention, and is what makes the SS7 network so robust.

Chapter

6

Message Transfer Part Level Three

There are two categories of functionality at this level: signaling message handling and signaling network management. These are level-three functions. Signaling message handling is used for routing messages to the appropriate link and determining if messages are addressed to the received node or are to be forwarded. Signaling network management is used to reroute traffic to other links when nodes become unavailable.

The Message Transfer Part (MTP) level three relies on the services of level two for delivery of all messages. The interface between the two levels consists of a set of primitives. Primitives allow for parameters to be sent to level two for routing over the network in the form of SS7 messages. At the same time, the primitives allow level two to send parameters to level three for message processing.

Message processing begins at level three. Level three must determine the destination of a message and the user of a message, and maintain the status of the network. Level three uses primitives to communicate to level-four users. Parameters are sent to level four through these primitives.

In the same fashion, primitives allow level four to send parameters down to level three for inclusion in a signal unit and transmission over the network. These primitives use the same structure as the primitives used between level two and level three.

Message-Handling Overview

Message handling includes three different functions: message discrimination, message routing, and message distribution. The interface to level two is the message discrimination function. Message discrimination determines who the destination signaling point is by reading the routing label of the message. If the destination point code is the same as the signaling point's self-identification (its own address), the received signal unit is given to message distribution.

The message-handling function uses the routing label found in all level-four messages to determine who the originator is and who the destination is. It should be noted that a signaling point may have more than one point code. In addition to its own self-identification, the signaling point can also possess a capability point code, or an alias point code. This allows a signaling point to be partitioned into separate entities. For example, it may be advantageous to assign a point code to an STP for gateway services, and another point code for global title services.

Message distribution is the mechanism used to deliver a message received by a signaling point when the *destination point code* (DPC) is its own. The distribution function must deliver the message to the proper user part or network management function.

The service indicator octet (SIO) is used by message distribution to determine who the user part is. The user (usually level four, but not always) can be any user part or network management. In the case of network management, the SIO will indicate whether the user is network management or network management testing.

Message routing is used to pass a message back to level two for routing over the network. If message discrimination determines that the message is not addressed to the receiving signaling point, it sends the message back to message routing. If level four has generated a message for transmission, it also is given to message routing.

Message discrimination overview

Discrimination applies to all received messages. Message discrimination uses the routing label of the Message Signal Unit (MSU) to determine the destination of a message. The routing label provides the origination address (origination point code—OPC), the destination address (destination point code), and the signaling link code (SLC).

Message discrimination reads the routing label, specifically the destination point code, to determine if the destination is the receiving node or if the message must be transferred to another node. If the message is intended for the receiving node, then the message is given to the message distribution function, which must determine which user part will receive the message (user part being the level-four function, such as ISDN User Part).

The user is determined by reading the service indicator octet field. This field is divided into two parts: the service indicator and the subservice field. The service indicator identifies the user part to receive the message. A user part does not have to be a level-four function. Network management and network management testing are level-three functions that use the information field of a Message Signal Unit.

The service indicator can also indicate whether special or regular testing messages are being sent by network management. This is also used by level-three network management, each requiring a unique procedure.

The service indicator octet subservice field identifies whether this message is from an international network or a national network. This part of the message is used to identify the type of point code structure used, so the discrimination function can determine how to read the routing label. If the value of the network indicator in the subservice field indicates an international network, the routing label uses a different structure than it would for a national message.

The national network indicator does not necessarily mean that this message came from an ANSI network. The national network indicator is used by ITU-TS to differentiate between various structures used within different countries. This allows each country to use its own point code structure within its own national network, while still conforming to the international standard set forth by the ITU-TS.

If the service indicator octet indicates that this is a national message, the SIO can also be used to indicate two different versions of a national structure. The "spare" or reserved code in the subservice field can be used to indicate a different version of the national structure.

For example, a country may have two unique point code structures used in different segments of its network. This could be indicated by using the reserved bits in the national indicator of the subservice field. In ANSI networks, these bits are used to indicate the priority of a message signal unit.

If used in a private network that does not access the Public Switched Telephone Network, the network indicator does not have to be used (since the network is closed and it will always access the same network). This field can then be used as an extension of the service indicator, providing additional user part identification. The implementation of this feature is network dependent, and it is not defined in the ANSI or Bellcore standards.

In the ANSI network, the national network indicator includes two spare bits. These spare bits are used to indicate the priority of the

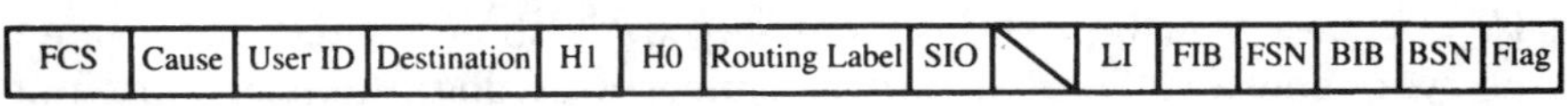

FCS	Cause	User ID	Destination	H1	H0	Routing Label	SIO		LI	FIB	FSN	BIB	BSN	Flag

Figure 6.1 This figure illustrates the components of the user part unavailable (UPU) message.

Message Signal Unit. Priorities are used by network management during periods of congestion and rerouting procedures. Recommended values for the priorities are given in the ANSI and Bellcore standards, but they can be network dependent.

If the message is not intended for the receiving node, then it is given to the message-routing function. The message-routing function must determine which link to choose to reach the destination node. In many cases, the next node to receive the message will not be the final destination. There may be other Signal Transfer Points (STPs) that will receive the message before it reaches its destination. A routing table is used to determine which is the shortest path to reach the destination. The objective in routing is to prevent multiple "hops" between nodes, introducing delay in the transmission. The routing table uses a priority mechanism for determining the most efficient route, that is, the route with the most direct path and the fewest hops.

If the message is for another node, and the message-routing function determines that it cannot reach the destination, network management must respond. Network management is a level-three function that is invoked when a signaling point failure is detected or a link failure is detected. We will discuss these messages later in the chapter.

If the point code in the routing label of a message is the point code of an alias, the message is given by message distribution to the global title translation function. If the global title translation function is not available at the signaling point, the message is given to message routing for transmission back onto the network.

The global title translation function is provided by Signal Transfer Points and can be located regionally or within several STPs.

Message distribution overview

If the destination point code is the address of the receiving signaling point, then the message is given to the distribution function. Message distribution must determine who the user is for any given message. This is determined by examining the service indicator octet service indicator field. As mentioned before, the user can be a level-four function or level-three network management.

In the event that a user is not available, a *user part unavailable* (UPU) message is sent by network management as an indication to the originating point that the message was discarded because the user part function was not equipped at the destination or was unavailable. The UPU message also provides a cause code, although this cause code is not very specific. There are three causes provided: the user part function was not equipped at the destination, the user part function was not accessible (i.e., out of service due to a processor outage), or the user part function could not be reached for an unknown reason. The UPU message is shown in Fig. 6.1.

Some user parts are processes that reside within a central processor at the signaling point. For example, the Signal Connection Control Part (SCCP) may provide global title translation, considered a user part in this case. When the processor at the signaling point becomes congested or unavailable, the user part unavailable message would be generated.

This does not indicate a problem with the signaling point, however. The signaling point may be able to carry on with other processing, such as levels two and three, or even level four. This, of course, depends on the architecture of the signaling point itself. When distributed processing is implemented, the impact of a user part unavailable message should be minimal from a signaling point perspective.

Message routing overview

If a message is received from another destination, then message discrimination passes the message on to message routing for transmission back onto the network. This is a common function of the Signal Transfer Point. An end signaling point, such as a Service Switching Point (SSP), would not likely receive messages that needed to be transferred, unless F-links were deployed. Level four uses the routing function to send messages originated by the point code out onto the network.

The routing function must first determine what the destination address is by reading the routing label and looking for the destination point code. In an ANSI network, the subservice field of the service indicator octet will indicate a national network, which means the point code will be a 24-bit field. If the subservice field indicates that the message came from an international network, then the point code will be a 32-bit field.

The international point code structure is very different from the ANSI structure. The first value indicates the zone of the address. The zone is a four-bit value, and can be used to address a country or a group of countries. The next field is the area identifier. This is an eight-bit field and identifies the network of the address. The third field is the individual signaling point identifier, and it identifies the actual member of the network. A member is any signaling point in the network.

It is important to note that the international point code structure is only valid at the international hierarchy. Once a particular gateway is reached, messages are converted into national formats, where the point code structure and protocol rules can change. This allows each country to deploy SS7 in its network to meet the needs of the individual country, rather than to define a standard that does not address everyone's needs. The ITU-TS distributes point codes for use in the international network.

In the United States, the national standard is ANSI. The ANSI standard is also used in the Bellcore networks, but it is modified for the specific requirements set forth by the Bell Operating Companies. The point code structure defined for usage in the United States is quite a bit larger than the ITU-TS version. Point codes in the ANSI network are distributed by Bellcore.

The first field in the point code is the network identifier and it is used to identify the individual network being addressed. This is an eight-bit field. The network identifier is reserved for only the largest of service providers. Those who receive a unique network identifier are also given usage of all numbers in the cluster and member fields.

Network identifiers 1 through 4 are reserved for medium networks. Network identifier 5 is reserved to identify small network clusters. There is a network cluster code associated with each state and territory in the United States and Canada. These are used to identify smaller networks. Medium networks are those that do not have enough signaling points to warrant the distribution of an entire network number. Presently, only network identifiers 1 and 2 have actually been assigned to service providers. The cluster field then identifies the individual networks.

The cluster field is also an eight-bit field and is used in two different ways. In the case of those networks that have their own network identifiers, the cluster number can be used to group signaling points together to provide more efficient routing (cluster addressing). In medium networks (network identifiers 1 to 4), the cluster field identifies the individual network.

Small networks (network identifier 5) use cluster value 1 or 2. The member codes are then divided into three code blocks and given to very small networks with only a few signaling points. The codes are assigned in blocks, depending on location and number of signaling points in the network. The member value of 0 is reserved for STPs.

Cluster addressing allows Signal Transfer Points to maintain small routing tables. All messages will contain a full point code, but in some networks partial routing is used. Partial routing only examines the network and cluster fields of the point code or, in some cases, only the network identifier. The message is then routed based upon that value only. The full point code does not appear in the signaling point's routing table.

When networks grow quite large, partial routing helps minimize the number of point code entries required at every signaling point. Every local Signal Transfer Point must have the full point code of all signaling points connecting to it directly (homing). If any other point codes are connected indirectly (through other signaling points), only the network identifier and cluster are entered into the routing table. The last field of the point code is the member code.

The message-routing function receives messages from level four when a newly created message is to be transmitted, and from message discrimination when a message is received but needs to be transferred to another signaling point. The message routing depends on the destination point code located in the routing label to determine the destination of the message, and to determine which link the message is to use.

There are two forms of routing that may be deployed within any signaling point: full point code routing and partial point code routing. In the case of full point code routing, the signaling point looks at the entire point code to determine how the message will be routed. This means the routing table for that signaling point must include all point codes for all connecting signaling points and end points in the network.

Even though a particular point code does not directly terminate with the routing signaling point, the point code of the remote signaling point must be resident in all routing tables. This means that large networks will require large routing tables in their signaling points.

The use of cluster routing reduces the number of point code entries in the routing table. With cluster routing, only the cluster code is read to determine the path for a message. This only applies to remote signaling points, however, and eventually full point code routing must be used to reach the final destination. The home Signal Transfer Point is usually the point at which full point code routing must be used.

All other Signal Transfer Points need only the cluster address to route a message. This is a significant reduction in the space required in the routing table. Memory requirements can get cumbersome in large networks, making partial routing a necessity.

When using cluster routing, the Signal Transfer Point can have the same cluster number as all those signaling points connecting to it, or it can have a unique cluster number. If it uses the same cluster number as its end points, then an alias point code must be assigned for global title functions resident within the STP. If the STP does not provide this function, then there is no need for an alias.

The use of a unique cluster for each signaling point code does allow the Signal Transfer Point to be directly addressed within any message and can be of benefit for network management. Whether the STP has a unique cluster number or uses the same cluster number as the other signaling points in this cluster group, the member code is usually zero, identifying it as an STP (this is optional, but widely used in many U.S. networks).

Another form of partial routing is network routing. This uses the same concept as cluster routing but reads only the network address. This is very useful at signaling points connecting to other networks. In some countries, there may be one common database for all networks in the country. In this case, network routing will allow signaling points to route to that database with only the network identifier.

Gateways are another likely candidate for network routing. Because a gateway connects to so many networks, routing tables can get cumbersome and difficult to maintain. The network administrator must be aware of all point codes it routes to. With network routing, only the network identifier need be known, saving administration and memory resources of the signaling point.

Link selection

The signaling link used to route the call is determined by the value of the *signaling link selection* (SLS) field located in the signaling link code (SLC) field of the routing label. The SLC code is a preassigned code representing a physical link. The SLC is not the link number, but a logical connection to a physical link.

All links are assigned a signaling link code and a terminal number. The terminal number is a logical connection to a physical link. The SLCs range from 0 to 127. This means that each link will have more than one SLC. When a link is deployed, it is up to the equipment software to reallocate the SLCs to accommodate the newly equipped link. When a link fails, software will again reallocate the SLCs so that the failed link no longer has an SLC.

Before transmission, level-three routing determines which link will be the next outgoing link. This is based on several factors. First, all links within a linkset must carry equal traffic (load sharing). This means no link will be carrying more than the others. The Message Signal Unit size must also be figured in the traffic capacity of every link, so that one link does not become burdened with large messages while another is sending only small messages.

After a link is selected, the signaling link code of the physical link is placed in the SLC field of the routing label and sent to level two for transmission. When the message is received at the remote signaling point, the bits in the SLC field are rotated one position to the right by level-three routing. Presently, only the five least significant bits are rotated, even though the standards now support an eight-bit SLC field. The three most significant bits are used for link selection in networks where eight-bit rotation is supported (this is optional today, and will slowly be implemented into all networks).

There are exceptions when bit rotation is not to be used. Call setup messages and database query messages are examples of messages that should not use bit rotation. These messages usually require more than one signal unit to be sent. These signal units must be received in order, or an error will occur. To prevent them from taking different routes and being received out of order, the messages relating to one transaction or to the same circuit connection always follow the same path (use the same links from one signaling point to another signaling point). Cross links (C-links) between mated Signal Transfer Point pairs do not use bit rotation.

There are also certain maintenance messages which must be routed over specific links. For example, a changeover order is always sent over an alternate link, but the acknowledgment of a changeback order can be routed over any available link. Level-three routing must also take these rules into account before selecting a link and rotating the signaling link selection bits.

Load sharing must be used whenever there is more than one link or one linkset. In the event that there is more than one linkset, the least significant bit (LSB) is used to select the linkset to be used. The remaining four bits then identify the link within the linkset. In the event that only one linkset exists, the least significant bit is used as part of the link identifier, but the most significant bit (bit 5) is left unused (maintaining a four-bit link selection code).

In the international network, load sharing is not used. The signaling link code (SLC) field is used to identify the voice *circuit identification code* (CIC) for all Telephone User Part (TUP) messages. For Data User Part (DUP) messages, the bearer identification code is identified in this field. Bit rotation is not used in international networks.

Normal Routing Procedures

Before a message is transmitted, the routing table must identify which link and linkset to route the message to (Fig. 6.2). The routing table is maintained by administration personnel. Every routing table must indicate the primary route to a destination point code and an alternate route(s). Each route is assigned a priority, which indicates whether it should be first choice, second choice, etc. Some equipment manufacturers use different techniques for identifying the priority for a route, but the basic concept is the same.

The rule is to always choose the most direct route to any destination. The most direct route will always be the route with the fewest "hops" through other signaling points. In the case of an end signaling point, if an F-link is available, it should be first choice if it terminates at the

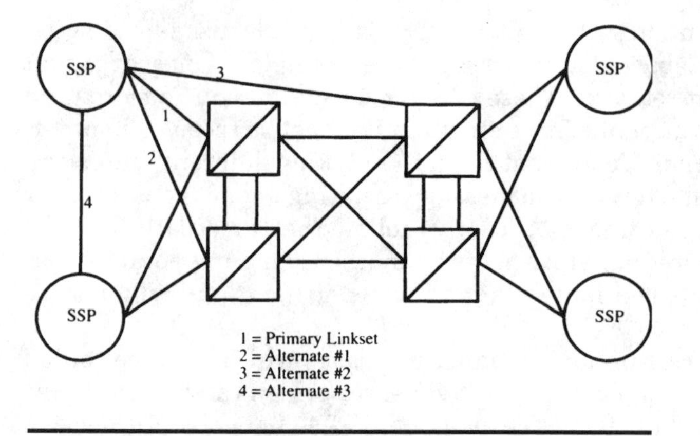

Figure 6.2 This figure shows a typical routing scheme for SS7 entities. The numbers indicate the choice, or weighting factor, assigned to each of the links. For example, the "one" indicates first-choice route, etc.

destination. If an E-link is available, it should be the second choice. A-links are last-choice routes. Of course, in many cases, the A-link may be the only route to a destination point code, in which case it will be the first choice.

For a Signal Transfer Point, if the destination is an end point (end points are Service Switching Points and Service Control Points), then an A-link should be first choice. If there are no available A-links to the destination point code, then an E-link should be used (if one exists). If the message must go through another Signal Transfer Point to reach the destination point code, then the STP which the destination "homes" to should be the next destination for the message. The home STP is that which connects directly with the signaling point (Fig. 6.3).

In the event that the home signal transfer point cannot be reached directly, then the message should be routed to a primary STP in a two-

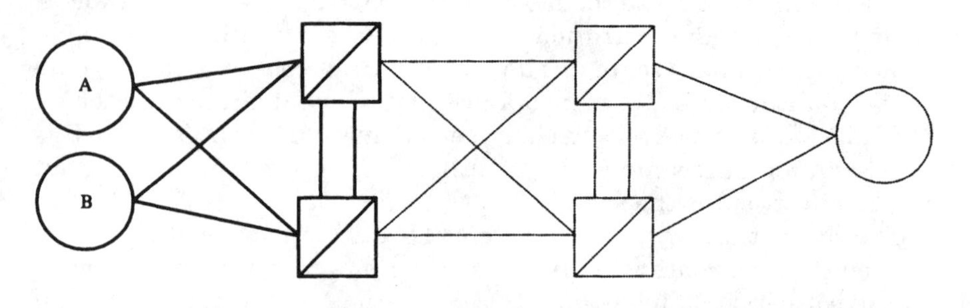

Figure 6.3 In this figure, both A and B have the same "home" STP. A home STP is one which is directly adjacent to the SSP.

level hierarchy. The primary STP of the destination point code should be the STP used to reach the home STP. In the event that the primary STP cannot be reached, then the primary STP of the message originator should be used.

Network management overview

Network management at level three provides the procedures and functionality to reroute traffic through alternate links and linksets or to control the flow of traffic to a specified destination point code. There are three separate network management functions used in SS7:

- Traffic management
- Link management
- Route management

Traffic management is used between two signaling points to divert traffic away from failed links. Traffic management messages are originated by the signaling point, which detects a problem in a link and is sent over an alternate link to inform its adjacent signaling point not to route messages over the affected link. Traffic management also triggers route management from the receiving signaling point to its adjacent nodes.

Traffic management messages are not propagated through the SS7 network. They are only point-to-point. The difference between traffic management and link management (which incorporates the use of level-two Link Status Signal Units) is the fact that the LSSU is carried over the signaling link that is in question.

This fundamental difference points out that the link must be capable of carrying level-two traffic, while not being able to carry level-three and -four traffic. In the event that a link cannot carry level-two traffic (such as in the case of a complete link failure), then level-three traffic management will be implemented to advise adjacent signaling points that the signaling link can no longer be used.

Link management consists of activation and deactivation procedures, as well as link restoration. These functions are used with level two to restore failed links back into service. The link management function is what triggers level-two link alignment procedures and guides the transitions through the various states of alignment.

Route management is used to divert traffic from a specific signaling point. Route management does not pertain to any one specific link as traffic management does, but to an entire signaling point. Route management is triggered by traffic management messages, and is sent to

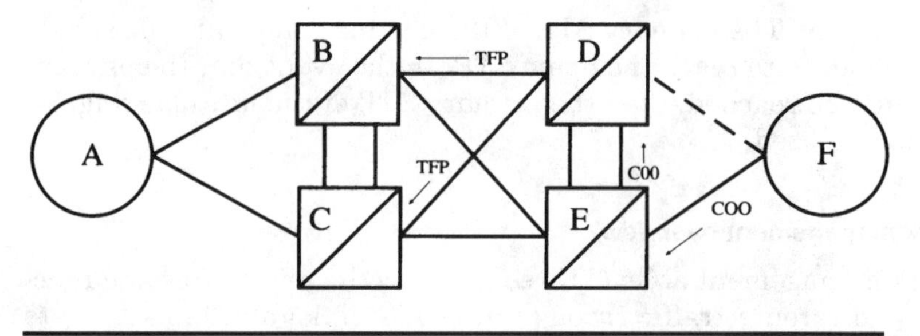

Figure 6.4 This figure depicts traffic management and route management.

the adjacent signaling points of any signaling point that receives a traffic management message.

For example, in Fig. 6.4, signaling point F has failed its link to signaling point D. Signaling point F sends a changeover message to signaling point D through signaling point E (traffic management). Signaling point E has no knowledge of what the traffic management message is about, nor does it care. It is transferring the message on to its final destination.

Signaling point D, upon receipt of the changeover message, will send a *transfer-prohibited* (TFP) message to its adjacent nodes. The TFP message indicates the destination point code, which is no longer accessible through signaling point D. All adjacent signaling points then invoke rerouting procedures to begin routing traffic destined for signaling point F through signaling point E, since this is the only available route. If there were other links available through D, then those links would be used to route traffic to F and the TFP would probably not be sent, since signaling point F can still be reached through D.

Link management is then invoked by signaling point F to restore the failed link. This triggers level two to begin the alignment procedure and controls the transitions from one state to the next through the use of timers. When the link has been restored by level two, level-three link management is notified, and traffic management sends a changeback message to restore traffic back to the newly restored link.

Upon receipt of a changeback message, signaling point D then invokes route management to alert adjacent nodes that traffic destined for signaling point F can now be routed back through signaling point D.

This is only one example of how these three network management functions work together. As we discuss each management function, we will discuss specific examples in greater detail.

Signaling Network Management Procedures

As we mentioned before, there are three types or categories of network management. Traffic management diverts messages away from failed links, link management is responsible for the activation and deactivation of signaling links, and route management is responsible for the rerouting of messages around failed signaling points. Route management also controls the flow of messages to a signaling point.

Network management also provides a layered approach to managing troubles in the network. Procedures are provided that deal with network congestion and outages, from the link level all the way up to the route level.

Level two provides the means to detect errors on individual links. Level two is not interested in problems outside the signaling point. Level two is only concerned about an individual link. Every link has this function, which is controlled by level-three link management.

When an error is encountered, level two reports the error to level three, which must then determine which procedures to invoke. Procedures begin at the lowest level, the link level, and work their way up to the route level.

Link management directs the procedures at the link level. These procedures do not have a direct impact on routing or the status of signaling points. They do trigger other network management events at level three, however.

Link management does have an impact on other functions, but indirectly. Traffic management is affected by link management, primarily because traffic management must divert traffic away from a link that link management has failed and removed from service. Traffic management ensures the orderly delivery of diverted traffic, providing for the transfer of unacknowledged messages to another link buffer and retransmission of messages on a different link.

Traffic management does not divert traffic away from a signaling point. The purpose of traffic management is to redirect traffic to a signaling point to different links. However, traffic management does impact routes and route-sets to specific destinations. If a particular route is used by another signaling point to reach a destination, and traffic management has diverted traffic away from that route, adjacent signaling points may have to invoke route management procedures.

Route management diverts traffic away from signaling points that have become unavailable or congested. The reasons vary, but regardless of the reason, traffic management and link management will be involved at the affected signaling point. In the meantime, all the signaling points around the affected signaling point are invoking route management procedures to prevent messages from becoming lost.

This layered approach to network management is what makes SS7 networks as robust as they are. Very few network troubles actually cause the network to fail. In fact, there are very few network outages at all in SS7. The protocol maintains a very high level of reliability, due mostly to the Message Transfer Part and the network management procedures discussed in this section.

In this section, we will review each of the messages used in network management and the procedures used. We will look at the structure of each message type and the parameters used.

Network management messages use the Message Signal Unit structure as shown in Fig. 6.5. The routing label is used for routing the message to the appropriate signaling point. Within the signaling information field (SIF) is information concerning the point code experiencing the failures or the link that has failed. In addition, status codes, priority codes, and other maintenance codes can be included.

Because of the nature of these messages, the information field will vary from one network management message to the next. As we discuss each type of network message, we will discuss the structure of the information field and how it is used.

Link management procedures

Link management provides the procedures necessary to manage the individual links within a signaling point. Link management does not view links from a linkset or even a route perspective, but from an individual perspective.

The link management procedures require the use of Message Transfer Part level-two functions or SAAL to report failures and link status to an adjacent signaling point. These messages are not propagated through the network. Link management is only concerned about the local end of a link.

There are three functions provided by link management:

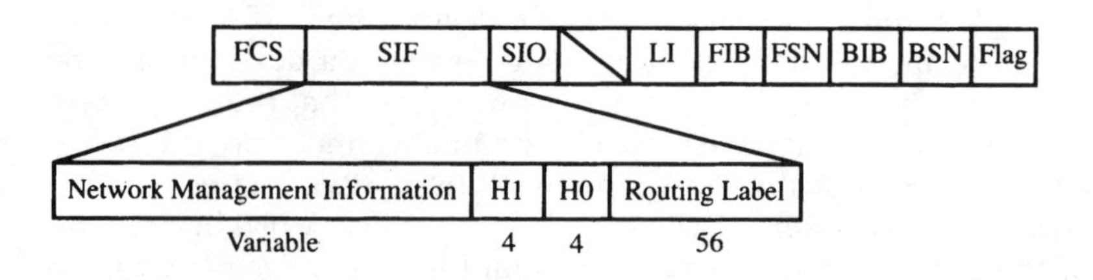

Figure 6.5 Network management messages use the Message Signal Unit (MSU). The H0/H1 field is the label which identifies the type of network management message being sent.

- Link activation
- Link restoration
- Link deactivation

These three functions are explained as follows in greater detail.

Link activation. When a link is first activated, level three directs level two to begin the alignment procedure (described in Chap. 5) and place the link in service. Before messages can actually be transmitted over the link, link management will also send test messages over the link, to ensure the integrity of the link.

Level two is used to inform the adjacent signaling point of the activities at level three. This is accomplished using the Link Status Signal Unit, described in Chap. 5. The LSSU identifies the status of the link, but does not instruct the adjacent signaling point on any procedures.

Once the link has been activated and is considered in service, a *signaling-link-test message* (SLTM) is generated and transmitted over the link. Upon acknowledgment of the SLTM (*signaling-link-test acknowledgment,* or SLTA), the link is restored to service and traffic is allowed over the link. In the event that the link has a failure (determined and reported by level two), link management invokes the restoration procedures.

Restoration involves the Link Status Signal Unit to inform the adjacent signaling point of the events taking place. The signaling point that detected the errors is responsible for invoking alignment procedures and notifying the adjacent signaling point about the status.

Level-three timers are used to control the procedures and ensure that the link does not get caught in an endless loop of alignment procedures. When a link has successfully passed the alignment procedures, level three generates a signaling-link-test message and transmits the test message over the network (as it did during activation). When an acknowledgment (SLTA) is received for the SLTM, the link is considered restored and level three changes the local status to available, in service.

Link deactivation is used when a link is found in error and needs to be placed into alignment. Link deactivation must first stop all traffic to the link, which will invoke traffic management procedures (diversion of traffic from the failed link). Link management then "disconnects" the link from its logical connection (signaling link code).

Every link has a logical connection. In any signaling point, there are a number of logical connections available for a link. All logical connections must be associated with a signaling link selection code. This code is a dynamic code assignment, which changes according to the link status.

Each link within a linkset is given a unique signaling link selection code, which is used by routing to determine which links a message should be routed over. This is directly related to the bit rotation procedure discussed earlier in the section about MTP level-three routing.

The bits rotated are the signaling link selection bits. A link may have more than one SLS code. All codes must have an assignment, because of the way the bit rotation is used to select a link. When any link becomes unavailable and must be placed out of service, link management must disconnect the link from its logical connection. This means the table will change. One can view this table as a mapping to every link in the system. Each signaling point must have a similar table in software that allows it to choose signaling links based on status.

Link management changes the table based on the local status (reported by level two) or the remote status (reported by level two via the Link Status Signal Unit). When a link becomes available again, it is reconnected to its logical connection.

Notice that all links are listed in numerical order. This ensures that the assignment is linear and not random, which allows both ends of the link to remain synchronous. Because the signaling link code is transmitted in maintenance messages, the code must be the same at both signaling points.

Link management can also drive a function referred to as automatic link allocation. Very few systems offer this feature, which allows links from one linkset to be disconnected (logically) by link management and reassigned to another linkset. This requires a link to be connected to the same signaling point as the failed link that it is replacing. This feature is usually invoked by a threshold value, which is typically administrable. When a linkset falls below the predefined threshold, a predefined link can be removed from its linkset and reassigned by link management to replace failed links in another linkset.

Automatic link allocation also provides for voice circuits to be used as links. The requirement is that the facility must be digital and of the same data rate as the failed link. This is not a concern, because most voice circuits in central offices today are DS0A circuits (interoffice trunks, not local trunks).

Link management is able to remove the voice circuit from its connection and reassign the voice circuit to a linkset. A predetermined signaling link selection code is assigned to the new link. When the failed link returns to service, the voice circuit is removed from the linkset and returned to service as a voice circuit.

This feature requires the voice circuit to be connected to the same signaling point as the failed link. This could be the case between two signaling points, provided the voice circuit was terminated in the same signaling point as the failed link. This forces the assumption that the signaling points must be end offices, which provide both voice circuits and SS7 circuits (Service Switching Points).

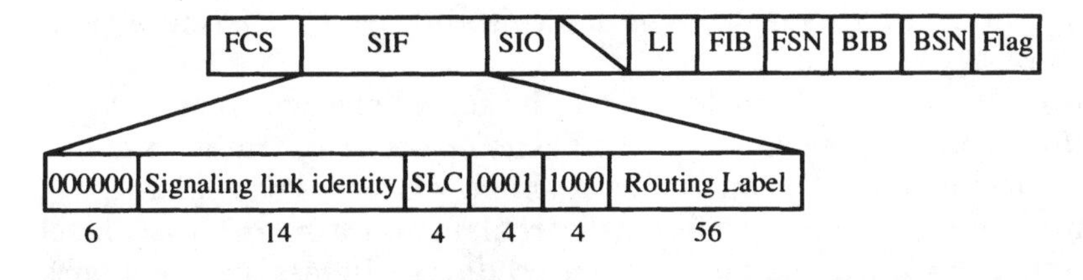

Figure 6.6 The signaling-data-link-connection (SDLC) message uses the network management structure, and consists of the components shown here.

To accommodate link management and automatic allocation, a signaling-data-link-connection message (Fig. 6.6) must be generated and sent to the adjacent signaling point to inform the adjacent signaling point of the new assignment. The adjacent signaling point must make the same assignment for the link to work.

In the signaling-data-link-connection message, the signaling link number is the physical link number (such as link one, two, or three), which in turn may be SLC 0, 6, or 12. When identifying the signaling link code, only the first code is provided. The other signaling link codes, even though they are assigned to the same link, are considered secondary and are not used in network management messages.

Once the signaling-data-link-connection message has been received, an acknowledgment must be sent to confirm that the adjacent signaling point has made the same assignment in its own database. This is accomplished using one of three messages: connection-successful, connection-not-successful, or connection-not-possible.

If the connection was successful, then traffic can begin on the newly allocated link. If connection was not successful (for any of the reasons provided), then the procedure is aborted.

In the unlikely event that two signaling points invoke automatic allocation simultaneously, the signaling point with the highest point code will be considered first, overriding the other signaling point.

To understand the impact that link management has on other portions of the signaling point, we must look at the other network management procedures. Keep in mind the layered approach discussed earlier in this section, and it will help you to understand the individual roles played by all these functions.

Traffic management procedures

Traffic management is used to divert traffic away from failed signaling links. It uses several messages which are sent using the Message

Signal Unit over adjacent links or adjacent linksets to adjacent signaling points.

Traffic management also deals with the source of congestion and provides flow control procedures as well. The objective of traffic management is to deal with the source of a problem whenever possible.

Traffic management also handles problems on a more direct level than other network management procedures. In essence, network management is offered at several layers, traffic management being the middle layer.

Traffic management provides the mechanisms for managing traffic diversion due to:

- Signaling link unavailability
- Signaling link availability
- Signaling route unavailability
- Signaling route availability
- Signaling point restricted
- Signaling point availability

Signaling link unavailability. In the event that a signaling link should become unavailable, be manually blocked, deactivated by network link management, or inhibited, traffic management provides for the diversion of all traffic normally routed over the affected link to alternate links within the same linkset or other linksets which can route to the same destination.

Signaling link availability. In the event that a failed, blocked, inhibited, or deactivated link should become available again, traffic management diverts messages back to the affected signaling link. Procedures are provided to ensure that messages are not lost and the transmission is controlled to ensure an orderly delivery of all buffered messages.

Signaling route unavailability. In the event that an entire route should become unavailable, forced rerouting is used to divert the traffic away from the affected route. A route is a linkset or a group of linksets with a common destination.

Signaling route availability. Controlled rerouting is used to divert traffic back to a previously unavailable route. It involves an entire linkset, rather than just an individual link. The diversion of traffic to another alternate linkset or route must be conducted in an orderly fashion to prevent messages from being lost.

Signaling route restricted. When a route becomes restricted, traffic must be diverted to a route of equal priority (or cost). In effect, this procedure invokes load sharing over two routes to prevent a route from becoming unavailable due to congestion (from multiple link failures, for example).

Signaling point availability. The Message Transfer Part restart procedure is used to divert traffic to a signaling point now made available. The MTP restart procedure is a new addition to the Bellcore standards and is not yet widely implemented. However, Bellcore added this ANSI procedure with the intent of implementing it in Bellcore networks in the future.

Traffic management message structure. The traffic management message provides the signaling link code of the failed link and, in some cases, the forward sequence number of the last good Message Signal Unit received on the failed link. This information is sent to adjacent signaling points so they can assume the traffic for the failed link and ensure that no messages are lost. In order to ensure that no messages are lost, the traffic management procedure includes a method for copying all Message Signal Units remaining in the transmission buffer of a failed link to the newly selected link. This procedure will be explained in further detail.

In all cases of traffic management, existing traffic on any one signaling link must not be interrupted. This means the procedures must allow normal traffic to continue while links are assuming the traffic of other failed links.

Changeover. The changeover message is used to divert traffic away from a failed link. Link Status Signal Units are sent by level two to indicate the status of the link throughout this procedure. This allows the two signaling points to maintain current status while the link is being realigned. When LSSUs are not being sent, Fill-In Signal Units are transmitted.

Figure 6.7 shows the contents of the changeover message. The forward sequence number is the FSN of the last Message Signal Unit received by the failed link. This serves as an acknowledgment for any unacknowledged signal units received on that link.

The signaling link code identifies the failed link. As we learned in our previous discussions about the SLC, this number is not necessarily a physical link number, but the logical link number assigned to the link. Refer to the section in this chapter regarding the routing label and the use of the SLC for details on this field.

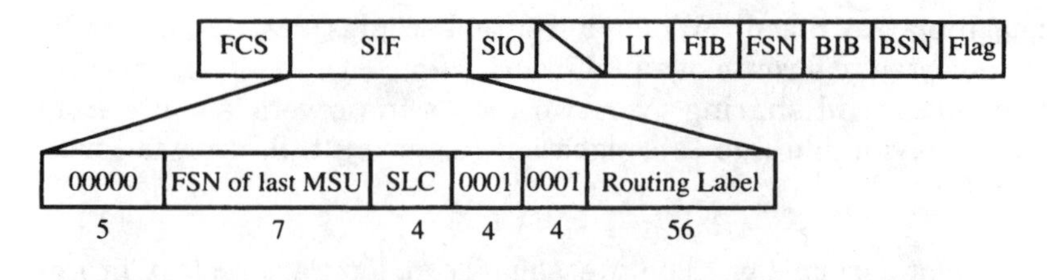

Figure 6.7 The changeover order (COO) consists of an SLC field, indicating the signaling link code of the failed link and the FSN of the last good MSU received on that link.

It should also be mentioned here that all changeover-related messages (changeback, acknowledgments, etc.) use the same structure, using the same fields. These related messages are discussed separately for clarity.

When a link fails (Fig. 6.8), there are still Message Signal Units in its transmit buffer which have not been acknowledged. Before these MSUs are discarded and the link realigned, they need to be dealt with in such a way that the messages will not be lost.

When level two detects a link failure, it first performs a buffer update. The messages in the transmit buffer of the failed link must be placed in the transmit buffer of the alternate link, so that they can be retransmitted in the event that an acknowledgment is not sent from the adjacent signaling point.

After the transmit buffer has been updated, the changeover message is sent over the alternate link to inform the adjacent signaling point that all messages to the affected node must be rerouted over the alternate link. The message will contain the signaling link code of the failed link, and the forward sequence number (for MTP L2) or the sequence number of the last sequenced data protocol data unit (for SAAL) of the last good Message Signal Unit received by the affected signaling point. The FSN in this message is used as an acknowledgment to the adjacent node so that it does not have to retransmit all MSUs in its transmit buffer.

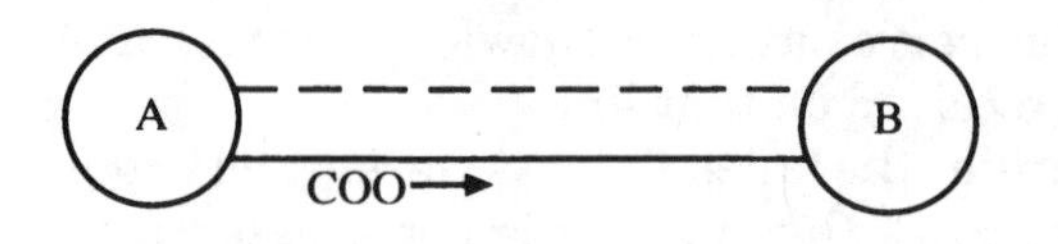

Figure 6.8 In this figure, the link indicated by dashes has failed. Signaling point A sends a COO to signaling point B, indicating the SLC code of the failed link.

The traffic is diverted to a link or links within the same linkset if there are any links available. However, there are times when there may not be any links available in the linkset. If there are no links available in the same linkset, another alternate linkset or linksets may be used. The destination must be the same for both linksets. In the event that there are no linksets available, routing management is triggered to reroute messages around the signaling point.

In the event that a Signal Transfer Point does not normally carry traffic for the affected signaling point (the changeover was to a linkset to a different signaling point), a transfer-prohibited message is sent by route management at the concerned signaling point. This is to prevent messages from being routed to the concerned signaling point and possibly causing congestion to occur (because the signaling point is now handling twice the traffic).

A transfer-prohibited message informs adjacent nodes that no traffic addressed to the affected signaling point should be sent to the concerned signaling point. The concerned signaling point is that which originated the changeover message and is no longer carrying traffic to the affected destination.

The affected destination is that which has the failed link. The affected point code is usually provided in network management messages. Note that messages being carried over the concerned point code are being sent from the signaling point that used to carry traffic to the destination, but lost its path. The concerned signaling point has a path, and is now "relaying" those messages, but should not be sent all traffic for this destination.

A good example of a situation like this is depicted in Fig. 6.9. A signaling point that is adjacent to the affected signaling point has received a

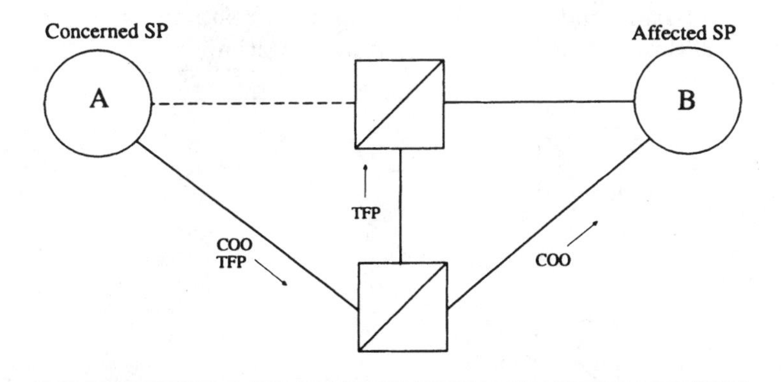

Figure 6.9 A COO may be sent through another signaling point, as depicted here. There is no other path for the message to get to signaling point B, so it must be sent through an alternate route. The TFP is sent to all adjacent signaling points to prevent messages from being sent through A to B.

changeover, but does not normally carry traffic for this destination; therefore, the concerned signaling point sends the transfer-prohibited message to any adjacent signaling points. When the changeover message is received by the concerned signaling point, a changeback acknowledgment is sent to indicate that the changeover was received and traffic is being diverted.

If there are no available links to the affected point code, then a time-controlled diversion procedure is implemented. This is also true if a processor outage is received, or if the signaling link is marked by the signaling point as inhibited, but is receiving traffic. The time-controlled diversion procedure prevents the signaling link from being failed and realignment from starting.

This is accomplished by setting level-three timer T1, "Delay to avoid message missequencing on changeover." When timer T1 expires, new traffic can then be transmitted on the alternate link. This helps prevent missequencing of messages, which may occur during a changeover initiated by the receipt of a processor outage or link-inhibited state.

Figure 6.10 shows when the use of time-controlled diversion may be necessary. Notice that signaling point A has lost its path to B, which is the Signal Transfer Point for signaling point D. The changeover procedure is diverting traffic through signaling point C, which is adjacent to the affected signaling point D.

If the signaling link becomes available, and the timer T1 has not yet expired, the time-controlled changeover is canceled and traffic may resume on the affected link. However, if the Message Transfer Part becomes unavailable at the affected point code, the MTP restart procedure is initiated.

MTP restart is used to reset all timers and counters used by the Message Transfer Part and to resynchronize the link. This means that all sequence numbering and error rate monitors are restarted as if for the first time. MTP restart is discussed in a later section.

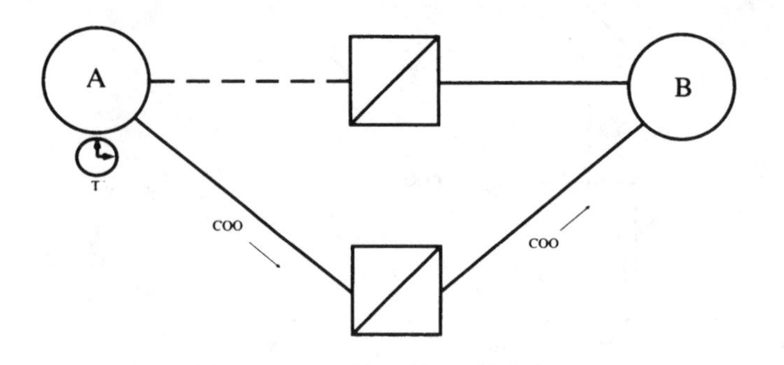

Figure 6.10 When timer T1 expires, the changeover order (COO) is sent to B. This is referred to as time-controlled diversion.

If there is a processor outage at the affected link, and the timer T1 expires ("Delay to avoid message missequencing on changeover"), all messages that are available for retransmission are discarded. The processor outage means that level four and possibly level three are not functional and the messages could not be processed anyway. The affected link will not have any recollection of these Message Signal Units ever being received, because its receive buffer will have been reset.

If there is no acknowledgment received within timer T2 ("Waiting for changeover acknowledgment"), retransmission on the alternate link begins. This means that the changeover procedure will start without the acknowledgment. Any new traffic will be transmitting on the new link(s). This prevents a bottleneck in the event that the changeover message or the changeover acknowledgment was lost.

If a changeover acknowledgment is received, but there was no changeover order sent, the acknowledgment is ignored and discarded. No further action is necessary.

Changeback. The changeback message is used when a failed link has been restored, and traffic may now resume over that link. The message structure, as seen in Fig. 6.11, consists of the signaling link code of the now-restored signaling link and the changeback code.

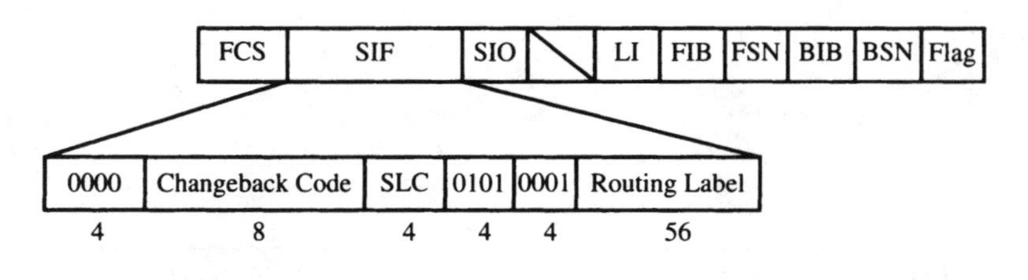

Figure 6.11 The changeback declaration (CBD) is sent to an adjacent signaling point to indicate that the link that previously failed has now returned to service.

The changeback code is a unique pattern assigned by the originator of the changeback message. The changeback code allows a signaling point to initiate the changeback procedure for a number of signaling links. When the acknowledgment is returned, the acknowledgment must also carry the unique changeback code. This allows for the discrimination between acknowledgments and allows the signaling points to begin sending traffic in relation to each of the individual changebacks.

Without this code, there would be no mechanism to allow for the orderly diversion of traffic over multiple signaling links. This is an issue only when there have been multiple link failures to one destination and all of the signaling links have been made available at the same time.

When the changeback is initiated, all transmission of Message Signal Units on the alternate link is stopped. The changeback message is sent over the alternate link to the affected signaling point to inform it that the changeback procedure has been stopped and transmission over the failed link will now resume (Fig. 6.12).

The affected signaling point must then send a changeback acknowledgment (Fig. 6.13). This will trigger the procedure. All Message Signal Units sent by level four or by other links destined to the affected signaling point are stored in a buffer until the changeback procedure is complete.

The changeback acknowledgment may be sent over any available link as long as the message is routed to the originating point of the changeback message. Once the acknowledgment has been received, the Message Signal Units that were stored in the changeback buffer are sent over the now-available signaling link. There is no need to transfer unacknowledged MSUs from the alternate link transmit buffer to the now-available links transmission buffer.

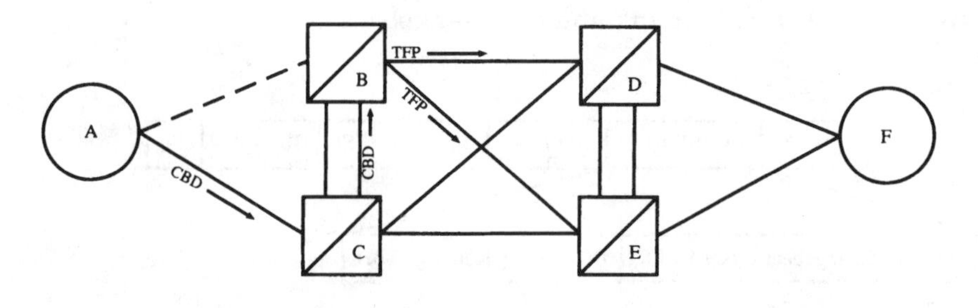

Figure 6.12 This figure shows the changeback declaration being sent to signaling point B by using an alternate link to C. If there had been an alternate link directly to B, it would have been the path used for the CBD.

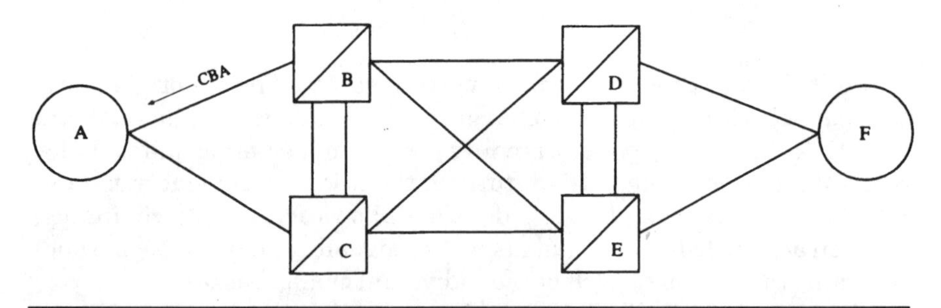

Figure 6.13 The changeback acknowledgment (CBA) is sent over any available link. In this case, the previously failed link is used for the CBA.

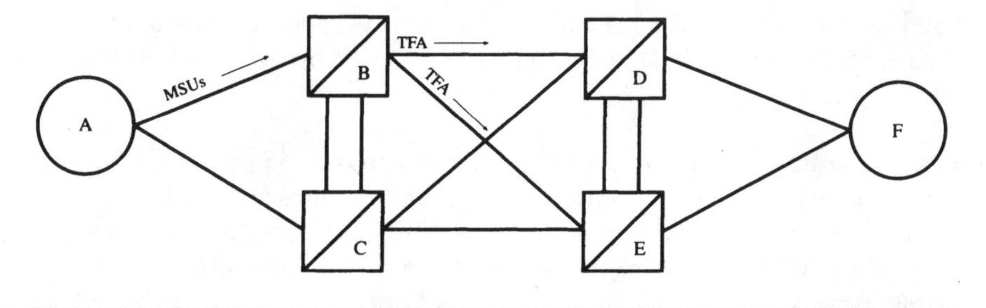

Figure 6.14 When MSUs are sent by the affected point code to STP B in this example, the transfer-allowed (TFA) is sent to all of B's adjacent signaling points.

In the event that the changeback is originated from a signaling point that sent a transfer-prohibited, a transfer-allowed is sent to adjacent signaling points, allowing messages to be routed to the signaling point (Fig. 6.14). Likewise, if the affected signaling point became isolated due to the failed link, the signaling point is now made available by route management.

If an acknowledgment is not received within timer T4, "Waiting for changeback acknowledgment—first attempt," the changeback declaration is repeated and timer T5, "Waiting for changeback acknowledgment—second attempt," is started. If timer T5 should expire before an acknowledgment is received, traffic is automatically started on the now-available signaling link. Maintenance functions within the signaling point are alerted in the event that there was an error in the acknowledgment transmission.

Emergency changeover. In the event that a changeover procedure is initiated but the transmit buffer cannot be read, an emergency changeover procedure is used. The emergency changeover does not provide the forward sequence number (MTP L2) or sequence number (SAAL) of the last good Message Signal Unit received, because the buffer has been cleared and that information is not available.

Level two begins sending the Link Status Signal Units on the failed link and Fill-In Signal Units when the LSSUs are not being sent. All new MSUs are diverted to the alternate link or linkset.

As was the case in the changeover procedure, the traffic can be diverted to multiple links or alternate linksets. Load sharing is invoked when there is existing traffic on these links to prevent a congestion condition on any one link.

In the event that there are no available paths to the affected signaling point on which changeover and emergency changeover messages may be transmitted, the time diversion procedure is invoked. As mentioned in the changeover procedure description, time-controlled diversion allows traffic to be diverted without failing the link (link status of "out-of-service").

Forced rerouting. The forced rerouting procedure is initiated in the event that a route to a specific destination becomes unavailable. The purpose is to reroute traffic around the concerned signaling point to the destination, without losing messages or causing any other route to become congested.

Figure 6.15 shows Message Signal Units destined for signaling point A. Signaling point A becomes unavailable through the route using signaling point C. Signaling point C sends a transfer-prohibited message to signaling points D and E to inform them that the route to signaling point A is inaccessible, and messages should be rerouted through an alternate route.

In Fig. 6.16, the alternate route is through signaling point B. All traffic to signaling point C (destined to signaling point A) is stopped. The forced rerouting buffers store all Message Signal Units with destina-

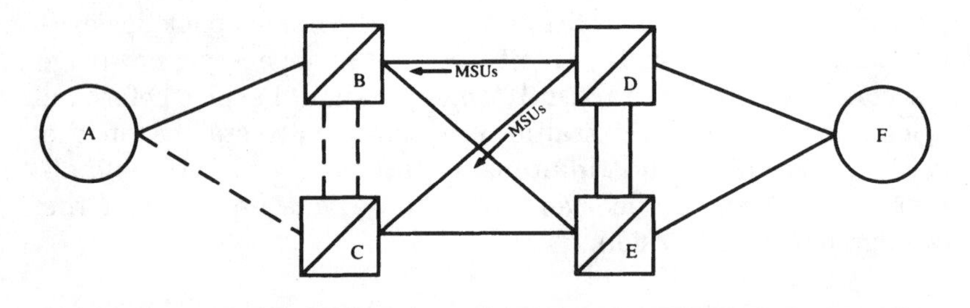

Figure 6.15 In this figure, MSUs are destined to A, through C. The link from C to A has failed. The links from C to B have failed as well, causing messages to be sent back to D and E (circular routing).

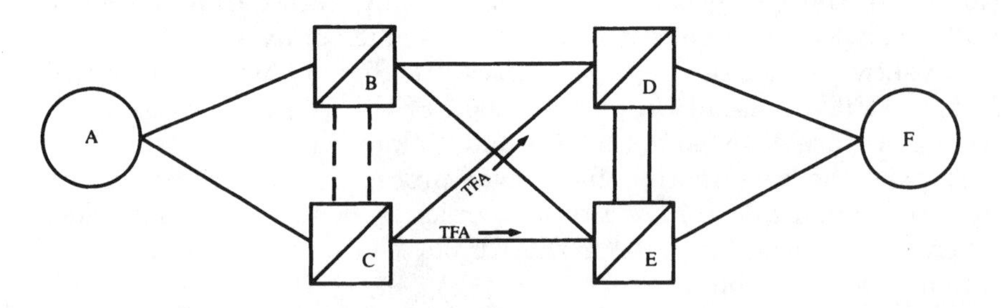

Figure 6.16 The link from A to C has been restored. C sends the transfer-allowed (TFA) to its adjacent nodes, allowing traffic to A.

tion of A. When the alternate route is determined (B in this example), all diverted traffic is transmitted to signaling point B. The contents of the forced rerouting buffer are sent first.

When the route to signaling point A through signaling point C becomes available again, signaling point C sends a transfer-allowed message to signaling points D and E (Fig. 6.17). Messages destined to signaling point A can now be rerouted back through signaling point C, or signaling point B, depending on the network configuration.

Any existing traffic on the link should not be interrupted in any way. If there is a lot of traffic on the alternate route, load sharing is used to spread the traffic over the links evenly. This procedure should not cause the alternate route to become inaccessible due to failure or congestion.

If there are no alternate routes available, traffic is blocked and messages stored in the forced reroute buffer are discarded. Flow control is used to inform user parts to stop sending traffic to the affected point code. A transfer-prohibited is also sent to the adjacent signaling points to stop traffic from being sent to the concerned signaling point (Fig. 6.18).

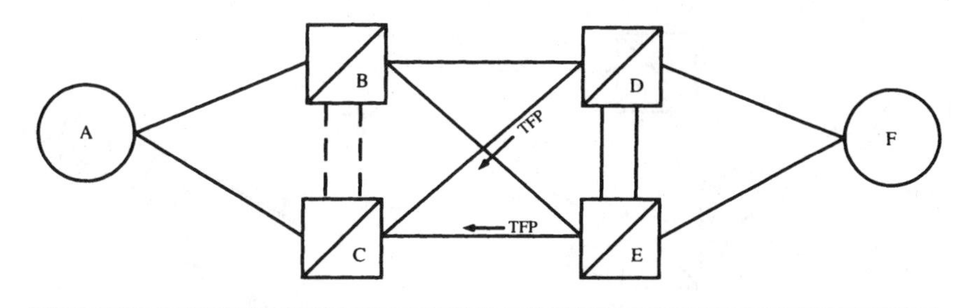

Figure 6.17 To prevent circular routing, D and E send TFPs to C. This prevents C from sending MSUs destined for A or B back to D and E.

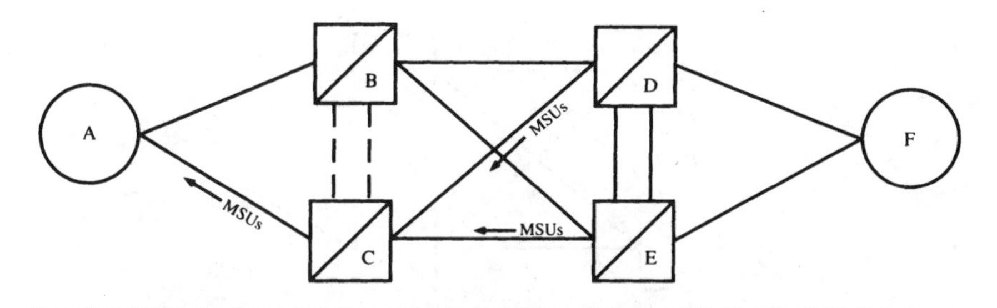

Figure 6.18 Once the transfer-prohibited (TFP) has been sent, MSUs in the buffers of D and E can be sent to A through C.

Controlled rerouting. The objective of the controlled rerouting procedure is to restore traffic to the most favorable route, in cases where a particular route was previously restricted. This is probably best explained by Fig. 6.19. In the figure, traffic is destined to signaling point A. The primary routes to signaling point A from signaling points D and E use signaling point C, with signaling point B used as an alternate.

The route from signaling point C to signaling point A fails. In addition to this route failing, the route from signaling point C to signaling point B fails (Fig. 6.20). This scenario in the previous example (forced rerouting) caused a transfer-prohibited message to be sent from signaling point C to signaling points D and E. This forces signaling points D and E to route all traffic destined for signaling point A through signaling point B.

The route from signaling point C then becomes available. All messages destined for signaling point A can now be routed through the primary route, signaling point C. To initiate this, a transfer-allowed message is sent to signaling points D and E (Fig. 6.21).

Signaling points D and E then initiate the controlled rerouting procedures. To prevent circular routing, signaling points D and E send a

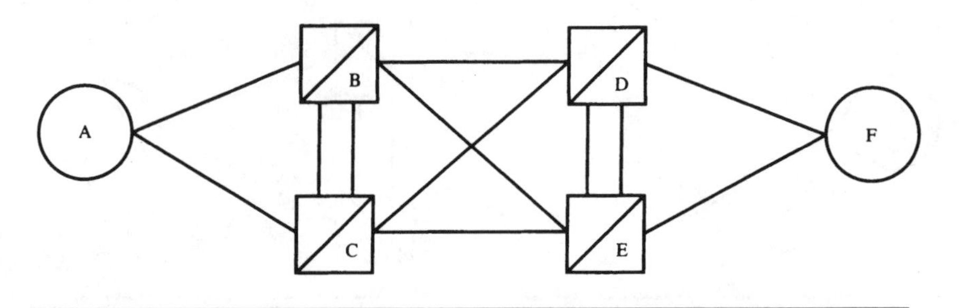

Figure 6.19 The links from C to A are primary linksets for routing to signaling point A.

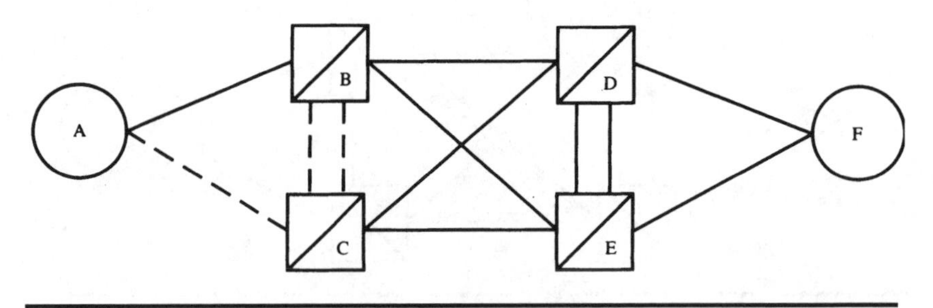

Figure 6.20 Links from C to A and C to B have failed.

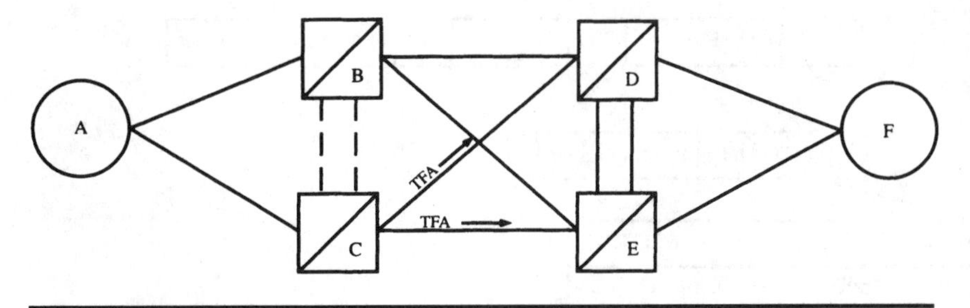

Figure 6.21 The link from C to A has now been restored. The transfer-allowed (TFA) is sent to D and E to allow traffic for A to be sent to C.

transfer-prohibited to signaling point B (Fig. 6.22). This prevents signaling point B from sending messages destined for signaling point A.

A timer T6, "Delay to avoid message missequencing on controlled rerouting," is set and, after its expiration, the controlled rerouting buffers at signaling points D and E are transmitted to signaling point C (Fig. 6.23).

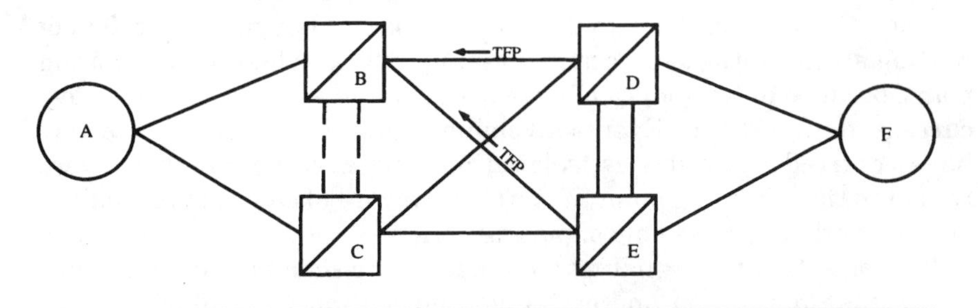

Figure 6.22 Signaling points D and E send TFPs to signaling point B, to prevent B from sending traffic destined to A back to D and E (circular routing).

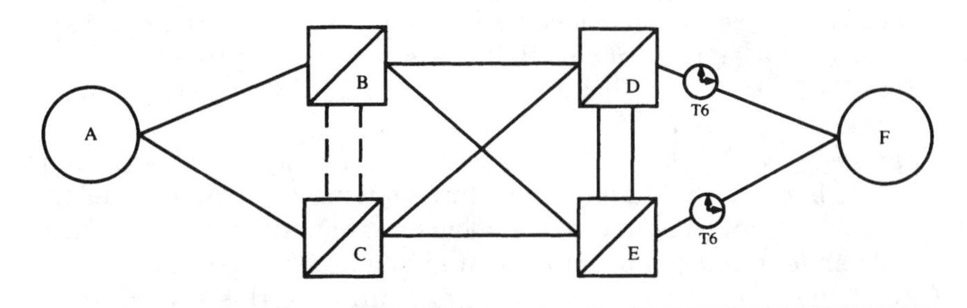

Figure 6.23 After T6 expires, MSUs resume on the links to signaling point C.

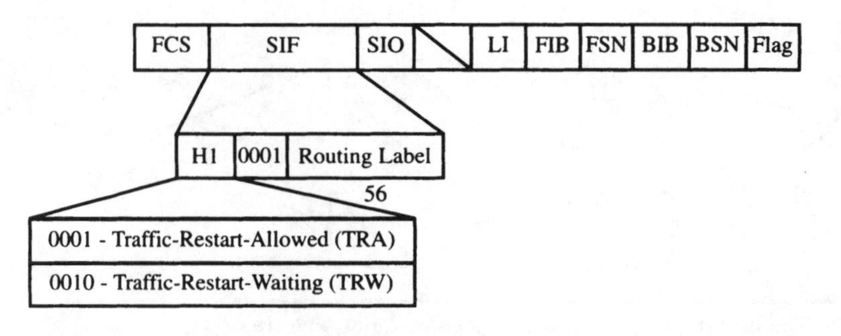

Figure 6.24 The traffic-restart-allowed (TRA) and the traffic-restart-waiting (TRW) messages are used with the MTP traffic restart procedures. There is no information provided other than the management message.

MTP restart. The MTP restart procedure (Fig. 6.24) is a new addition to the Bellcore standards, even though it has been in the ANSI publications for awhile. The purpose of this procedure is to protect the network and a signaling point in the event that a signaling point becomes isolated for a period of time and then becomes available.

When this occurs, the signaling point must exchange a great deal of routing status with its adjacent signaling points. The failed signaling point, because of the length of time it was out of service, may not have current routing status. Events involving adjacent signaling points may have occurred while it was isolated and, because no messages were routed to the signaling point, it cannot be aware of congested signaling points or other inaccessible signaling points.

To ensure that the signaling point has ample time to update its routing status information and exchange information with all of its adjacent signaling points, the MTP restart procedure is implemented with a set of timers. The timers ensure that the signaling point has ample time to retrieve routing status on each linkset or route.

There really are no set rules for when an MTP restart procedure should begin. One suggestion by Bellcore is to start the procedure upon receipt of a Link Status Signal Unit with a value of processor outage (SIPO).

There are two levels of MTP restart: a full restart and a partial restart. It is up to the management functions at level three to determine which is implemented and under what conditions. This is network dependent and can be implemented in a number of conditions.

One of the suggestions by Bellcore for initiating this procedure is that, when a route becomes available after a remote processor outage, the route initiates a restart procedure. This procedure at the remote end of the route involves the use of timers (T25 and T28), which are initiated upon receipt of the first link status of in-service (determined by level two).

When a partial restart is initiated at the isolated signaling point, it sends out a *traffic-restart-allowed* (TRA) message to its adjacent signaling points. The remote signaling points can then initiate their own restart procedures on the direct routes to this signaling point, as specified in the ANSI and Bellcore standards. This procedure involves the use of timers T25 and T28 (as previously mentioned).

If a full restart is required, then timer T27, "Minimum duration of unavailability for full restart," is started. This timer is used to ensure that all adjacent signaling points see the unavailability of the routes leading to this signaling point and have ample time to respond. During the duration of T27, the isolated signaling point sends level-two processor outage messages to its adjacent signaling points using the Link Status Signal Unit.

After expiration of timer T27, the signaling point then attempts to place predetermined links back into service. The primary link of each linkset should be the first to be restored. To expedite the alignment procedure on these links, the emergency alignment procedure is suggested for the first links within the linksets.

Once these links have been brought back into service, route management messages can be exchanged. The route management messages will enable the isolated signaling point to determine current status of all direct routes. While the in-service links are exchanging route status information, the other links can begin their alignment procedures.

When the first link in a linkset goes into service, timer T22, "Timer at restarting signaling point waiting for signaling links to become available," and timer T26, "Timer at restarting signaling point waiting to repeat traffic restart waiting message," are started. Timer T22 is stopped when sufficient links have become available ("sufficient links" is a network-dependent parameter). If T26 expires, a *traffic-restart-waiting* (TRW) message is sent to all adjacent signaling points. Timer T26 is then restarted. This timer (T26) ensures that enough time is allowed for the procedure to be completed. Timer T26 is stopped upon the expiration of several other timers, depending on the events taking place.

There are many other timers and events that take place during the MTP restart procedure. The intent here is to give you an idea of what this procedure tries to accomplish. Without this procedure, each link is left to its own accord, and the alignment procedure is begun on a link-by-link basis.

When the links are left to their own devices to get restored, there is no orderly fashion in which routes are reinstated. This procedure brings more order to the alignment procedure and the procedures which take place when an entire signaling point has been isolated.

Management inhibiting. Management inhibiting is used by link management to block a signaling link from level four. The status of the link does not change at level two. The purpose of this procedure is to allow personnel to send test messages over the inhibited link, or to allow link management to send test messages over the link without interference from any of the user parts.

However, inhibiting a link is not permitted if the signaling point is under a congestion status or if there are no other links available. If the link being inhibited is the last available link, the procedure is denied. If a signaling point should suddenly become isolated (due to other link failures) or if all other links within the same linkset as the inhibited link should become unavailable, inhibiting is canceled and the link is returned to normal service.

If no other links fail and the inhibit procedure is uninterrupted, only the originator can uninhibit the link. The link inhibition can be initiated either through a command entered at a system terminal or by network management.

The link is inhibited by sending an inhibit request to the remote signaling point. This informs the remote signaling point that the originator wishes to inhibit the link and the remote signaling point should mark the link as inhibited. If, for any of the reasons here mentioned, the link cannot be inhibited, the request is denied.

To ensure that the link shows as inhibited at both ends of the link, both signaling points periodically send test messages to check the status of the link at the adjacent signaling point. This is accomplished through the inhibit test message.

During the time the link is marked as inhibited, the local signaling point sends a local inhibit test message. This is to ensure that the link status in the remote signaling point is still shown as inhibited. If the remote does not acknowledge the local inhibit test message, the procedure begins again, with an inhibit request. The local signaling point must first force an uninhibit of the link before inhibiting the link can start.

Likewise, the remote signaling point will also periodically send a remote inhibit test message. If the remote inhibit test message does not get an acknowledgment, the link status at the remote signaling point is changed to available (through a forced uninhibit) and traffic is allowed on the link. These two test messages ensure that both ends of the link are aligned properly and show the same status. A signaling point is allowed two attempts at inhibiting a link.

When the inhibit link message is received by a signaling point, the receiving signaling point initiates a changeover, diverting traffic away from the link onto other links within the same linkset or other linksets. If the signaling point receives an inhibit message, the link is marked as remotely inhibited.

The originator of the inhibit marks the link as locally inhibited. A changeover procedure is initiated to divert traffic to other links. The local signaling point uses the time diversion changeover procedure.

If the local signaling point has not received an acknowledgment to its inhibit message within timer T14, "Waiting for inhibit acknowledgment," the procedure is started again. Two consecutive attempts are allowed. If the signaling point is still unsuccessful after two attempts, the inhibit procedure is aborted and the link remains available for traffic.

Inhibiting a link is usually a manual procedure, used by maintenance personnel for testing the reliability of a link. Test messages with specific patterns can then be exchanged over the inhibited link and checked for accuracy. The signaling point has the responsibility for ensuring that the inhibited link does not remain inhibited after testing is complete (caused by lost uninhibit messages never received by remote signaling points) and that the inhibited link is not the last available link to another signaling point.

Flow control. Flow control at level three is used to control the flow of user part messages from the source. Much about flow control is implementation dependent. The procedures implemented for congestion or unavailable user parts are dependent on the manufacturers of SS7 equipment. The standards only define the need for such procedures and make suggestions as to how they can be addressed.

The intent of these procedures is to deal with congestion at the source, where the messages are being generated. This, of course, is at level four—user parts. The protocol has no interaction, really, at this level, because these are internal functions. However, the protocol does trigger these internal functions.

If a transfer-prohibited message is received for a particular destination, level three will interact through network management with the protocol and will direct level four as well. Communications from the protocol to these internal functions are accomplished through the use of primitives, which were discussed in Chap. 5.

The primitives offer a structured communications format to interact with other levels. In the case of flow control, level three must be able to notify level four of a congestion condition at another signaling point. The result is a reduction in traffic being generated for the affected signaling point.

The advantage to this is twofold. Only the user part experiencing the congestion is affected, rather than the entire signaling point. The congestion flow control is directed at a specific user part (such as ISUP) rather than an entire signaling point. This allows other traffic to continue without impedance.

The other advantage is that congestion is dealt with at the source, rather than by trying to redirect messages around the congestion. If a particular node is causing congestion, the amount of traffic generated by that node is throttled, instead of redirecting all that traffic to another destination.

There are procedures invoked by route management at level three which also deal with congestion conditions; however, they deal with signaling point congestion. Route management does not communicate with the source, the user parts, directly. Traffic management deals with the source.

The Message Transfer Part uses traffic management flow control to deal with traffic destined to a user part that has become unavailable. This is from the perspective of the receiver, rather than that of the source.

If a Message Signal Unit is received by the Message Transfer Part, and the message discrimination function has determined that the MSU is addressed to the local signaling point, the message is sent to message distribution to be given to level four. However, if message distribution is unable to give the message to level four because of a congestion condition, or because the user part at level four is not available, traffic management MTP flow control is invoked to inform the originator of the problem (Fig. 6.25).

An example of how this could occur is best explained using the feature of global title translation. Global title translation is performed by a Signal Transfer Point when it receives a Signaling Connection Control Part message with a called party address that contains only digits.

These digits are referred to as *global title digits.* The digits are typically nonroutable (i.e., 800 number or 900 number). The Signaling Connection Control Part must deliver the message to a user part

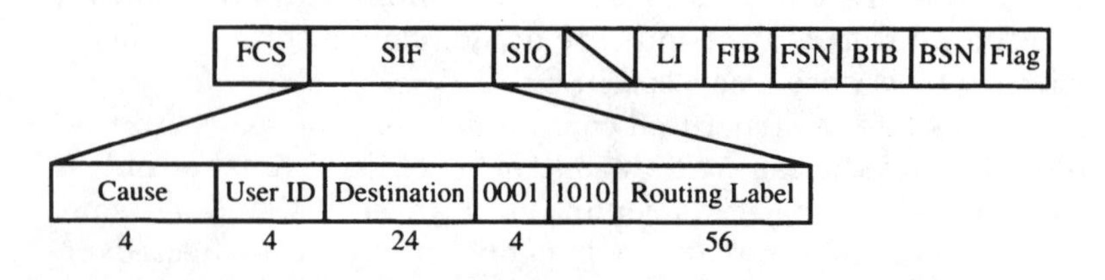

Figure 6.25 The user part unavailable (UPU) indicates that a user part is not accessible at the specified destination. The user ID is the same value as that provided in the SIO field.

(SCCP in this case) at the receiving signaling point for the global title translation function. If the processor dedicated to that function has failed, there are no resources available to perform the global title translation.

The signaling point in this case would return a user part unavailable message to the originator of the message indicating that the message processing could not be completed because the necessary resources are not available to perform the function.

In Fig. 6.25, the user part unavailable message provides the destination point code (the point code of the failed user part), the user part that is not available (SCCP, ISUP, etc.), and the cause. The MTP user code is the same code used in the service indicator octet of the Message Signal Unit.

There are only three causes today: reason unknown, an unequipped remote user part, or the remote user part is inaccessible. The condition for each of these causes is purely implementation dependent and may have different meanings in different networks.

In the event that traffic management should create a routing problem to any other signaling point, the routing management function at adjacent signaling points will be invoked. Likewise, if a signaling point that has invoked traffic management has lost a route to a destination, it will also invoke routing management to resolve routing issues. Route management is described in the following section.

Routing management procedures

Routing management is used by a signaling point to notify its adjacent signaling points of a routing problem. The routing problem is usually attributed to the loss of a signaling link or linksets, which together comprise a route.

The purpose of routing management is to redirect traffic around the failed route. This is accomplished through the use of transfer messages, which identify the failed destination by point code and instruct receiving signaling points on how to react (Fig. 6.26).

When a network is using cluster addressing, cluster routing management can also be invoked. In cluster addressing, each Signal Transfer Point is assigned a unique cluster address. All of the signaling points that "home" to that STP are assigned the same cluster address, with unique member addresses (Fig. 6.27).

This allows routing management cluster messages to be sent to an STP and distributed among all of its cluster members. The advantage of cluster routing management is that one message can be sent to address an entire group of signaling points, rather than individual signaling points.

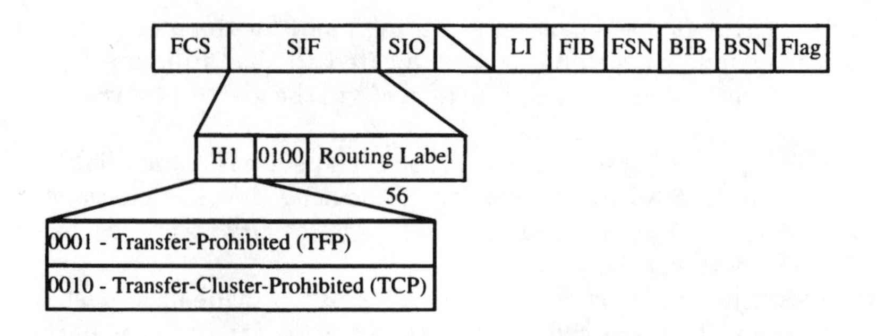

Figure 6.26 Both the transfer-prohibited (TFP) and the transfer-cluster-prohibited (TCP) work in much the same way. The TFP identifies a single entity, while the TCP identifies a cluster of entities.

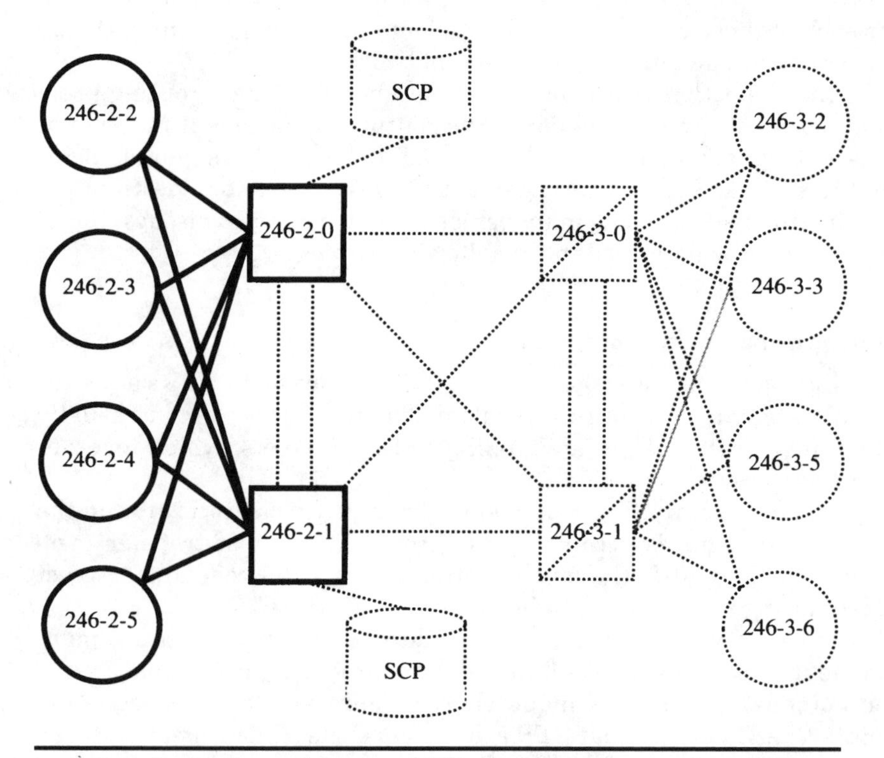

Figure 6.27 In this figure, the group of SSPs that share the same home STP have the same cluster address. Network management messages can then be sent from the home STP concerning status of the entire cluster. Signaling traffic can also be routed by cluster address rather than by the entire point code.

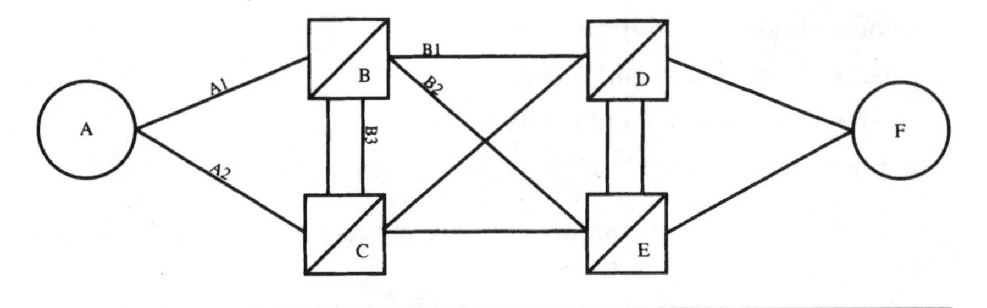

Figure 6.28 In this example, A1 is a primary route to destination F, while A2 is the alternate route. For signaling point B, B1 is the primary route to F and B2 is the secondary route, while B3 is an alternate route.

In addition to cluster routing, all networks must use some prioritizing on their routes. This is typically done by weighting each route in order of efficiency (Fig. 6.28). For example, all signaling points will have a primary route. The primary route (one for every destination) is the fastest path to the destination. Therefore, it is considered the most efficient.

In addition to the primary route, there should also be a secondary route, which is not the most favorable path but provides an alternative in the event that the primary path should fail. This is not the best path, but it will get messages to the same destination.

Additional routes are weighted in similar fashion. The more paths provided, the more reliable the network. The objective is to provide diverse routes to the same destination so that, in the event that a major failure should occur, messages can still be routed to their destinations.

When a route fails, a routing management message is sent to adjacent signaling points to advise them that the originating signaling point can no longer reach a destination through its routes, and an alternative should be selected ("Do not send to me, because I can't get there"). For normal routing management, the following messages are used:

- Transfer-prohibited (TFP)
- Transfer-allowed (TFA)
- Transfer-restricted (TFR)
- Transfer-controlled (TFC)
- Signaling-route-set-test (RST)
- Signaling-route-set-congestion-test (RCT)

During cluster routing management, the following messages are used:

- Transfer-cluster-prohibited (TCP)
- Transfer-cluster-allowed (TCA)
- Transfer-cluster-restricted (TCR)
- Cluster-route-set-test (CRST)

These messages and their accompanying procedures are explained as follows.

Transfer-prohibited (TFP). The transfer-prohibited message is sent by a signaling point when it determines that it can no longer reach a destination (adjacent signaling point). The reason for the isolation is due to the loss of an entire route, which connects directly to the affected destination.

The message will provide the destination point code for the affected signaling point. This allows adjacent signaling points to determine which alternate route to select to reach the affected destination.

In Fig. 6.29, signaling point D can no longer route messages to signaling point F. A transfer-prohibited is sent to signaling points B and C to advise them to select an alternate route. When signaling points B and C receive the TFP message, they stop transmission of all Message Signal Units to the concerned signaling point (D) with the address of the affected signaling point (F) until an alternate route is determined. Once the alternate route has been determined, MSUs are transmitted via the alternate route to the affected destination (F). Any traffic generated at signaling point D destined for signaling point F is sent to signaling point B or C for routing to signaling point F.

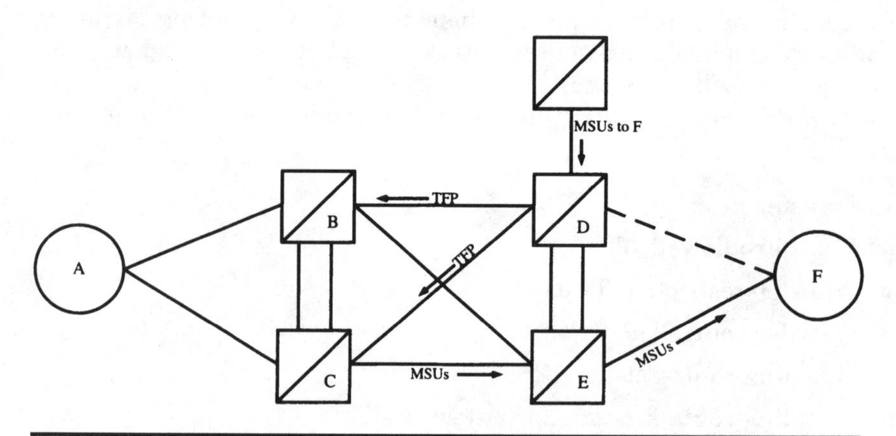

Figure 6.29 In this figure, D can no longer send MSUs to F. A TFP is sent to B and C, to prevent them from sending traffic for F to D. If D gets any traffic from other signaling points, the traffic is sent to C for routing to F.

This is a very simple example. Depending on the network configuration, this procedure can involve very little rerouting or a lot of rerouting. The objective in any routing plan is to keep all routing as direct and as simple as possible.

When the failed route becomes available again, a transfer-allowed message is sent to all adjacent signaling points by signaling point D, and normal routing is resumed.

Transfer-cluster-prohibited (TCP). Like the transfer-prohibited message, the transfer-cluster-prohibited message is sent by a signaling point to a cluster of signaling points. Each Signal Transfer Point is assigned a unique cluster address, while all signaling points which home to the STP have the same cluster address as the STP, but use unique member addresses.

As seen in Fig. 6.30, the home STP sends a transfer-cluster-prohibited message concerning all of the signaling points which is "home" to the STP. This allows one message to be sent regarding the entire cluster, rather than numerous messages sent by each signaling point in the cluster.

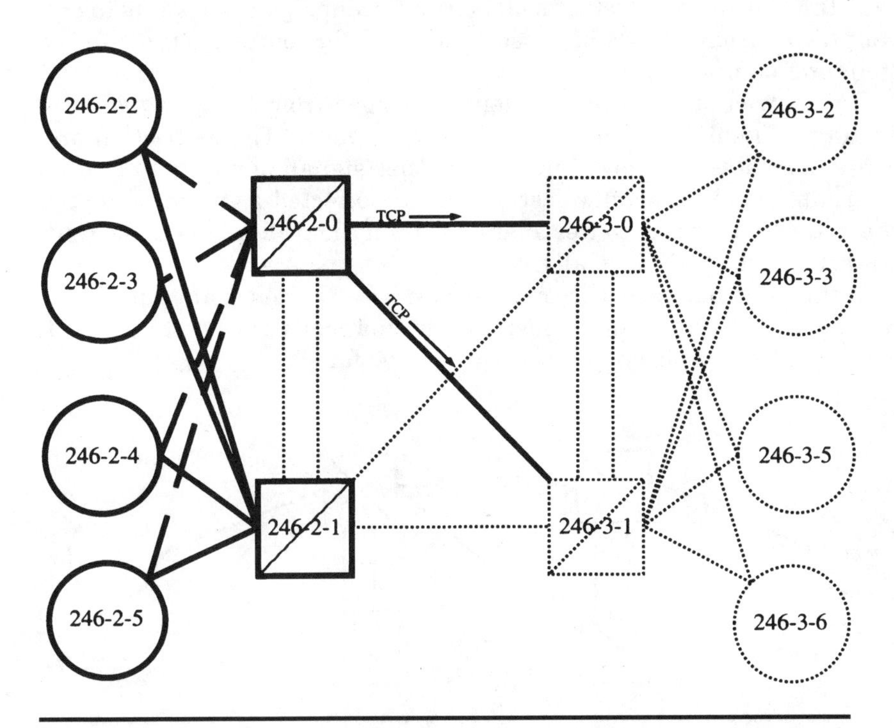

Figure 6.30 In this example, the STP 246-2-0 has failed and cannot reach any of the signaling points that home to it. A transfer-cluster-prohibited (TCP) message is sent to indicate this condition and cause routing to be forced to STP 246-2-1.

Transfer-allowed (TFA). The transfer-allowed is used when a route becomes available again. This is sent by the originator of a transfer-prohibited message to indicate that traffic may be sent once again to the affected signaling point.

The message structure is the same as the transfer-prohibited. When received, the transfer-allowed should trigger a changeback to occur on the concerned link.

Transfer-restricted (TFR). The transfer-restricted is sent by a Signal Transfer Point when it is determined that messages to a particular destination should no longer be sent to the STP for routing to the affected signaling point (Fig. 6.31). The transfer-restricted is always sent to adjacent signaling points.

In the event that a signaling link to the affected destination experiences a long-term failure (such as a processor outage), the Signal Transfer Point receives a changeover from the affected signaling point, ordering all traffic to be diverted away from the link at fault.

The Signal Transfer Point may then determine that it is necessary to send the transfer-restricted to all of its adjacent signaling points to prevent traffic from being addressed to the STP for routing to the affected signaling point.

The restriction does not prevent messages from being transferred from the Signal Transfer Point entirely, however. The restriction prevents normal traffic flow, and forces other signaling points to find an alternate route for traffic destined to the affected signaling point. If there are no alternate routes available, then the traffic can still be routed through the STP normally.

In this previous case, the transfer-restricted is sent using the *broadcast method.* The broadcast method automatically transmits the TFR when the link has been determined to have failed.

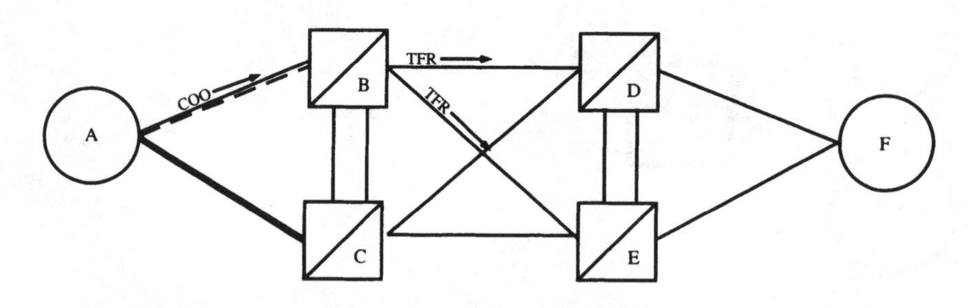

Figure 6.31 The transfer-restricted (TFR) message is used to restrict traffic flow. In this example, a link from A to B has failed, leaving only one link available. A changeover has occurred, and a TFR has been sent by the STP to all of its adjacent signaling points.

When a significant number of links in a linkset fail, a transfer-restricted is sent to adjacent signaling points using the *response method.* The response method sends the TFR when a Message Signal Unit is received on a link for transfer to the affected signaling point.

The criterion for sending a TFR message under these conditions is implementation specific. The objective is to restrict traffic to the signal transfer point on the linkset that has experienced link failures before congestion occurs. Congestion can occur at both the link level and the user part level.

If a signaling point was previously prohibited and links become available, the signaling point would be considered restricted and transfer-restricted messages would be sent to all adjacent signaling points until the signaling point was 100 percent in service again. This means that all signaling links to the affected destination would have to be in service.

There are cases when traffic on an alternate route would be diverted via the controlled rerouting procedure to the restricted route. This should occur only when the restricted route has a higher priority than the alternate route. The priority of a route is set by an administration command at each signaling point.

In the event that both routes should be of equal priority and both are restricted, then load sharing should be implemented to distribute traffic evenly over both routes. This ensures that links are carrying an even amount of traffic, and one link does not become burdened with all the traffic from the failed route.

Transfer-cluster-restricted (TCR). When several routes within a cluster fail or become congested, the transfer-cluster-restricted message is used (Fig. 6.32). This message indicates to an adjacent Signal Transfer Point that the concerned cluster should not be routed any messages if possible.

In Fig. 6.33, the signaling points attached to Signal Transfer Point A have the same cluster address as Signal Transfer Point A. This means that Signal Transfer Point A is their "home" STP. In the event that one

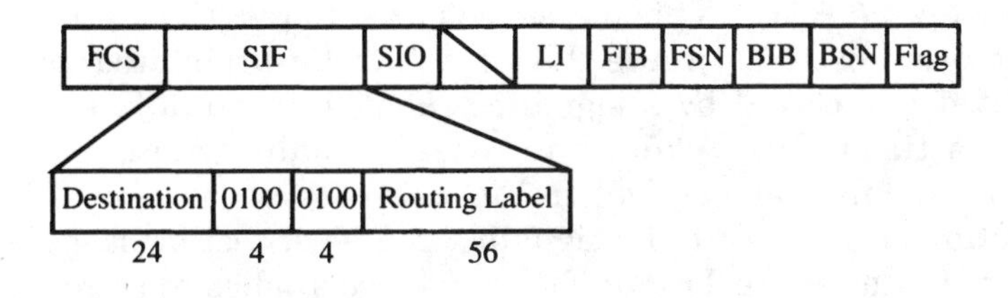

Figure 6.32 This is the message structure of a transfer-cluster-restricted (TCR) message.

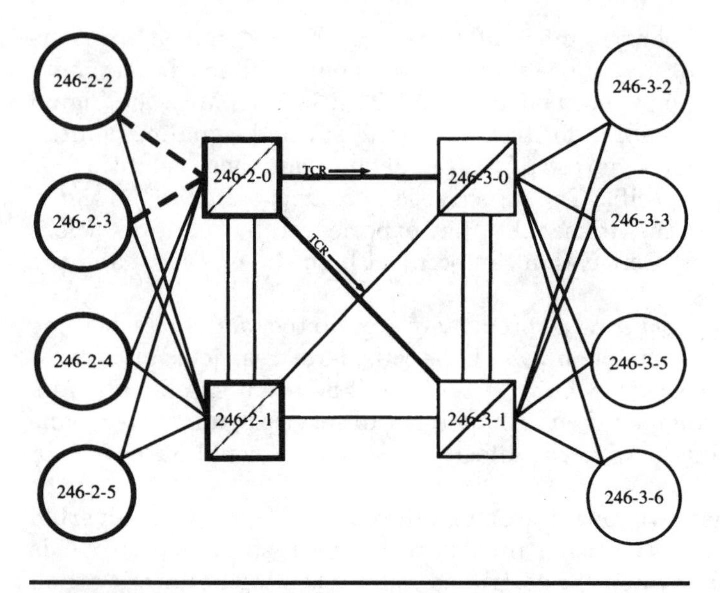

Figure 6.33 Two signaling links to STP 246-2-0 have failed. The failure of these signaling links triggered a transfer-cluster-restricted (TCR) to be sent to the adjacent STPs.

or more routes within this grouping of signaling points should become unavailable, Signal Transfer Point A would send a transfer-cluster-prohibited to any adjacent STPs. This allows control over the traffic to the signaling points attached to Signal Transfer Point A and any of its concerned signaling points. The same rules apply to the transfer-cluster-restricted as the transfer-restricted.

It should be noted that not all networks use cluster routing. Cluster routing is a network management feature and is not implemented for normal message routing. However, partial point code routing can be used for normal routing of all messages through the network. If partial point code routing is offered within a network, then, most likely, cluster routing is also implemented, since the two are closely related.

Signaling-route-set-test (SRST). This message is used to test the status of any prohibited or restricted route. When a transfer-prohibited or transfer-restricted is received by a signaling point from an adjacent signaling point, a timer T10, "Waiting to repeat signaling-route-set-test message," is automatically activated (Fig. 6.34).

At the expiration of timer T10, the signaling-route-set-test message is sent to the originator of the TFP or TFR (this also applies to transfer-cluster-prohibited and transfer-cluster-restricted messages). When the SRST is sent, the timer T10 is reset.

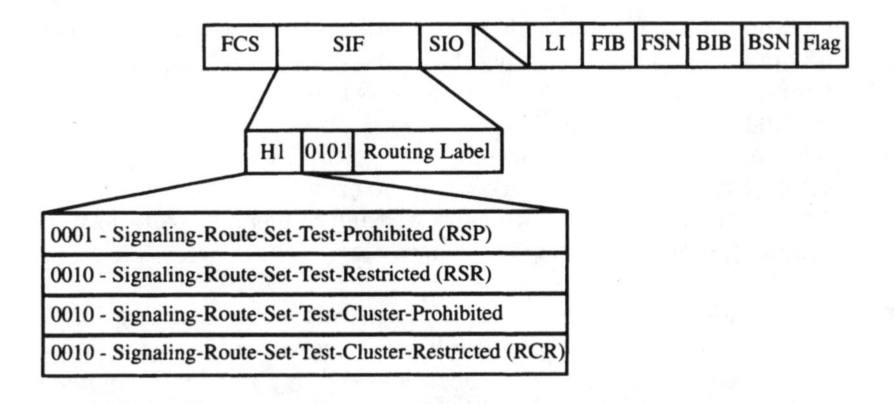

Figure 6.34 The message structure of signaling-route-set-test (SRST) messages is shown in this figure.

The signaling-route-set-test message is retransmitted after every expiration of timer T10, until a transfer-allowed has been received by the testing signaling point. This procedure is used to ensure that a prohibited or restricted signaling point does not get stuck in that condition indefinitely. The message contains the status information (from the perspective of the originator) for the concerned signaling point, as well as the heading code. No other information is necessary.

Another use for this message is when a link becomes available, but traffic is not restarting. Notification is given by level two (Link Status Signal Unit) that the link has started the proving period. The adjacent signaling point receives this status and also begins the proving period. After the proving period has ended and no other indications of failure are received, the link should be considered available.

However, only the signaling point that initiated the proving period can restart traffic on the link. The adjacent signaling point typically waits until receipt of a Message Signal Unit. The adjacent signaling point should then send the signaling-route-set-test message to determine if the route has become available. The message would then be transmitted every T10 until either an MSU is received or the status is indicated by level two.

In the event that a signaling-route-set-test message is sent to a previously restricted cluster, the receiving Signal Transfer Point (considered the home STP for all other signaling points in that cluster) will compare the actual status of the cluster with that indicated in the message. If the status of the cluster is not prohibited or restricted, the home STP will send a transfer-cluster-allowed back to the originator of the SRST message. The originator of the SRST then updates its status indicator for the cluster as available, and allows traffic to be routed to the concerned STP. This procedure prevents a cluster from erroneously being marked by a signaling point as unavailable or restricted when a transfer-allowed (TFA or TCA) message is lost.

If the cluster is not available and there are any signaling points within the cluster that are in danger of becoming congested, then the transfer-cluster-restricted message is sent to the originator of the signaling-route-set-test message. This prevents further congestion from forcing the signaling point into a busy condition.

Transfer-controlled (TFC). The transfer-controlled message (Fig. 6.35) is sent by a Signal Transfer Point when it receives a Message Signal Unit destined for a route that has been marked by the STP as congested. The TFC is addressed back to the originator of the MSU.

In Fig. 6.36, a signaling point has sent a message signal unit to a Signal Transfer Point, to be routed over any available route to the destination addressed in the routing label. The STP determines that the route for that destination is congested and there are no other routes to the destination.

When the Message Signal Unit is received on a signaling link, level-three routing must determine which route to send the message out on.

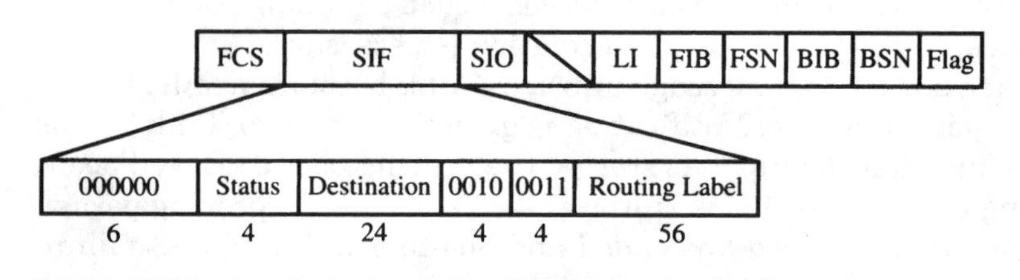

Figure 6.35 The structure of the transfer-controlled (TFC) message.

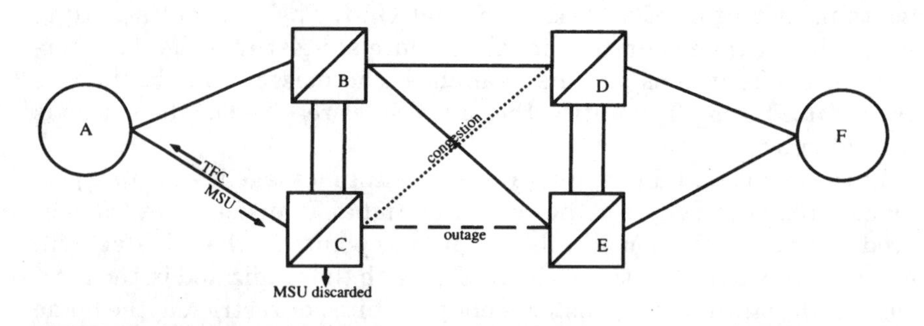

Figure 6.36 The link between C and E has failed in this example. As a result, the link from C to D has become congested. MSUs sent by A to C are discarded, and a transfer-controlled (TFC) is sent by C to A.

If the only route is congested, level-three network management will discard the message and send the transfer-controlled back to the originator of the MSU. The TFC should be generated by the link on which the message came in, rather than using the same processor showing the congestion. In other words, if resources are showing a congestion status, sending them additional work compounds the problem. Implementation of this procedure will differ from system to system, but the overall objective is the same. Reduce the traffic to the congested route.

The reduction is accomplished by returning a congestion status code in the transfer-controlled message. The congestion status indicates the priority level a message must possess before it will be routed over the congested route. Only messages of a higher priority than indicated in the congestion status field will be allowed on the congested route.

The priority is determined by the individual signaling point. Each signaling point must assign a congestion status level to a route. The status is a two-bit code, providing for a status level of 1, 2, or 3 (0 indicates no congestion). When a Message Signal Unit is received by an STP and is routed to the congested route, the network indicator field of the service indicator octet is examined to determine what priority has been assigned to the received message.

Priorities are assigned by the originating signaling point. This coding is implementation dependent. It can be an administrable value, or automatically assigned based on configuration of the signaling point. Regardless, in ANSI networks, this field determines whether or not a message will be allowed to pass through a congested route, based on the congestion level of the route.

If we examine the network indicator field, there are two spare bits associated with the national network indicator. They can have values of 0, 1, 2, or 3 (which correspond with the congestion-level values). The Message Signal Unit must have a priority value equal to or greater than the current congestion level of the congested route.

In addition to indicating the affected destination point code (the affected point code is the one that is adjacent to the Signal Transfer Point and the congested route) is the current congestion status of the route. This allows the signaling point that originated traffic towards the destination to determine which type of messages it can send. If it has any messages to transmit that have a priority less than the congestion status indicated in the transfer-controlled, then the messages are not sent. This prevents the STP from receiving unnecessary messages that will have to be discarded.

Messages received by the Signal Transfer Point for a congested route are not processed and are not returned. They are simply discarded, and the transfer-controlled message is sent in the backward direction to indicate that they have been discarded. The only time the TFC is sent is when a message has been discarded because of a congested route. If the Message Signal Unit has a priority equal to or higher than the congestion level, the message is allowed to pass through the route to its destination and no TFC message is created.

It should also be noted that if a signaling point received a transfer-controlled message, and it determined that another route is available to the same destination, than it can choose another route to the destination. In this case, the signaling point would mark the route on which it received the TFC message as congested, and would invoke routing management procedures.

The transfer-controlled message is sent on a regular basis to update the originating signaling point of the congestion status. If timer T15 expires within the originator of a Message Signal Unit, and no TFCs have been received within timer T15, the signaling-route-set-congestion-test procedure is invoked to determine if the route is still congested.

The signaling-route-set-congestion-test procedure is explained in full detail in the next section. The concept is to send a message (the test message) with a priority value one less than what the originator thinks the congestion status is. If the route is still congested, the receiving STP will send a new transfer-controlled message indicating the congestion level.

If the signaling-route-set-congestion-test message gets through and a timer T16 expires, then the route is considered congested but at a lower congestion level than the test message. The signaling-route-set-congestion-test message is sent again with a lower priority, and the procedure is repeated until the route is found to be at level 0 (no congestion).

Signaling-route-set-congestion-test (RCT). In the event that a signaling point receives the transfer-controlled message, it needs to periodically verify that the indicated route is still under TFC procedures. There are no indications sent by the originator of the TFC message that the route is no longer under TFC procedures.

If we look at the big picture of what is happening during this procedure, we can see multiple tiers of network management. Procedures have been invoked at the link level, which only involves two signaling points directly adjacent to one another.

Traffic management is diverting traffic away from the affected signaling link, but, again, only between two adjacent signaling points. This sometimes may trigger the routing management procedure, which involves adjacent signaling points of an STP undergoing any one of the previous procedures.

The transfer-controlled procedure, on the other hand, is sent to any signaling point which originates a Message Signal Unit and is received by the concerned signaling point. Without keeping track of every TFC message sent and the destination of each message, it is not feasible for a signaling point to inform another signaling point that it is not adjacent to of the changed congestion status during the TFC procedure. For this reason, the responsibility is placed on the receiver of the TFC message to continuously monitor the congestion status of a route to determine when the status has changed. This is accomplished through the use of the RCT message.

The receiver of the transfer-controlled must send out the signaling-route-set-congestion-test message at the expiration of level-three timer T15. This timer is set when the TFC message is first received. When the signaling-route-set-congestion-test message is sent, the timer is reset.

The message is sent using the structure shown in Fig. 6.37. The test message provides only the H0/H1 heading code, indicating the type of test message. The Message Signal Unit, however, carries the priority level of the concerned congested link. In other words, the congested route is considered congested at a specific level, which is determined by the concerned signaling points.

This congestion level, as discussed before, corresponds to the priority of a Message Signal Unit. In ANSI networks, the priority of an MSU is indicated in the service indicator octet subservice field.

The idea is that the Message Signal Unit should be sent at the same priority level as the congestion status of the concerned route. If the test message is passed through the concerned signaling point towards the affected destination, then, obviously, the status of the route has changed. However, there is no indication as to what the current congestion level is (Fig. 6.38).

If the congestion status has not changed, and the Message Signal Unit carrying the signaling-route-set-congestion-test message is received, the message will be discarded by the receiving Signal Transfer Point and

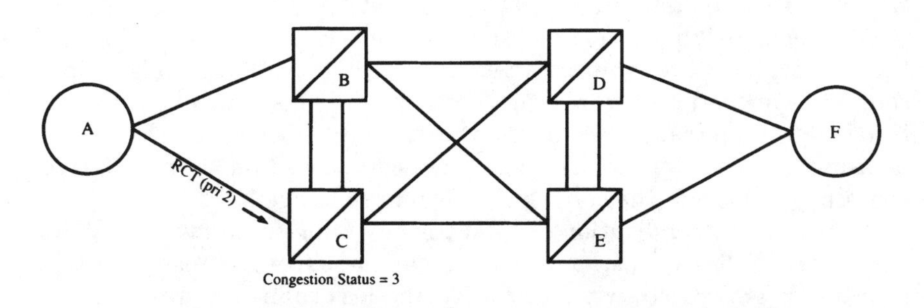

Figure 6.37 In this example, signaling point C is under congestion. Signaling point A has already been notified by a transfer-controlled (TFC) message of the congestion status. The signaling-route-set-test (SRCT) message is sent with a priority one less than the actual status of the route.

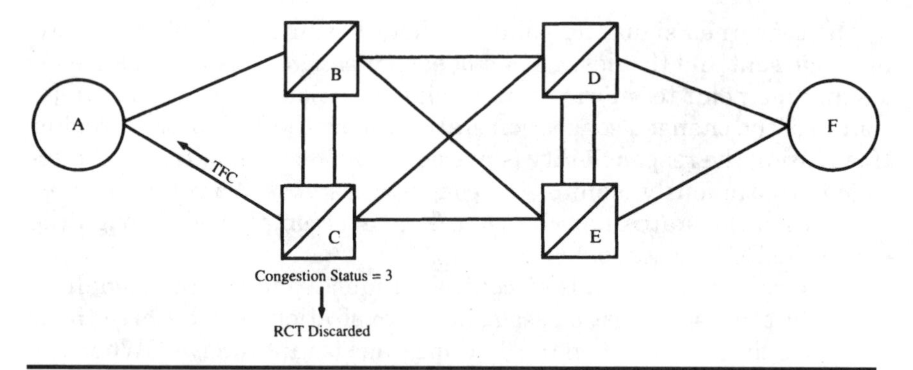

Figure 6.38 The SRCT was received, but because the congestion level has not changed, the SRCT was discarded. A TFC is returned to the originator (A) indicating the current congestion level.

a transfer-controlled message will be returned to the originator of the test message indicating the present congestion status level towards the affected destination.

Remember that the Message Signal Unit must have a priority equal to or more than the congestion status level. With every transmission of the signaling-route-set-congestion-test message, the MSU priority is set one less than the perceived value of the concerned route. For example, if the congestion status is determined to be at level three, only MSUs with a priority of three will be passed through. The message is sent with a priority that is one less than what the originator thinks the congestion level is, based on the last transfer-controlled message received.

If the congestion level has not changed, the test message is discarded, and a transfer-controlled message is returned. If the congestion status has changed, the Message Signal Unit is passed on to the destination. There is no indication or acknowledgment that the message was received and routed to the affected signaling point.

To determine when a test message has been allowed to pass through the concerned route, the originating signaling point of a signaling-route-set-congestion-test message sets timer T16. At the expiration of timer T16, if no transfer-controlled message has been received, it can be assumed that the test message has been passed on to the affected signaling point, and the congestion level has changed.

There is still no indication of what the new congestion level is. When the timer T16 expires, another signaling-route-set-congestion-test message is generated, with a priority one less than the previous test message. This new test message is sent once every T15, until the congestion status has changed, and the timer T16 expires with no receipt of a transfer-controlled message (Fig. 6.39).

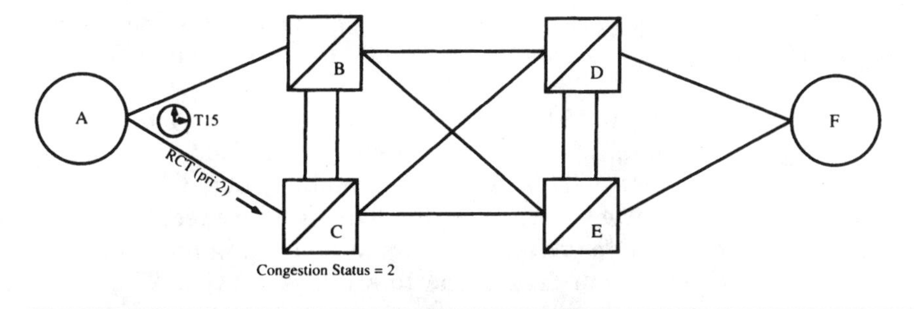

Figure 6.39 When the congestion level has decreased, eventually an SRCT will be accepted. The timer T15 triggers the transmission of SRCT messages.

This procedure continues until the congestion status has abated and the route is considered at congestion level zero. The congestion level is not defined by Bellcore, but by each individual signaling point. This means that the congestion levels are implementation dependent, and there are no rules as to what constitutes a level-three congestion status versus a level-one congestion status.

The ANSI and Bellcore standards do indicate what type of messages should be allowed during the various congestion levels. These can be found in Bellcore publication T1.111.5. The rules state that any network management messages sent by level three should always have a priority of three. This, of course, is the highest priority available, and ensures that network management messages always reach the affected destinations, despite the congestion status.

Any messages that are related to future connections, that is, voice circuit connections which have not yet been established, must have less priority than those messages related to existing connections. This is done to ensure that present connections are maintained properly, and any messages related to present connections can be routed through the network. Any new connections would then be based on the network's ability to route those requests through the congested route.

These messages should carry a priority of zero or one, being of a much lower priority than the existing one. This prevents new connections from being established during congestion periods, requiring more processing than what is available. This, of course, only impacts the congested route and does not prevent these connections from being established using other available routes, if any are available.

Any message sent in response to another previously received message, such as an acknowledgment, should be of the same priority as the request. For example, if a request was sent to a signaling point for information relating to an existing connection, the response should have the same priority as the request. This also ensures proper maintenance of existing connections and prevents messages from being lost and affecting these connections.

Any large messages should also be given low priorities. A large message requires additional processing resources, which adds to the congestion level of the route. This should be avoided whenever possible. What constitutes a "large" message is implementation specific.

The entire concept is to prevent additional processing requirements on a congested route from taking the link out of service. The route needs to be able to finish processing the connections it is already servicing, and should be able to return to a normal state within a reasonable time if network management is successful in controlling the traffic flow to this route.

Network Maintenance Procedures

The procedures previously described are used by the network to maintain the reliability of the SS7 network. Usually, there is very little that can be done by service personnel during any of these procedures, because the network is maintaining the status and invoking these procedures autonomously. There are occasions, however, when the network may not be successful in returning routes and links to service, requiring the intervention of service personnel.

This section will explain the big picture in network management, showing a variety of events taking place at all levels. The objective of this section is to show all the events that may take place during a failure or congestion, so that readers may have a better understanding of what is taking place and what to watch for in their own networks.

It is not the purpose of this section to prescribe any procedures that should be taken in the event that a failure or congestion is experienced, because this will be dependent on your own company's maintenance procedures and the equipment used with the network. Understanding what is happening will assist you, however, in determining the next course of action.

Congestion management

At the physical layer, maintenance will depend on the type of interface used. For example, if the interface for a link is a V.35, the testing procedures will be different than for a DS0 interface. Regardless of the interface type, this is usually the best place to start troubleshooting any transmission troubles.

The use of various test tools can aid in troubleshooting a link and can help determine what type of tests need to be run. The types of test equipment vary, depending on the type of interface and the sophistication required. A transmission analyzer looks at the transmission from the perspective of the physical layer and does not perform any protocol analyzing.

When a high-level view is needed, a protocol analyzer can aid in detecting both transmission problems and protocol problems. A protocol analyzer will decode all layers of the protocol, and allows a maintenance technician to determine at which level the error occurred. Since most troubles in the SS7 network are at the physical layer, the protocol analyzer will not be of much use to someone in the field or in the central office. The protocol analyzer is best suited for the remote maintenance center, where messages can be monitored from and to all points in the network.

At the various exchanges, a transmission tester is best suited for troubleshooting link troubles under the direction of the remote maintenance center. When a transmission test is performed, the technician is testing the facility for its ability to send and receive bit streams to the other end of the link without error.

No addressing or other protocol functions are needed with this type of testing because the test is being performed between two entities. The purpose of such a test is to check the integrity of a signaling link. The SS7 protocol will also be checking the link integrity whenever the link is active.

In the event that the protocol takes a link out of service, maintenance personnel may be required to test the link and isolate the fault. This is not always the case, however, since most failures on digital facilities such as DS0s are clock oriented and can be resolved through the diagnostics of the protocol.

The procedures used for testing at this level are company dependent. The only suggestion here is to check each link before placing it into service, to alleviate any obvious problems that may occur when turning up a signaling point for the first time. Once the link has been placed into service, there should be a minimal amount of problems.

The SS7 protocol, as we have already seen, uses a hierarchical approach to network management. As we have discussed all through this chapter, there are three levels of network management:

- Link management
- Traffic management
- Route management

Even though we have already discussed these procedures in full detail, we haven't looked at the whole picture during a link failure or

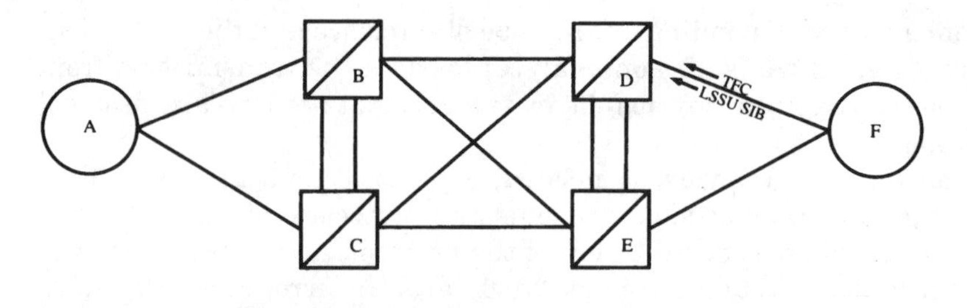

Figure 6.40 This figure shows a couple of activities. Congestion has occurred on the linkset from D to F. In this figure, the single line represents multiple links. A TFC is sent to indicate congestion. Meanwhile, one of the other links has become busy, prompting the LSSU with status indication of busy (SIB).

during congestion of a route. Let's look first at a congested route and what events could take place.

In Fig. 6.40, the signaling point D has been notified of a busy condition on one or more of the links towards signaling point F. One of the links at signaling point F has experienced a busy condition. This occurs when too much traffic is sent over one link, and the processor used by that link cannot handle all of the traffic. When this occurs, the link management software initiates a Link Status Signal Unit with a condition of busy. The LSSU is sent by level two under the direction of level-three link management. The LSSU is sent every T5 (level-two timer), for a period of T6.

During the period in which this link is under congestion, it can still send Message Signal Units, but it will not accept any from its adjacent signaling point. In fact, signaling point F, in this example, will hold all acknowledgments and negative acknowledgments until the congestion subsides.

Timer T6 prevents the link from remaining in a busy condition for too long. When this timer expires, the link is removed from service and the alignment procedure is started. In this example, the link was still in congestion mode when T6 timed out. The link was failed and the level-three link management software began the alignment procedure (recovery).

Because there are only two links to signaling point F, the failure of one of these links has created a problem. Traffic must now be diverted away from the failed link onto the adjacent link within the same linkset. However, this link is already near capacity and adding the load of another link will possibly create a congestion condition on this link as well.

Level-three link management places the link out of service by initiating the Link Status Signal Unit (LSSU) of "out-of-service" towards signaling point D. This is done to ensure that signaling point D does not send any traffic onto the failed link. The link may be very capable of sending traffic, but signaling point F cannot process it. In fact, the link, in this case, must be able to carry traffic, because the LSSU is sent on the same link for which it represents status.

This brings up a good point. When a link is failed, the link is not always at fault. When we speak of a link failure, the link processor at either end is included as part of the link. So even though the link itself is perfectly fine, the processor at either end or the interface card at either end could be at fault.

The LSSU is used at level two to inform the adjacent signaling point of the status of the link at the other end, so that both signaling points can be in synch during the diagnostics phase. Once the link has been removed from service (by transmission of the LSSU "out-of-service"), the signaling point F sends another LSSU with a status of out-of-alignment. This indicates that the link is no longer aligned and cannot be used to carry messages, other than level-two messages.

The next Link Status Signal Unit to be sent carries the status of "normal alignment," indicating that the processor at signaling point F has begun the normal alignment procedure and signaling point D should do the same.

During the proving period, Fill-In Signal Units are sent from signaling point F to signaling point D. These FISUs are monitored for errors using the alignment error rate monitor (AERM) during the proving period, which is usually around two to three seconds.

If there are more than four errors during the proving period, the link is failed and the alignment procedure begins all over again, beginning with the Link Status Signal Unit of out-of-service. This procedure will continue until the link has been returned to service, or until level-three link management removes it from service entirely.

While the Fill-In Signal Units are being sent between the two adjacent signaling points, traffic must be diverted away from the failed link. This actually begins before the link begins the alignment procedure. This is initiated by level-three traffic management. The purpose is to notify the adjacent signaling point that all traffic that was destined for the failed link must be rerouted to a new link. The traffic management message will also instruct signaling point D which link to use.

Normally, this would not be necessary. However, both ends of the link have a transmission and receiving buffer. These buffers hold transmitted and received Message Signal Units until an acknowledgment is sent or received. In the case of the transmission buffer, these MSUs have been sent but have not yet received any acknowledgments.

The traffic management at level three will first instruct signaling point F to move the contents of the failed link's transmit buffer to another link. The other link is either a link within the same linkset, or of another route to the same destination. Once the Message Signal Units have been moved to the new buffer, traffic management initiates the changeover procedure.

The changeover order is sent on the newly selected link, indicating the signaling link code of the failed link. This means that the link code table within the signaling point must also be modified. Every signaling point maintains a signaling link code table, which identifies the link code assignment for every link in the system. Remember that the link code is a logical assignment and does not necessarily correspond to the physical link code.

The signaling link code (SLC) allows every link to have multiple codes. When a link is failed, the link must be removed from this table so that it is not selected by the routing function of level three. All of the signaling link codes are then reassigned accordingly.

The receiver of the changeover message must mark the indicated link as failed, remove it from its own signaling link code table, and transfer the contents of the transmit buffer to the link on which the changeover order was received. When this has been accomplished, the link alignment procedure is able to start on the failed link. Signaling point D sends a changeover acknowledgment to signaling point F to indicate completion of the changeover procedure within itself.

All unacknowledged Message Signal Units can now be re-sent from the transmission buffer, and newly generated MSUs can be sent via the new link as well. This will continue until the failed link has been returned to service.

At this point, network management has controlled the traffic between two adjacent signaling points and initiated a recovery procedure to return a link to service. No other signaling points have been informed of these activities. Unless the failed link creates congestion of the entire route, there is no need for further action. Let's see what would happen, however, if the condition did not change and congestion was to occur on the route.

In Fig. 6.41, signaling point D has been diverting traffic under the direction from signaling point F to another link. However, signaling point D only has two links in its route to signaling point F. With one link carrying all of the traffic, signaling point D has become congested over that route. This does not mean that signaling point D is under congestion and cannot accept messages, but that messages destined for signaling point F cannot be sent through signaling point D without some method of controlling the throughput of these messages.

There are two methods used for controlling messages in this situation. Route management is primarily responsible for rerouting traffic away from a congested signaling point (or signaling point with conges-

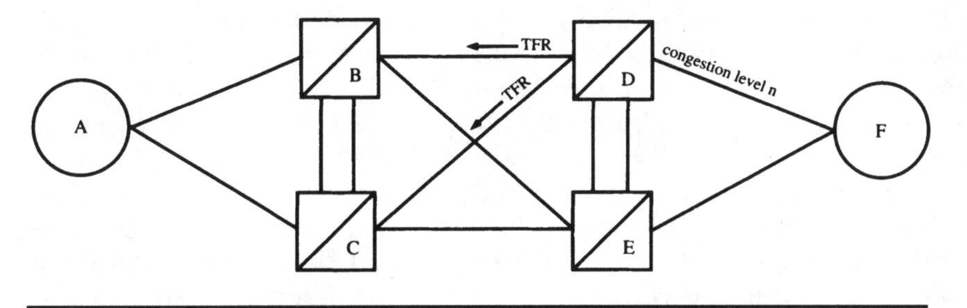

Figure 6.41 In this example, a link to signaling point F has become congested. Since this link is part of a route to F, the route congestion procedure is invoked.

tion on one of its routes). At the same time, route management can also throttle the amount and type of Message Signal Units sent to a specific destination. We will look at both of these procedures.

When signaling point D determines that the route has reached a predetermined threshold (usually determined by some configurable percentage, assigned at deployment time), the signaling point sends a transfer-restricted message to all adjacent signaling points. The TFR indicates a congestion condition to a given destination. Only the destination point code is given in this message. The receiving signaling points (signaling points B and C in this example) must then determine which alternate routes they will use to route messages to destination F.

Remember that signaling point D is not the affected signaling point. Signaling point D is simply a relay station for messages destined to signaling point F. The signaling point is not congested—just its route to signaling point F. Only messages destined for F are affected. All other messages can be sent to signaling point D with no impact, since they are not routed over the congested route.

The transfer-restricted indicates that messages destined to signaling point F should not be routed through signaling point D. This does not stop signaling point D from receiving messages destined for F. If no other alternate routes are available, messages can still be routed through signaling point D to signaling point F. This could compound the problem of congestion if too many messages were being routed to D.

When signaling point D receives a Message Signal Unit addressed to signaling point F, it must examine the congestion level of the affected link. We are no longer talking about the failed link, since it is still tied up in the alignment procedure. The link that all traffic was diverted to has now become the problem. Signaling point D must determine the congestion level (0, 1, 2, or 3) for the link and notify the originator of any MSUs.

Congestion levels are also implementation dependent, and can be configurable parameters based on network performance and number of links to a given destination. In this example, we will say that the signaling link has reached a congestion level of two. Level three is the most severe.

Signaling point D will use the congestion level of the link to determine which type of messages to allow through. For example, if the congestion level is level two, only Message Signal Units with a priority of two or above will be permitted to route through signaling point D to signaling point F. All other MSUs are discarded, and a transfer-controlled is sent to their originator.

The transfer-controlled message is sent by signaling point D to the originator of the received Message Signal Unit. Only MSUs received and discarded can trigger the TFC message. The TFC message carries with it the current congestion level and the destination of signaling point F.

The receiver of this message then stops generation of all Message Signal Units with a priority less than two. As discussed in the section about transfer-controlled procedures, there are certain types of messages which are allowed during this congestion status, and each type of message receives a particular priority. To review those priorities, refer to the discussion on the TFC procedure.

To determine when the route has become available again, the receiver of the transfer-controlled message must periodically send the signaling-route-set-congestion-test message. This message contains a priority of one in our example (since the TFC indicated that congestion status at level two). This message is re-sent every T15, until the congestion abates.

We can now see that network management has been performed at various levels: the link level, the traffic level, and even the message-origination level. As the severity increases, so does the role of network management. In Fig. 6.42, the failed link has been returned to service, and the congestion is subsiding.

Let's start by looking at the link again. The failed link has been through the alignment procedure and has successfully passed the alignment procedure. Before the link is returned to normal service, level three sends a signaling link test message (SLTM). This message is also configured at deployment time and carries with it a predetermined test pattern. The purpose is to ensure the ability of the link to carry level-four traffic and not just Fill-In Signal Units.

Once the signaling link test message has been sent and successfully received, the link is returned to service. Level two does not send any more Link Status Signal Units, but does allow Message Signal Units

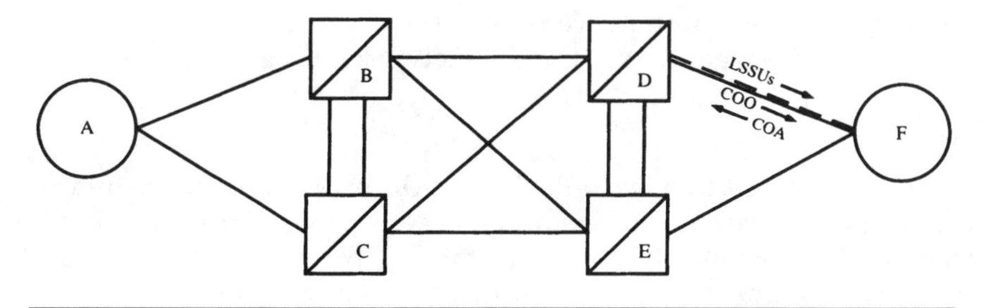

Figure 6.42 After testing a previously failed link by sending an SLTM, signaling point F sends a changeback declaration (CBD), which is then acknowledged (CBA). Traffic is returned to normal.

through the link. Actually, level three is the driver here and initiates the transmission of MSUs again. Before this can happen, a few procedures must be initiated internally.

First, the link must be reassigned its signaling link code, so that level-three routing can select this link for transmission. Once the link has been restored in the link code table, level-three traffic management sends a changeback declaration (CBD) message to signaling point D. The buffers do not need to be transferred this time, because normal processing can continue on both links.

To indicate the success of the changeback declaration, signaling point D sends a changeback acknowledgment on any available link. No other action is necessary. The failed link has now been restored and traffic has been diverted back to the link. Both links are now operational. This means that the congestion condition should also be corrected.

As the congestion level subsides, the signaling link begins processing messages of a lower priority. No indication is sent by signaling point D of the new congestion level. However, any receivers of a transfer-controlled are still sending messages. This is their only means of determining the current status of the link.

When the congestion level has abated, and the route is now free of congestion, the congestion level should be at level zero. Any signaling points sending messages will find they no longer receive a transfer-controlled when they send a test message with a priority of zero. They then mark the destination in their routing tables as available and begin normal transmission of signaling again.

Throughout this whole series of events, there has been no human intervention. In fact, this whole scenario may have taken only a few minutes to complete. By the time the events were detected by personnel at the remote maintenance center, the condition could have been repaired. This is what makes the SS7 network such a robust network.

Failure management

In the preceding section, we talked about the procedures that would be used to correct a congestion condition. Now let us talk about the events which could occur during a network failure. We will start at the link level and watch as the failure migrates to an entire route.

In Fig. 6.43, signaling point D has detected a failure at one of the links to signaling point F. This failure appears to be at the processor level. Level two is operational, level three is functional, but level four cannot be reached by level-three message discrimination.

This failure affects only one link in a two-link linkset. The first event to occur is the initiation of the Link Status Signal Unit by level-three link management. Level-three link management instructs level two to send the LSSU with a status of "processor outage" to signaling link F. Signaling link F thinks all is well until it receives the LSSU on the failed link. When it receives this message, it holds all Message Signal Units to prevent transmission over the failed link. Level two has done its job.

Level-three link management must now begin the recovery procedure on the link. This entails marking the link as out of service, which initiates changing the signaling link code and beginning the changeover procedure.

Level two sends the Link Status Signal Unit with a status of "processor outage" to signaling point F, indicating the processor failure at signaling point D. This should be followed by the LSSU of normal alignment. However, before normal alignment can begin, the traffic must be diverted away from the failed link.

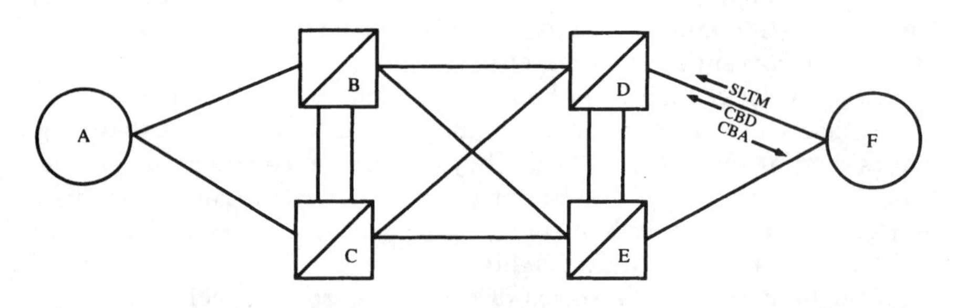

Figure 6.43 In this example, one link has failed in a two-link linkset. Level-two LSSUs are sent on the failed link, while the changeover procedure begins on the other link.

The changeover procedure is invoked by level-three traffic management. All unacknowledged messages in the transmit buffer of the link at signaling point D are transferred to the new link (in this example, the adjacent link within the same linkset).

When this procedure has been completed, the changeover order is sent to signaling point F providing the failed signaling link code. The changeover order, as seen in the illustration, is sent on the new link. When received, signaling point F transfers the Message Signal Units still in the transmit buffer of the failed link to the new link. The MSUs are then retransmitted over the new link in both directions.

To acknowledge completion of the buffer transfer and receipt of the changeover order, signaling point F sends the changeover acknowledgment message to signaling point D. This signifies that all is completed and Message Signal Units can now be diverted to the new link.

Now that the messages have been retransmitted and the buffers have been transferred, the link management recovery procedure can begin. This entails sending the LSSU of normal alignment over the failed link and beginning the alignment procedure. All timers and counters are set to zero, and the alignment procedure begins.

During the diverting of traffic, the processor outage problem has migrated to the new link. This means that both links to signaling point F are now inaccessible. Levels two and three are operational, but level four is not. This requires the link management procedures to be reported on the other signaling link.

The failure of both links has now isolated signaling point F from the rest of the network. While there is still another route available (through signaling point E), failure of E could cause a major outage.

The link management procedure on the last link of the linkset works a little differently than it did before. Now the emergency alignment procedure is used. This is virtually the same as the normal alignment procedure, except the time is much shorter (0.6 seconds) and only one error is allowed during the proving period.

Traffic management must now divert traffic away from the failed link and choose another alternate link. But this time, the links in this linkset are all out of service. This means that link management must choose a link within another linkset going to the same destination.

In this example, the other linkset is to signaling point C. The changeover order is sent to signaling point C on the linkset from D to C. All Message Signal Units received by D for signaling point F will now be routed through C. Since we have already discussed what events occur during the changeover order, we will not go through them again here.

Traffic is now diverted away from the failed links and rerouted to the linkset from D to C. However, signaling point D is still receiving messages destined for signaling point F. To stop messages from being sent to D for F, signaling point D route management sends a transfer-prohibited message to signaling points B and C. This will prevent either signaling point from sending an MSU destined for signaling point F through D.

Both signaling points B and C must now search for an alternate route. If there are no alternate routes, then the transfer-prohibited message is sent to their adjacent signaling points to indicate that they can no longer reach the destination of signaling point F.

Notice that signaling point B does not need to send a changeover order to signaling point C. In essence, signaling point B is diverting traffic away from one route to an alternate route. This is done through routing table states.

In the event that either signaling point B or C becomes congested, they may enter into a transfer-controlled procedure. Hopefully, this will not occur, if both signaling points have ample links in their alternate routes.

Now let's see what happens when the links all return to service. The failed links between signaling points D and F are restored. This means they have successfully passed the alignment procedure, and they are capable of processing level-four messages again. This is determined by level-three message distribution.

The links must first be assigned the signaling link codes so level-three routing may select them during the routing function. Once this has been completed, Message Signal Units can be sent over the affected links.

The changeback declaration message must be sent between signaling points D and C. This is to indicate to signaling point C that the traffic should now be diverted back to their old routes, and all Message Signal Units destined for signaling point F can now be routed through signaling point D. At the same time, since the route to signaling point F through D is now accessible, a transfer-allowed message is sent by signaling point D to its adjacent signaling points (B and C) to indicate the accessibility of the route.

To prevent signaling point D from routing messages destined to signaling point F through signaling point C, both signaling points B and C may send a transfer-prohibited message to signaling point D to prevent circular routing. Circular routing could occur if the route from signaling point C to F suddenly became unavailable. Messages would then be routed from D to C and then back up to signaling point B.

Once again, very little or no human intervention is required in such procedures. Remote maintenance personnel may redirect traffic through alternate networks, especially if a two-tiered network is used. In networks of this nature, the second tier allows routing through regionally located signaling points to route around clusters or regions that may be experiencing difficulty.

The rules for network outages in the Bellcore networks are stringent. Any one interface should not be "down" more than three minutes per year. The user interface is the access from level four to level three. This means that a processor outage (which indicates the failure of the interface to level four) should not render the user part inaccessible for longer than three minutes.

A network access unit should not be down more than two minutes per year. This includes the access to signaling points from the central office, such as the Service Switching Point. These must remain accessible all the time. Failure at any of the access points (end nodes) means telephone calls cannot be made.

Chapter

7

General Description of SCCP Functions

The Signaling Connection Control Part (SCCP) is much like X.25 in the services it provides. The only really significant feature added by SCCP is global title translation. Many applications within the SS7 network rely on this routing feature. We will discuss global title translation in full detail a little later.

The Signaling Connection Control Part is a protocol used for accessing databases and other entities within the network. As part of level four, SCCP relies on the services of the Message Transfer Part (MTP), just like the other level-four protocols in the SS7 network. The primary difference between the two services is in the addressing scheme and routing. SCCP also provides connection-oriented and connectionless services, whereas the MTP is strictly datagram.

The Signaling Connection Control Part provides both connection-oriented and connectionless services, with connectionless services currently supported in ANSI networks. However, if one examines the procedures closely, the protocol uses a number of procedures and parameters to emulate connection-oriented services. In fact, SCCP has the ability to maintain a dialog with another network entity, just as if there was a virtual connection between the two resources. SCCP does not, however, establish a connection before beginning the dialog—hence, the reason it is considered connectionless. Still, the service it delivers is very much like connection-oriented service, which is why I classify it as a connection-oriented "like" service.

When we look at the SS7 protocol stack, there is an indication that the ISDN User Part (ISUP) also uses the Signal Connection Control Part services to deliver messages in the network. The ITU, ANSI, and

Bellcore standards all provide procedures for ISUP services over SCCP, but this has not yet been implemented. The thought is to allow SCCP to deliver end-to-end signaling capability for ISUP messages, rather than node-to-node via the Message Transfer Part.

It is unclear whether this capability will ever be implemented. One would think that with Advanced Intelligent Network (AIN) features, this may be a useful feature, but even this standard has not been fully defined to date.

When we examine the services provided by the Signaling Connection Control Part (SCCP) in contrast to the Message Transfer Part (MTP), we can begin to understand the fundamental differences between the two. MTP provides routing, sequencing, and flow control. However, SCCP also provides these functions. The difference lies in the user of these services.

The Signaling Connection Control Part relies on the Message Transfer Part to route its payload from one node to another node. This means that SCCP must provide enough information to MTP that these functions can be carried out. SCCP provides the same functions, but the user is either the Transaction Capabilities Application Part (TCAP) or the ISDN User Part. This means that the functions used at SCCP will be somewhat different from those used by MTP. Let's look at the basic services that SCCP provides, and you can refer to Chap. 6 on MTP to compare the two.

Services of SCCP

The Signaling Connection Control Part is divided into five classes of service, or *protocol classes.* Each protocol class defines which level of service SCCP is to provide. The five classes of service are:

- Class 0—basic connectionless
- Class 1—sequenced connectionless
- Class 2—basic connection-oriented
- Class 3—flow control connection-oriented
- Class 4—error recovery and flow control connection-oriented

The first two classes, class 0 and class 1, support the connectionless environment, which is all that is used in today's networks. Classes 2, 3, and 4 are used for connection-oriented services and, even though well-defined, they are not used in today's network.

Class 0 services provide for the basic transport of TCAP messages (the payload does not need to be TCAP, although this is the most commonly used today). There are no procedures used to segment data or provide any sequencing of data. Typically, class 0 is used to deliver non-critical messages (such as database queries) when guarantee of delivery is minimal.

Class 1 is used whenever more than one SCCP message exists for a transaction. A transaction is something that takes place at the Transaction Capabilities Application Part level. The best way to define a transaction is as a function that is being requested by the application level.

For example, if there is a feature to be invoked at a remote switch, TCAP would send a message with several components. Each of the components would direct a particular portion of the switch to provide a function. All of the functions combined cause the feature to be invoked. The entire process is considered a transaction.

When this occurs, the TCAP message may be too big to fit in one SCCP message. SCCP must then provide segmentation of the data and possibly sequencing as well. The segmentation function divides the TCAP header and the application entity data into smaller segments, which can then easily fit within multiple SCCP messages. The TCAP header and the application-level information are left intact before the segmentation.

In other words, the TCAP header is not duplicated for each of the SCCP messages. Rather, it is segmented, as is, right into the payload portion of the first SCCP message. The sequencing function is used to ensure that the SCCP messages are received in the same order in which they were sent.

The routing function at level three examines the protocol class parameter to determine if the signaling link selection (SLS) filed should remain the same or be rotated. If messages are part of the same transaction, the SLS is not rotated, ensuring the messages are received in sequence. If rotation of the SLS field is allowed, associated messages may be received out of sequence.

To ensure that the messages that are part of the same transaction are delivered in sequence, the routing function at level three examines the protocol class parameter to determine if the signaling link selection (SLS) field should remain the same or be rotated. If rotation is allowed, then a message may be allowed to travel through a different path, whereas if the SLS field is not rotated (bit rotation), the messages may be received out of sequence.

Protocol class 2 provides a basic connection-oriented service. A connection must first be established between the two entities and, once established, a two-way dialog may take place. In the case of SS7, the same physical link may be used to carry multiple connections at the same time. This is the same concept as multiplexing several telephone conversations onto one physical facility.

To maintain an order between the various "connections," a reference number is established at both ends as an index to each of the particular connections. This allows either end to determine which virtual connection a message is destined to, much in the same way that X.25 uses virtual channels for a connection.

Protocol class 3 provides the same services as class 2, but adds the function of flow control and expedited data. Message loss and missequencing can also be detected and reported to the opposite end of the connection by way of an SCCP management (SCMG) message. When such an event occurs, class 3 has the capacity to reset the connection and restart transmission of the upper level messages again.

Protocol class 4 adds error recovery to the class 3 functions. This has presently been removed from Bellcore standards, but is still defined in the ANSI standards. Error recovery supports retransmission of errored messages.

All of these protocol classes involve services of SCCP. Primitives are used to convey information between the levels, whether between the user part and SCCP, or between SCCP and MTP.

Figure 7.1 illustrates the various functions provided by SCCP. Through the use of primitives, all user parts that use the services of SCCP must discern whether they need connection-oriented or connectionless services. A separate interface is maintained for either one of these services.

The various routing functions of SCCP are handled by the SCCP routing control (SCRC) function, which interfaces directly with MTP. This includes any global title translation if the signaling point is equipped with that function. Remember that not all signaling points are equipped with global title translation.

The signaling point may also provide the reverse option of global title, which is translating the calling party address from point code and subsystem number into nothing but global title. This is especially useful for secure networks, where a message is being routed outside of the network and only the global title needs to be given to prevent outside networks from learning the point codes of internal signaling points. This is a function which would be provided by a gateway Signal Transfer Point.

MTP sends all received messages to the SCCP routing control for discrimination and distribution. If you remember our discussion from Chap. 6, you will remember that MTP also has a message discrimination function as well as a distribution function. The SCRC function is much like MTP in that it also must discriminate messages at the SCCP level and distribute to a higher level if addressed to the local application.

The called party address field provides the addressing information necessary for the SCRC to route a particular message to the correct destination. The SCRC is not concerned about which signaling point to

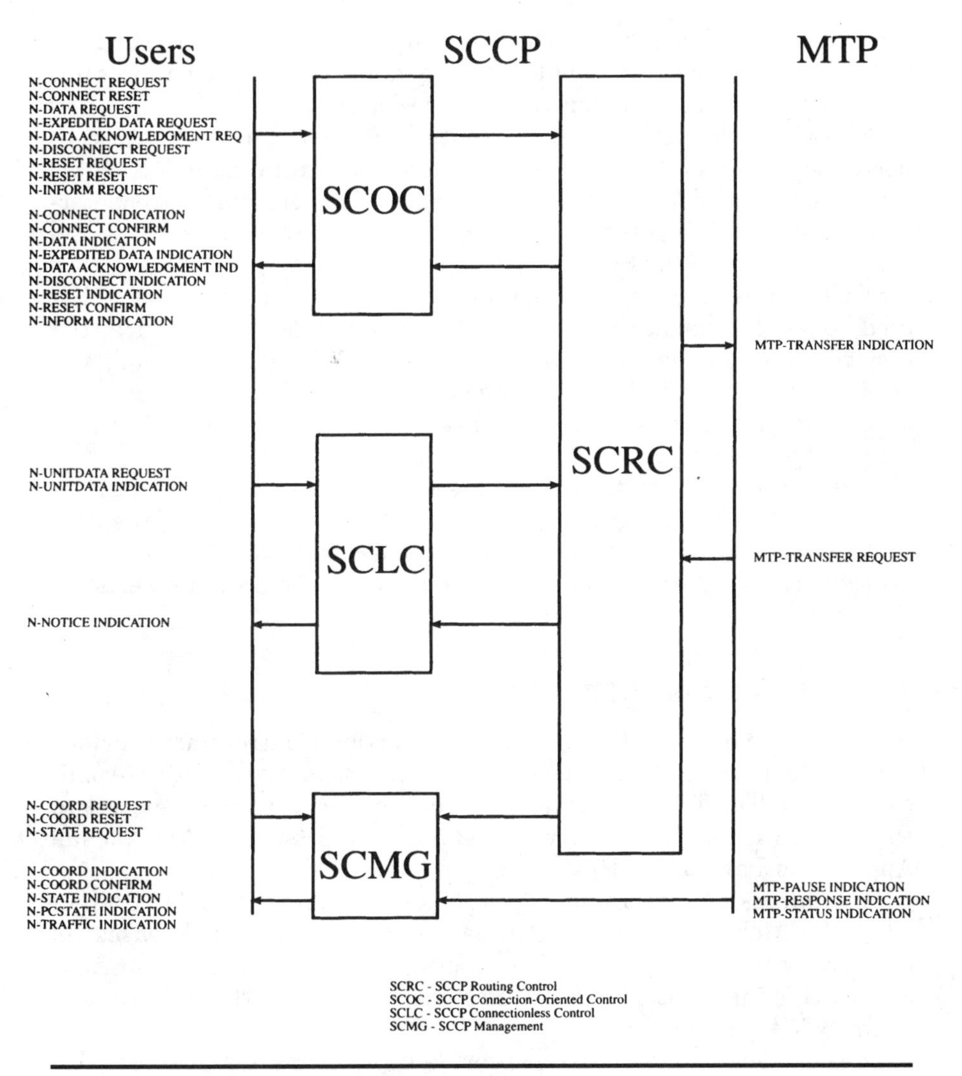

Figure 7.1 This figure illustrates the four functions defined in SCCP and the primitives used to interface to other protocols in SS7.

route the message to, because this is the function of MTP. The SCCP level is concerned only about its users, which are at the application level. Therefore, the destination point code is not used by the SCRC function. It exists simply for the sake of the MTP. The subsystem number is the only information (besides the global title, which is passed to the subsystem) that is needed at this point for routing by the SCRC.

The message distribution function routes the message to the correct user part, as defined in the called address field. Later, we will define the values of the various subsystems which have been defined by both Bellcore and ANSI and identify which applications they represent. Routing is discussed further in the next section.

The SCCP connection-oriented control (SCOC) and the SCCP connectionless control (SCLC) are both used to control functions such as segmentation and sequencing of messages. The user part must determine which service is required for the particular transmission. At the present time, only the SCLC function is used.

SCCP management is notified any time a message cannot be delivered to a user part. In the case of the connectionless service, the received message is actually returned to the originator using the Unitdata Service (UDTS) or Extended Unitdata Service (XUDTS) message structure. In either one of these formats, the payload is also returned to the originator.

In the case of connection-oriented service, the connection request is denied and SCCP management is notified. There is no return of the data sent. SCMG may reset the connection, if one has been established, or simply return a connection reject message if no connection previously existed.

Routing Services of SCCP

As mentioned earlier, the Signaling Connection Control Part provides end-to-end routing functions to the Transaction Capabilities Application Part. The TCAP protocol, in turn, provides transport services to the application entity with which it interfaces (such as the Mobile Application Part, or MAP).

The addressing requirements are somewhat different for TCAP than what the Message Transfer Part is able to provide. MTP bases its addressing (the point code in the routing label) on information from SCCP. This information is found in a called/calling party address field in the SCCP message.

The called/calling party fields provide three forms of addresses. The first consists of digits. These digits, referred to as global title, are usually what the subscriber (calling party) dialed in the case of an 800 number call. However, they do not have to be digits dialed. They can also be the mobile identification number (MIN) used in the cellular network to identify a cellular subscriber. At any rate, whatever the address is (dialed digits or some other form of digits), SCCP provides this information to MTP in the form of a primitive. This allows MTP level three to determine which signaling point to route the SCCP message to. If the destination point code is part of the SCCP called party address, this information is given to the MTP of routing.

If this is the only information found in the called party address, then nothing else is included in the primitive to MTP. However, if the called party address includes global title and subsystem number, then this information is given to MTP for transmission only. MTP simply includes this information as part of the SCCP header and sends it to the destination. If MTP does not get this information, then these fields are filled with zeros.

Typically, when routing an ISUP message, level-three MTP knows what the end destination will be for a particular circuit. But remember that MTP needs to know only about signaling points that have a direct trunk connection to their location. This is because of the nature of the signaling taking place with ISDN User Part services. The whole basis for this protocol is to establish a physical connection between the originating signaling point and an adjacent signaling point for a circuit connection (telephone call).

In TCAP transactions, the intent is different. We are no longer interested in circuit-related transactions. The purpose of TCAP is to provide a means for information exchange between two network entities, whether they are a database or another switch. This requires different routing procedures.

Global title translation

Most end-office switches will not know the actual address of all of the databases in the network, especially if the database is located in another company's network (the exception is when the network is very small and there are only a few point codes). Large networks need to consolidate resources and maintain smaller routing tables. This means that the Message Transfer Part must be able to route based on global title digits (such as an 800 number).

The Message Transfer Part will route the message to a Signal Transfer Point, which is capable of providing a more detailed address, point code, and subsystem number for the actual database that will provide the 800 number routing information. This application is referred to as *global title translation.*

Not every STP in the network needs to have this capability. In fact, global title translation is usually a function which is centralized within the network. The global title digits are given to the SCCP function of global title translation, so that SCCP routing can determine the point code and subsystem number the message needs to be routed to for the information. This feature applies to both 800/900-type numbers as well as to cellular mobile identification numbers. The ability to provide this type of routing at the SCCP level allows MTP routing tables within individual signaling points to remain condensed, without the need to know every point code in the network.

When using the global title translation feature, messages from a Service Switching Point (which is always the originator of such messages) are routed to a Signal Transfer Point, which must then translate the SCCP address fields into the point code/subsystem number combination. The message is then given back to the Message Transfer Part with the additional address information so that it may be routed by MTP level three to its final destination.

This method of routing also allows networks to accept messages from other networks, without disclosing their own internal point codes. For example, a company providing a database feature to other carriers may provide the point code of a gateway Signal Transfer Point. The purpose of the gateway STP is to control who has access to the network (using a security feature called *gateway screening*) and also to provide global title translation.

The carriers route SCCP messages to the database service using only global title digits. It is then up to the database service to determine which of their databases will receive the messages by performing global title translation on the global title digits and routing the message through their own network using the point code/subsystem number combination. This method is currently being used in Canada and some other countries with centralized databases, and will likely be used in the United States for cellular databases.

I should mention here, by the way, that the point code in the called party address field is that of the Service Control Point (SCP), which is providing an interface to the actual database. While it is possible to have a database resident within an SCP, these devices are usually front-end database servers used to route SCCP messages to actual database systems located on an X.25 network or TCP/IP network.

The subsystem number is actually the address of the database itself and is necessary for getting the TCAP message to the database. Each Service Control Point may interface to a number of databases, each with its own subsystem number.

We can now see the fundamental differences between SCCP routing and MTP routing. SCCP is providing a more flexible routing scheme and actually provides three addresses: the global title, the point code, and the subsystem number. MTP routes to the point code of a signaling point, which has a direct voice facility connection or signaling points within a cluster. It may also address signaling points that provide centralized services, such as global title translation.

Flow Control

Another fundamental difference between the Message Transfer Part (MTP) and the Signaling Connection Control Part is flow control. The

MTP network management procedures of level three provide flow control to a signaling point. In the event that a signaling point experiences a failure at the link level or the processor level, level-three management provides the procedures for routing around the failure.

The Signaling Connection Control Part also provides flow control, but at a different level. As previously mentioned, SCCP interfaces to the TCAP and ISUP protocols, which, in turn, provide services to an application entity.

The flow control in the SCCP management procedures provides management of message flow to a user part rather than a signaling point. In the event that a particular user part (today it is only TCAP) becomes congested, then the SCCP management function throttles the traffic destined to that user part. This has no effect on the traffic destined to the signaling point, unless configuration of the signaling point warrants rerouting.

There are procedures that allow for a database function or application entity function to be duplicated or replicated at another mated signaling point. If this is the case, then through configuration of the signaling point, the messages destined for a congested user part may be redirected to another signaling point.

This is not necessarily a feature of the protocol, other than as an indication as to how the message should be handled (the indication is through the multiplicity indicator, located in the SCCP management messages).

The multiplicity indicator indicates whether there are duplicated subsystems or not. There are no handling procedures other than the indicator. If the indicator says that the subsystem is solitary, then messages to a congested user part can be returned to the sender by SCCP management. If the subsystem is duplicated, then the signaling point configuration can determine the method to be used for handling messages. They may be rerouted to the duplicate subsystem, or they may be returned. This, of course, depends on the implementation at the signaling point.

In Bellcore networks, messages routed to a congested or failed user part must be rerouted to the duplicate subsystem, only if the duplicate subsystem is available and not under congestion itself. Otherwise, the message is discarded and a Unitdata Service or Extended Unitdata Service (XUDTS) message with a return cause is sent to the message originator.

This is also supported in the ANSI standards, but has not made its way into the standards publications as of the date of this publication. There is mention of the requirements, however, in Bellcore STP Generic Requirements TA-NWT-000082, Issue 5, May 1992 Revision 1, July 1992. This document cites the need for rerouting SCCP messages per the ANSI T1S1.3 standards group, but indicates that the requirement has not yet been added to the other standards publications and has, therefore, been included in the STP generic requirements publication.

Flow control procedures

Flow control is provided only in connection-oriented procedures. The purpose is to manage the number of data units sent on a particular connection. The unique thing about SCCP connection-oriented services is the ability to use one transaction to manage several connection sections.

The *credit* parameter in the connection request and connection confirm is used to establish the window size for a connection section. There may be multiple connection sections within one signaling point at any given time. The connection section is the equivalent of a virtual connection within an entity.

During the connection phase, the connection request message sends a credit parameter with the requested window size. This is based on the number of the messages that need to be sent. The credit may be negotiated between the two entities.

When the connection confirmation is returned, the credit parameter indicates the window size that was granted, based on the available resources of the destination signaling point. The actual credit granted may be larger or smaller than what was originally requested.

The sequence numbering then determines when additional messages may be sent. If the window size has already been met, then no other messages may be sent until messages sent can be acknowledged and processed. Once they have been removed from the receiving signaling points buffer, additional messages may be sent.

The sequence numbers in the SCCP messages are used to notify both entities when messages have been acknowledged. This works the same way as message sequence numbering at the MTP level.

Another unique point about flow control compared to MTP is the fact that flow control at SCCP is used to control the flow of messages to a user part and not a signaling point. MTP flow control controls the traffic destined to a signaling point, whereas SCCP controls traffic to an application.

If the resources available to a particular user part (which serves as the communications interface to applications) become congested, then SCCP flow control is used to throttle the messages to the affected subsystem.

Connection-Oriented Services

The Signaling Connection Control Part supports connection-oriented services for the Transaction Capabilities Application Part and the

ISDN User Part. However, none of these uses connection-oriented services in networks today. It is important to remember that the protocol defines the procedures and functionality for a great many services, but only a fraction are actually implemented to date. For the sake of this book, we consider the possibility of all of these procedures and functions being used at some point in time.

This is a particularly interesting development, since the mechanisms used in the connectionless service emulate a connection-oriented protocol. Part of the reason for using connectionless versus connection-oriented services lies in the resources required to support hundreds of connections at any given time.

Because of the nature of the SS7 network, to establish a virtual connection with an application entity for every transaction that takes place in the network and maintain that connection through the entirety of the transaction would require far too many resources. If the same information could be sent to the application without having to establish a connection, then the results would be the same, but with fewer resources.

Nevertheless, connection-oriented services are defined and described here because there may come a time when these services will be needed. Besides, no discussion could be complete without discussing all the capabilities of this network, instead of what is implemented.

Connection-oriented services provide for two types of connections: permanent and temporary. Permanent connections are established for operations, maintenance, and administration functions. These connections must be maintained permanently so that continuous real-time information can be exchanged between the network entities and the operations centers. In today's networks, this is not used, since connection-oriented services are not supported.

Temporary connections are used for all other services and, as the name implies, are established on a temporary or as-needed basis. These must be established when data transfer is requested and released when data transfer is complete. This works just like a telephone call in the Public Switched Telephone Network.

For example, in the case of remote control of another telephone switch, there may be a need to establish a temporary connection before transmitting the actual data. This requires the services of connection-oriented SCCP, which will send a connection request to the distant end and establish a connection with the resources required at the remote switch. Once the connection has been established, the actual data, or control information, is sent over the established connection (which is actually a virtual connection, or session, if you think in mainframe terms).

When the transaction is complete and there is no other control information to be sent, the connection can be released. This means that whatever resources that were required for the transaction are now made available for other entities. Temporary connections, if used today, would be the most commonly used connection.

As mentioned already, connection-oriented services are not presently supported in ANSI or Bellcore networks, nor in any of the ITU networks known by this author. This may change in light of the many new Advanced Intelligent Network features being deployed throughout the world. To date, there have not been any features defined that require connection-oriented services.

Even though connection-oriented services are not supported, the connectionless services do emulate many of the features of a connection-oriented protocol. Index numbers are used throughout the message in the TCAP portion to provide references back to previous transmissions, much in the way that virtual circuits are used in X.25. These index numbers are of no significance to the SCCP protocol. It is important to understand how these work and how TCAP works in general, so that one may understand the services of the SCCP. You may want to read Chap. 8, on TCAP, first or review this chapter after reading about TCAP.

Connection-oriented services consist of several phases, which represent the various activities that take place during transmission of data between two entities. Some of these phases are represented during connectionless services as well.

The main disadvantage of connection-oriented services is the number of resources required for a connection-oriented transaction. For example, for two entities to exchange data regarding a mobile telephone subscriber (such as location data), a connection would be established first, virtually dedicating resources on both ends for the transaction. The data could then be transferred and the resources released.

The problem comes when there are many of these transactions occurring throughout the network. In cellular networks, a mobile subscriber's phone may send a location update message every three to five minutes. You can imagine the number of such messages which must travel through a cellular network supporting 10,000 to 15,000 subscribers! This is one of the reasons connectionless is so widely used in SS7 networks today. Nevertheless, let's look at the procedures of the connection-oriented protocol and discuss the functions.

Connection-oriented procedures

As mentioned earlier, there are several phases that take place during connection-oriented transmission. The first phase is the *connection establishment phase.* This phase is when the resources of both entities, the receiver and the originator, are dedicated to the transaction.

This is then followed by the *data transfer phase,* which is when the data is actually being exchanged between the two entities. During this phase, no other entities are allowed to send anything to the resource that has been assigned to this transaction. Notice we said "resource." Switches and computers have multiple processors and have the capability of handling many transactions at once. This is due to the usage of distributed processing.

Following the data transfer phase is the *release phase.* This consists of a volley of messages exchanged between the two entities, releasing the resources which were dedicated for the particular transaction.

To better understand the various phases and the exchange of messages that takes place, let's look a little closer at each of the three phases.

Connection establishment. To understand the events that take place, we must first understand the routing of SCCP messages. A discussion of SCCP routing can be found in previous sections; here we need to talk about the differences between routing in connection-oriented and connectionless procedures.

In connection-oriented messages, the addressing is somewhat different than that of connectionless. The address information for all SCCP messages is located in two fields: the called party address and the calling party address.

In connectionless services, the called and calling party addresses represent the originating and destination point codes for the *message.* However, in connection-oriented services, each node involved in the connection establishes a connection between itself and the next intermediate node.

A good example of how this works is the voice network. A central office usually does not have a direct trunk connection to the final destination of the call, unless it is a local call. To reach the final destination, it must route through any number of intermediate switches.

During the call setup, a message is sent from the originator to the next intermediate node involved with the connection. The connection is then established between these two entities. Any messages regarding that connection are directed to the intermediate node and not to the final destination.

The intermediate node is then responsible for establishing a connection between itself and another intermediate node or the final destination if possible. This requires messages that are originated from the intermediate node to the next node in the connection. This same scheme is used for SCCP routing when connection-oriented services are to be used.

However, with connectionless services, the message can be addressed directly to the destination and, regardless of the number of intermediate nodes in the path, the message is routed according to the final address. No connection is established and the message addressing consists of the message originator and the final destination of the message.

In connection-oriented services, each node will have to establish two connections. The first connection is for the incoming message. The second connection is for the outgoing message. Both of these connections must be correlated to one another. This is referred to as a connection-section in the protocol. The entity that establishes the connections must be responsible for maintaining an association between the incoming and the outgoing ports used for this connection.

Other messages can be received over these facilities, addressed to other entities. This makes this connection-section a logical, or virtual, connection, rather than a physical connection.

To establish a connection between two entities, the originator must send a *connection request* (CR) message. This is sent to the first intermediate node, or the final destination if there is a direct connection available (the destination is an adjacent node to the originator). The CR contains the information necessary to define the parameters of the connection. This includes the Quality of Service (QoS) (defined by the protocol class) and the addresses.

The data parameter may contain bearer information to be exchanged with the application with which a connection must be established. A maximum of 130 octets may be sent in the data field.

To ensure minimal delay caused by routing through an excessive number of intermediate nodes, a hop counter is maintained. This hop counter is configured per network and decrements as it passes through each node. When the counter reaches zero, the message is in error, the connection is aborted, and an error message is returned with a cause code to the originator.

When the connection request is received by the destination, a *connection confirmed* (CC) message is returned. The CC is acknowledgment that the resources necessary to maintain such a connection are available and are now reserved for the transaction. In the event that the resources are not available, then a *connection refusal* (CREF) message is returned, along with a refusal cause code.

When a connection has been established, a local reference number (Fig. 7.2) is assigned to that connection. This works much the same way as a virtual channel in the X.25 protocol. Other messages may be received over the physical link, even though this link has been established for a connection, but any messages bearing the same address and reference numbers previously established will be handled in the same way as if the link was a permanent physical connection. You may

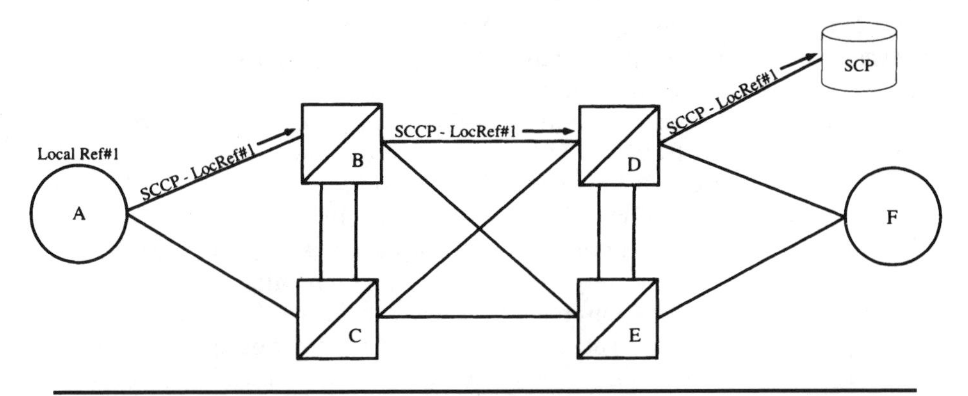

Figure 7.2 This figure illustrates an SCCP message being sent to an SCP with a local reference value of 1. The local reference number is only of significance to A, which uses this number much in the same way as logical channels are used in X.25. Any responses to this SCCP message will carry the same local reference number.

think of this as a switched virtual circuit, which will be released upon termination of the transaction.

The reference number is assigned to both the incoming connection and the outgoing connection, but both independent of one another. They are referred to as the *source local reference number* (incoming) and the *destination local reference number* (outgoing).

All messages that come in to the signaling point on a connection are then given the source local reference number of that signaling point. When a message is returned to that entity, it may provide the reference number so that the signaling point then knows to which connection the message is relevant.

Reference numbers are of local significance only. In other words, if the reference number of a connection at the originating signaling point were sent to the destination signaling point, the destination would not know what the number was, because it was created and corresponds to a connection back at the originating signaling point.

The local reference number is used by the remote node as an address. It becomes the destination reference number for all subsequent messages. This address is not found in the connection establishment phase, because it has not yet been created. During the data transfer phase, all data transfer messages will contain a destination local reference number, which is assigned by the remote entity.

When a connection has been released, the reference number assigned to the connections section related to that connection is "frozen." This means that the number may not be reassigned for a certain time. A timer is set to determine when the reference number may be "unfrozen," and available for reassignment to a new connection. This procedure is necessary to prevent an incoming message associated with a previous connection section from being routed to an incorrect reference number.

Also a part of the connection establishment phase is the negotiation for the Quality of Service. The receiver may choose a different QoS based on present conditions within that node and the resources available. The QoS is directly related to the protocol class parameter.

If we review for a moment the valid protocol classes, this means that an originator may propose protocol class 3, which calls for connection-oriented services with sequenced delivery and flow control. The receiver may instead assign protocol class 2, which provides the sequenced delivery without flow control. The connection confirm (CC) message is used to indicate which protocol class will be used.

If flow control is provided, the window size of the receiver may also be negotiated. This is accomplished using the credit field in the connection request message and the connection confirm message.

When the connection request is sent, the credit field is sent with the proposed window size for the connection. The receiver may then assign a different (lower) window size if it wishes. This is sent using the connection confirm message credit parameter. The credit parameter is also related to the Quality of Service of a message. This is assigned to a connection throughout the lifetime of the connection.

Data transfer phase. When a message containing data is sent over the established connection, the called party address is used by the SCCP routing control function to determine which connection section to route the message over. All subsequent data messages with this called party address will be routed over the same connection section.

In the event that there is an incoming and outgoing connection section (represented by the source local reference number and the destination local reference number), then the SCCP routing control must determine which of the outgoing ports is associated with the source local reference number and routes the message to that destination.

The message, when received by the final destination, is given by the Message Transfer Part to the SCCP routing control to determine the address of the message. If it is determined that the address is of the signaling point which has received it, then the data is given to the user part (TCAP or ISUP) via a primitive. Remember that primitives are used whenever a lower level function is passing information up to a higher level function. The primitive is the interface through software that communicates with the other signaling point entities.

Sequence numbering is provided (if the protocol class warrants) by using the sequence number parameters in the data form 1 (DT1) and the data form 2 (DT2) messages. The sequence numbers act as an acknowledgment for previously received messages as well. This works much the same way as in other protocols and as sequencing works at the Message Transfer Part (MTP) level.

The originator of data may send up to 127 data messages through a connections section in one direction. Another 127 messages may be sent through the same connection section in the opposite direction. At any time, the receiver may change the size of the window, depending on resources and other parameters, by changing the value of the credit parameter. The credit parameter indicates the number of messages that may be sent in any one direction.

For example, if the credit parameter has a value of 7, only seven messages may be sent in one direction (towards the sender of the credit parameter of 7). When the receiver of this credit has sent seven messages, it must wait until it receives a message acknowledging receipt of the previously sent seven messages. It may then continue transmission.

If the credit size has been changed, then the sender is limited to the number of messages it may send or, if the number has been increased, the sender may be able to send more messages. This dynamic window size allows the protocol to control the number of messages sent to an entity according to real-time events, rather than using a fixed parameter that is not sensitive to events that may take place during transmission (such as congestion).

Data acknowledgment messages may be sent even when there is no data to be transmitted. This allows for the acknowledgment of received messages and allows transmission to continue, even when transmission is only in one direction.

Some messages may exceed the capacity of the SCCP envelope. When this occurs, the data must be segmented into multiple packets before transmission. Keep in mind that the level above SCCP (the user part) includes header information. This header information is not included in every single data segment. Rather, the data and the header information are sent in their entirety to SCCP routing and, when segmented, the first set of bits is encapsulated and transmitted.

The remaining bits are then encapsulated and transmitted, with no knowledge of header information. The header information of the user part will then be received along with some of the data in the first segment. The header information will be in its entirety. The remaining data will be received in subsequent SCCP messages. Segmentation is not necessary if the message is equal to or less than 255 octets.

To ensure that data has been received, an *expedited data* message may be sent instead. The expedited data message allows up to 32 octets of user data to be sent to the destination. No further data may be sent, however, until an acknowledgment has been received. Once an acknowledgment has been received, then additional expedited data messages may be sent.

This is only true on one connection section. Multiple expedited data messages may be sent to a node (signaling point) on different connection sections, but only one at a time may be sent on any one connection section. This ensures that the transmitted data is indeed received without error. Flow control can be accomplished by withholding the expedited data message acknowledgment. Expedited data only applies to protocol classes 3 and 4.

In the event that two entities are no longer in sync with each other, a reset can be initiated. The reset changes the sequence numbering back to zero at both ends of the connection, and changes the window size back to the initial window size from when the connection was originally established. The credit field is also reset back to zero.

When a reset message is received, the receiver also resets its sequence numbers back to zero. Any data messages received after the reset request are discarded. A confirmation must be sent to confirm that the reset has been initiated before data can begin transmitting again. Once the confirmation has been sent, the originator of the reset begins sending data using sequence numbers beginning with 1.

Release phase. Once data transmission has been completed, and there are no further transmissions necessary, a release may be initiated. The release may be initiated by either node at any time during the life of the connection. However, there are measures which are taken at the user part to ensure that a connection is not prematurely released before an entity has completed its transactions.

The Transaction Capabilities Application Part uses a permission parameter, which is inherent within specific message types, indicating whether or not an entity has permission to disconnect a connection or not. Permission is not granted when there is additional data to be sent in association with a particular transaction.

As previously mentioned, the release may be initiated by either node, and by the user part or SCCP. When SCCP requests a release, it usually indicates a problem with the connection. SCCP may request a release or a pause on the connection, or it may also request a release without permission to reestablish the connection.

The release cause parameter indicates the reason for the release and will also implicate the originator of the release. A release must be acknowledged by a release complete before the connection is considered available for another transmission.

Connectionless Services

Connectionless transfer of data using the services of SCCP requires the use of the unitdata and the extended unitdata message structures.

These message structures provide all the information necessary for data to be transferred to a remote entity and to be processed by that remote entity.

There are two protocol classes that support connectionless services: protocol class 0 and protocol class 1. When protocol class 0 is used, there is no guarantee that the subsequent data will arrive in the same order in which it was transmitted. That is because the signaling link selection field in the routing label of the MTP header may be rotated at each node. When this occurs, a message may travel a different route than its associated messages.

Due to cross-delay, messages are received out of sequence when this occurs. To prevent this from happening, protocol class 1 can be specified. Any message with a protocol class of 1 indicates that the signaling link selection field should not be rotated, and any other messages received for the same destination are transmitted over the same SLS as the previous associated messages. This ensures that messages that are associated with one another follow the same path and do not get delivered out of sequence.

There is really only one phase during connectionless procedures: data transfer. There is no connection establishment, because a connection is not necessary. Data is encapsulated into a message envelope with all of the information necessary for the receiving entity to process the information as it was received.

Subsequent data may be sent in the same fashion, as long as there is enough information for the receiving entity to process the data. Obviously, connectionless services do not require the same resources as connection-oriented services and, thus, have found more favor in SS7 networks today. Let's look a little closer at the procedures used to transfer data with a connectionless protocol.

Connectionless procedures

The application service element wishing to send data to a remote entity requests connectionless services from SCCP. The application service element can be TCAP or some other transport mechanism used by an application entity. The application entity could be the Operations, Maintenance, and Administration Part (OMAP) or the Mobile Application Part (MAP).

The method used to transport data between two entities is transparent to the application entity and is the responsibility of the application service element, such as TCAP. All routing is handled by the SCCP routing control (SCRC), which provides routing parameters to the Message Transfer Part.

In large networks, it is not feasible for every node to know the address of every other node. For this reason, the SCCP routing control function can provide additional translation services. This is known as global title translation and is explained in detail in the preceding section on SCCP routing.

When there is too much data to fit into one SCCP envelope, or when it is determined that the message should be segmented (this can be determined by network management based on network conditions), the extended unitdata (XUDT) is used. The data is then divided into equal-length segments. The rule is that the first segment should be sized in such a way that the total message length is less than or equal to the size of the first segment multiplied by the number of segments being sent. This is to prevent buffers from becoming unmanageable.

All subsequent messages are configured with the same address information, and protocol class 1 is selected to ensure in-sequence delivery of all messages. The segmentation parameter in the XUDT message is set to indicate that there are additional segments to be sent, and the segment number field indicates how many additional segments are yet to be transmitted.

In the event that a message is received in error, an error message is returned to the originator of the message. The error message is in the form of the unitdata service or extended unitdata service messages. The data is also returned, along with a return cause code, which indicates why the message is being returned.

Unlike connection-oriented services, connectionless services do not assign any logical reference numbers to transmissions, because there is no connection to be established. Therefore, tracking is not important. The only objective is to get the data to its destination. At the user part level, there are many schemes used to ensure that messages are received without error and, at least in the case of TCAP, indexing schemes to keep track of associated data.

SCCP management

SCCP management (SCMG) is used to maintain the integrity of SCCP services. While the Message Transfer Part maintains link integrity and alerts adjacent signaling points of congestion at another signaling point, SCCP management is concerned about the status of a subsystem or application entity.

To accomplish this, SCCP management is divided into three tasks: signaling point status, subsystem status, and traffic management. Management messages use the unitdata message structure found in connectionless SCCP.

To maintain the status of the signaling point, SCCP management relies on information from the Message Transfer Part, which is sent to SCMG through primitives. These primitives consist of the MTP-Pause,

MTP-Resume, and MTP-Status primitives and their parameters. Subsystem information is gathered by SCCP management through primitives from the subsystem directly to SCCP management.

Probably the most advantageous feature of SCCP management is the ability to route messages away from a failed or congested subsystem to a mated subsystem in another location. This ensures that services are not lost when a subsystem fails, and guards against failures due to subsystem congestion.

This requires that subsystems be "replicated," or duplicated and placed in different geographical locations. When this has been adhered to, the diversity of the network increases, and the reliability increases. The protocol can now control message flow to the replicated databases.

There are several roles which a database can play within the protocol structure. One is as the dominant subsystem. The dominant subsystem will hold a higher priority than its replicated subsystems. All replicated subsystems possess the same subsystem number, so it is up to the signaling points to determine how to route to the subsystems. This is decided at configuration time, and each signaling point is configured to handle SCCP traffic according to these rules.

Each replicated subsystem may have a priority, with the subsystem with the highest priority receiving the bulk of the traffic. The assignment of the dominant subsystem may change dynamically or may be a fixed configuration, depending on the network. One thought is to allow the priority to change based on current load. This provides for a more dynamic routing scheme, but could prove difficult to implement.

A solitary subsystem is one that is not replicated and must, therefore, handle all traffic. In the event that this subsystem fails, traffic may be routed to another network or stopped altogether. This is the least favorable in any network, because it increases the single point of failure and decreases reliability in the network.

Another scheme is to have a primary subsystem that receives all traffic until a failure occurs, in which case, all traffic is routed to an alternate. The alternate would then handle all SCCP traffic until a failure occurred, in which case it would be marked as inaccessible and all traffic would be routed back to the other alternate.

This uses a standard master/slave relationship, but it is not favorable, because there is no use of the other subsystem until a failure occurs. This means that the other subsystem is sitting idle and not being utilized in any capacity. If there is something wrong with this subsystem (in terms of being able to handle messages), you will not likely find out until it goes online and begins handling messages, which is a little too late.

SCCP management also uses a concept referred to as the "concerned" point code. There are really two point codes that are affected by SCCP management. One is the affected point code, which is the failed or congested entity. The other is the point code that utilizes the service of the affected point code. The concerned point code must be notified when there is a status change at the affected point code, so that it knows how to route SCCP messages. The concerned point code is updated about an affected point code's status using SCCP management messages and connectionless services of SCCP.

SCCP management messages are sent to "adjacent" signaling points (adjacent in the logical sense only) to alter the translation functions located within those signaling points. By altering the translation function, when a protocol class 0 message is received, messages can be routed to other replicated subsystems. The action to be taken depends on the type of SCCP management intervention.

Signaling point status management

Signaling point status management is concerned with the status of a Service Control Point (SCP). If the SCP becomes congested or should fail, then the subsystems adjacent to the SCP cannot be reached. Traffic must then be diverted to replicated subsystems.

This requires a series of management messages that provide the status of a signaling point (SCP) and subsystem combination. These messages are not to be confused with the messages used by network management at level three, although there are some similarities. Level three is more concerned with all signaling points within the network, rather than just a select type.

A signaling point prohibited procedure indicates that the affected signaling point has been prohibited and cannot receive any traffic. The signaling point will be a Service Control Point, rather than a Service Switching Point or Signal Transfer Point, because this is the only entity SCCP management is really concerned about.

When a prohibited message has been received, the receiving node changes its translations to route traffic to the replicated subsystems, if any exist, according to the configuration of the replicated subsystems (dominant role, alternate role, or solitary).

When the signaling point (SCP) is considered as "allowed," the translation tables are once again modified according to the roles of the replicated subsystems. Traffic is then allowed to be directed towards the affected signaling point, and subsystem status tests may be invoked.

All translation changes are made at "adjacent" nodes. Adjacency refers to a logical adjacency. This means that a path exists from one signaling point to the affected signaling point. An example would be a Signal Transfer Point that provides global title translation. This entity needs to be kept apprised of signaling point status, because it will have to change its routing tables and translation tables based on the status of the Service Control Points and their subsystems.

Subsystem status management

The purpose of subsystem status management is to monitor the status of individual subsystems within a signaling point. A Service Control Point may have multiple subsystems. If the signaling point becomes congested, or fails, then none of those subsystems can be reached (signaling point status management). If only one of the subsystems becomes congested or fails, then subsystem status management redirects traffic from that one subsystem to other replicated subsystems.

This should point out the fundamental difference between signaling point status management and subsystem status management. One is concerned with the status of the Service Control Point while the other is concerned with the status of the subsystems located within or adjacent to the SCP.

Translation tables must also be changed to allow routing to be diverted away from the failed or congested subsystem and routed to replicated subsystems, depending on their roles in the network (solitary, dominant, or alternate).

The same status is provided for subsystems as for signaling points. A subsystem is either prohibited or allowed. There is no restricted mode as used by level-three management. If a subsystem is unable to handle traffic due to congestion or failure, then traffic is immediately diverted to replicated systems. Throttling of traffic cannot be tolerated, due to the nature of the transactions taking place at a subsystem.

When a subsystem is marked as prohibited, a status test is used to audit the prohibited subsystem and ensure the status is correct. The test is invoked by the Service Control Point so that it may keep track of the status of all of its subsystems. If, for any reason, the subsystem status has changed to *allowed* and the subsequent management messages were not received, a subsystem could remain prohibited in the status tables of the adjacent SCP, while the actual status of the subsystem is allowed.

In addition to testing the status of a subsystem, a Service Control Point may also invoke a broadcast which allows the SCP to inform other local subsystems (concerned subsystems) of the status of other signaling points or subsystems. This is reserved for local subsystems, which means they have a direct adjacency to the SCP and can be communicated to through the use of primitives, rather than protocol messages.

Other signaling points can be notified through the use of a broadcast procedure for signaling points. This procedure is used to inform concerned signaling points of status changes concerning subsystems. Only concerned signaling points are sent status updates. A concerned signaling point is that which regularly routes messages to the Service Control Point/subsystem combination. Usually, this is a Signal Transfer Point, which provides the global title translation function for the rest of the network.

Now you can see one of the distinct advantages behind using a centralized Signal Transfer Point with global title translation, rather than spreading the routing function through all of the nodes. SCCP management, as well as routing, can be simplified if only a few of these entities provide this functionality.

There is also a procedure which allows for the calculation of traffic mixes, although much of this is still under study. Traffic mix information can be provided as an option in ANSI networks and can prove useful to some databases for the purpose of network management and network monitoring.

The traffic mix indication informs end databases as to the type of SCCP traffic being routed: normal SCCP traffic or "backup." Normal SCCP traffic is that which would normally have been routed to the subsystem without network management intervention. Backup traffic is all messages routed to the subsystem as the result of an SCCP management function. This indicates that the receiving subsystem is a replicate subsystem and is receiving traffic from another subsystem that is prohibited.

As mentioned earlier, a subsystem is either prohibited or allowed. There are no procedures currently defined for flow control to a subsystem. Because of the nature of the transactions that take place at a subsystem, flow control may not be a viable option.

SCCP Message Structure

The Signaling Connection Control Part is divided into several sections, as we saw earlier. In this section, we will define the various fields and values for these fields, and identify where they can be found in the SCCP message. These fields are defined according to their location in the SCCP message. There are three parts to the message: the mandatory fixed part, mandatory variable part, and optional part (Fig. 7.3).

The mandatory fixed part consists of those parameters that are mandatory for the particular message. Each message will have different parameters which will be fixed in length (according to the message type). The message type may have one octet or may have several octets of parameters. The field will not vary, however, and, because of the message type, it can be determined how large these parameters will be.

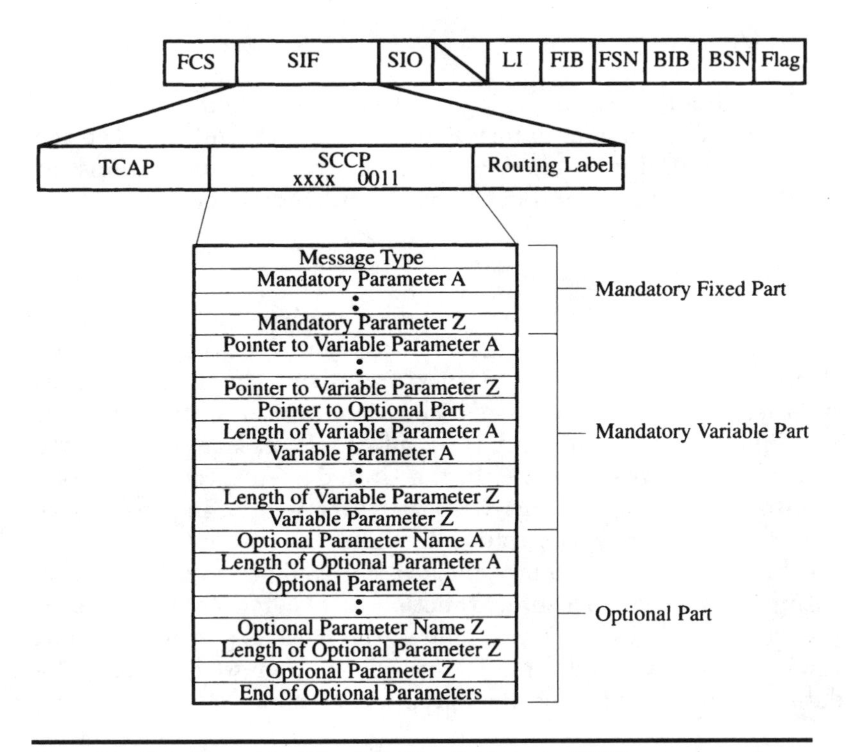

Figure 7.3 This figure identifies all of the fields in an SCCP message and its location in relation to the rest of the protocol.

The mandatory variable part consists of those parameters that are required of a particular message type, but are not of a fixed length. A good example of a mandatory variable part is a called or calling party address. This field uses length indicators to identify the length of each individual parameter field and the beginning of the parameter. Also in the mandatory variable part is a pointer that identifies the beginning of the optional part. This field consists of the binary representation of the binary offset. An offset is the octet count from the beginning of the pointer to the optional part indicator.

Each of the parameters found in the mandatory variable part is preceded by a length indicator and the parameter indicator. A pointer at the beginning of the mandatory variable part points to the octet of each of the individual parameters.

The optional part also uses length indicators after every parameter. The parameter is identified by a one-octet indicator, which provides a unique pattern for every parameter type.

Following the optional parameter is the length indicator, which provides the length for the entire parameter, excluding the parameter indicator. The length indicator is then followed by the parameter itself.

Not all SCCP messages will use all of these fields. Some SCCP messages will only use the mandatory fixed part; others will only use the mandatory fixed and variable parts. Following are the definitions for the various parameters found in the SCCP message structure.

Mandatory fixed part

The mandatory fixed part succeeds the message type code, which identifies what type of SCCP message is being sent. As discussed in the introduction, there are two types of services provided by the SCCP: connection-oriented and connectionless. The connection-oriented classes of service require several different types of SCCP messages, while the connectionless requires only one. In the United States, connection-oriented classes of service are not supported. We will define them here anyway, since there may come a use for these services in the future.

Even though there are no present uses for connection-oriented services using SCCP, there are many functions of SCCP that emulate connection-oriented services. We will look at the various parameters that are used for connection-oriented emulation when we examine the mandatory variable fields and the optional parameters.

Mandatory variable part

The parameters found in this field will depend on the type of message being sent. Each message has different requirements and may or may not require variable parameters. Not all SCCP messages will use the mandatory variable part.

The "mandatory" indicates that this field may be required for specific message types. The "variable" indicates that the length of these param-eters is not fixed in nature and will be variable. This includes addresses and other parameters, which will vary from message to message.

Optional part

The optional part is always a variable-length field, and may or may not be used with specific message types. "Optional" indicates that the field is not required for a specific parameter, but may be used to provide additional information relating to a transaction.

Length indicators are used before every parameter in this field to delineate between the various parameters. By providing these length indicators, the receiving signaling point can determine where the beginning and the end of a parameter is without the use of pointers.

Message Types

The first field of the mandatory fixed part is the message type. This field is found in all Signaling Connection Control Part messages. The message type will determine which parameters will be used in the mandatory variable part and the optional part.

The mandatory fixed part will be followed by variable and optional fields in some situations. This depends on the message type. The various parameters will provide additional information, again depending on the message type (the parameters are not shown in their entirety in this section).

This section describes the message types that are supported for connection-oriented and connectionless services. These message types and their parameters are shown as follows with explanations for the message types and their functions. The parameters are explained in their entirety in the next section.

End of Optional	Hop Counter	Data	Clg Party	Credit	Cld Party	Proto Class	Src Loc Ref	Msg Type
8	24	24 to 3120	32+	24	24+	8	24	8

Connection Request (CR)	**0 0 0 0**	**0 0 0 1**
Mandatory Fixed Part		
Source Local Reference	0 0 0 0	0 0 1 0
Protocol Class	0 0 0 0	0 1 0 1
Mandatory Variable Part		
Called Party Address	0 0 0 0	0 0 1 1
Optional Part		
Credit	0 0 0 0	1 0 0 1
Calling Party Address	0 0 0 0	0 1 0 0
Data	0 0 0 0	1 1 1 1
SCCP Hop Counter	0 0 0 1	0 0 0 1
End of Optional Parameters	0 0 0 0	0 0 0 0

End of Optional	Data	Cld Party	Credit	Proto Class	Src Loc Ref	Dst Loc Ref	Msg Type
8	24 to 3120	32+	24	8	24	24	8

Connection Confirm (CC)	**0 0 0 0**	**0 0 1 0**
Mandatory Fixed Part		
Destination Local Reference	0 0 0 0	0 0 0 1
Source Local Reference	0 0 0 0	0 0 1 0
Protocol Class	0 0 0 0	0 1 0 1
Optional Part		
Credit	0 0 0 0	1 0 0 1
Called Party Address	0 0 0 0	0 0 1 1
Data	0 0 0 0	1 1 1 1
End of Optional Parameters	0 0 0 0	0 0 0 0

End of Optional	Data	Cld Party	Cause	Dst Loc Ref	Msg Type
8	24 to 3120		8	24	8

Connection Refused (CREF)	**0 0 0 0**	**0 0 1 1**
Mandatory Fixed Part		
Destination Local Reference	0 0 0 0	0 0 0 1
Refusal Cause	0 0 0 0	1 1 1 0
Optional Part		
Called Party Address	0 0 0 0	0 0 1 1
Data	0 0 0 0	1 1 1 1
End of Optional Parameters	0 0 0 0	0 0 0 0

End of Optional	Data	Cause	Src Loc Ref	Dst Loc Ref	Msg Type
8	24 to 3120	8	24	24	8

Released (RLSD)	**0 0 0 0**	**0 1 0 1**
Mandatory Fixed Part		
Destination Local Reference	0 0 0 0	0 0 0 1
Source Local Reference	0 0 0 0	0 0 1 0
Release Cause	0 0 0 0	1 0 1 0
Optional Part		
Data	0 0 0 0	1 1 1 1
End of Optional Parameters	0 0 0 0	0 0 0 0

Src Loc Ref	Dst Loc Ref	Msg Type
24	24	8

Release Complete (RLC)	**0 0 0 0**	**0 1 0 1**
Mandatory Fixed Part		
Destination Local Reference	0 0 0 0	0 0 0 1
Source Local Reference	0 0 0 0	0 0 1 0

Data	Seg/Reassembly	Dst Loc Ref	Msg Type
16 to 2048	8	24	8

Data Form 1 (DT1)	**0 0 0 0**	**0 1 1 0**
Mandatory Fixed Part		
Destination Local Reference	0 0 0 0	0 0 0 1
Sequencing/Reassembling	0 0 0 0	0 1 1 0
Mandatory Variable Part		
Data	0 0 0 0	1 1 1 1

Data	Seq/Segment	Dst Loc Ref	Msg Type
16 to 2048	16	24	8

Data Form 2 (DT2)	**0 0 0 0**	**0 1 1 1**
Mandatory Fixed Part		
Destination Local Reference	0 0 0 0	0 0 0 1
Sequencing/Segmenting	0 0 0 0	1 0 0 0
Mandatory Variable Part		
Data	0 0 0 0	1 1 1 1

Credit	Rcv Seq #	Dst Loc Ref	Msg Type
8	8	24	8

Data Acknowledgment (AK)	**0 0 0 0**	**1 0 0 0**
Mandatory Fixed Part		
Destination Local Reference	0 0 0 0	0 0 0 1
Receive Sequence Number	0 0 0 0	0 1 1 1
Credit	0 0 0 0	1 0 0 1

Data	Clg Party	Cld Party	Proto Class	Msg Type
16 to 2032	2+	3+	8	8

Unitdata (UDT)	**0 0 0 0**	**1 0 0 1**
Mandatory Fixed Part		
Protocol Class	0 0 0 0	0 1 0 1
Mandatory Variable Part		
Called Party Address	0 0 0 0	0 0 1 1
Calling Party Address	0 0 0 0	0 1 0 0
Data	0 0 0 0	1 1 1 1

End of Optional	ISNI	Segment	Data	Clg Party	Cld Party	Hop Cntr	Proto Class	Msg Type
8	24 to 144	48	16 to 2032	2+	3+	8	8	8

Extended Unitdata (XUDT)	**0 0 0 1**	**0 0 0 1**
Fixed Mandatory Part		
Protocol Class	0 0 0 0	0 1 0 1
SCCP Hop Counter	0 0 0 1	0 0 0 1
Mandatory Variable Part		
Called Party Address	0 0 0 0	0 0 1 1
Calling Party Address	0 0 0 0	0 1 0 0
Data	0 0 0 0	1 1 1 1
Optional Part		
Intermediate Signaling Network Identification (ISNI)	1 1 1 1	1 0 1 0
Segmentation	0 0 0 1	0 0 0 0
End of Optional Parameters	0 0 0 0	0 0 0 0

Data	Clg Party	Cld Party	Return Cause	Msg Type
16 to 2032	2+	3+	8	8

Unitdata Service Message (UDTS)	**0 0 0 0**	**1 0 1 0**
Mandatory Fixed Part		
Return Cause	0 0 0 0	1 0 1 1
Mandatory Variable Part		
Called Party Address	0 0 0 0	0 0 1 1
Calling Party Address	0 0 0 0	0 1 0 0
Data	0 0 0 0	1 1 1 1

End of Optional	ISNI	Segment	Data	Clg Party	Cld Party	Hop Cntr	Rtn Cause	Msg Type
8	24 to 144	48	16 to 2032	2+	3+	8	8	8

Extended Unitdata Service Message (XUDTS)	**0 0 0 1**	**0 0 1 0**
Mandatory Fixed Part		
Return Cause	0 0 0 0	1 0 1 1
SCCP Hop Counter	0 0 0 1	0 0 0 1
Mandatory Variable Part		
Called Party Address	0 0 0 0	0 0 1 1
Calling Party Address	0 0 0 0	0 1 0 0
Data	0 0 0 0	1 1 1 1
Mandatory Variable Part		
Intermediate Signaling Network Identification (ISNI)	1 1 1 1	1 0 1 0
Segmentation	0 0 0 1	0 0 0 0
End of Optional Parameters	0 0 0 0	0 0 0 0

Data	Dst Loc Ref	Msg Type
16 to 264	24	8

Expedited Data Message (ED)	**0 0 0 0**	**1 0 1 1**
Mandatory Fixed Part		
Destination Local Reference	0 0 0 0	0 0 0 1
Data	0 0 0 0	1 1 1 1

Dst Loc Ref	Msg Type
24	8

Expedited Data Acknowledgment Message (EA)	**0 0 0 0**	**1 1 0 0**
Mandatory Fixed Part		
Destination Local Reference	0 0 0 0	0 0 0 1

Reset Cause	Src Loc Ref	Dst Loc Ref	Msg Type
8	24	24	8

Reset Request Message (RSR)	**0**	**0**	**0**	**0**	**1**	**1**	**0**	**1**
Mandatory Fixed Part								
Destination Local Reference	0	0	0	0	0	0	0	1
Source Local Reference	0	0	0	0	0	0	1	0
Reset Cause	0	0	0	0	1	1	0	0

Src Loc Ref	Dst Loc Ref	Msg Type
24	24	8

Reset Confirmation Message (RSC)	**0**	**0**	**0**	**0**	**1**	**1**	**1**	**0**
Mandatory Fixed Part								
Destination Local Reference	0	0	0	0	0	0	0	1
Source Local Reference	0	0	0	0	0	0	1	0

Err Cause	Dst Loc Ref	Msg Type
8	24	8

Error Message (ERR)	**0**	**0**	**0**	**0**	**1**	**1**	**1**	**1**
Mandatory Fixed Part								
Destination Local Reference	0	0	0	0	0	0	0	1
Error Cause	0	0	0	0	1	1	0	1

Credit	Seq/Segment	Proto Class	Src Loc Ref	Dst Loc Ref	Msg Type
8	16	8	24	24	8

Inactivity Test Message (IT)	**0**	**0**	**0**	**1**	**0**	**0**	**0**	**0**
Mandatory Fixed Part								
Destination Local Reference	0	0	0	0	0	0	0	1
Source Local Reference	0	0	0	0	0	0	1	0
Protocol Class	0	0	0	0	0	1	0	1
Sequencing/Segmenting	0	0	0	0	1	0	0	0
Credit	0	0	0	0	1	0	0	1

SCCP Management

The drawings below show the fields for an SCCP management message and the size of those fields. Each field is preceded by an indicator, which identifies the type of parameter to follow.

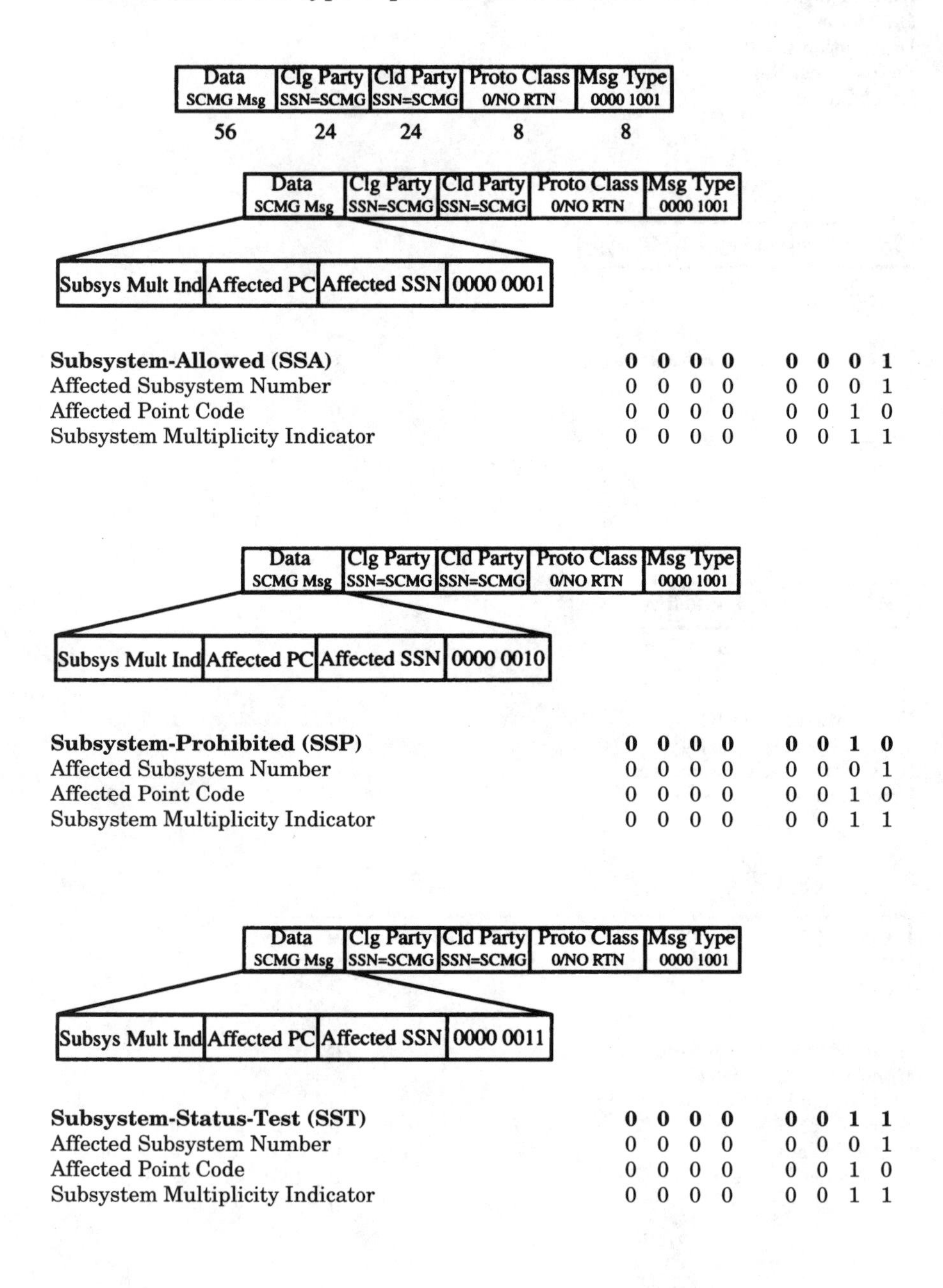

Subsystem-Allowed (SSA)	**0 0 0 0**	**0 0 0 1**
Affected Subsystem Number	0 0 0 0	0 0 0 1
Affected Point Code	0 0 0 0	0 0 1 0
Subsystem Multiplicity Indicator	0 0 0 0	0 0 1 1

Subsystem-Prohibited (SSP)	**0 0 0 0**	**0 0 1 0**
Affected Subsystem Number	0 0 0 0	0 0 0 1
Affected Point Code	0 0 0 0	0 0 1 0
Subsystem Multiplicity Indicator	0 0 0 0	0 0 1 1

Subsystem-Status-Test (SST)	**0 0 0 0**	**0 0 1 1**
Affected Subsystem Number	0 0 0 0	0 0 0 1
Affected Point Code	0 0 0 0	0 0 1 0
Subsystem Multiplicity Indicator	0 0 0 0	0 0 1 1

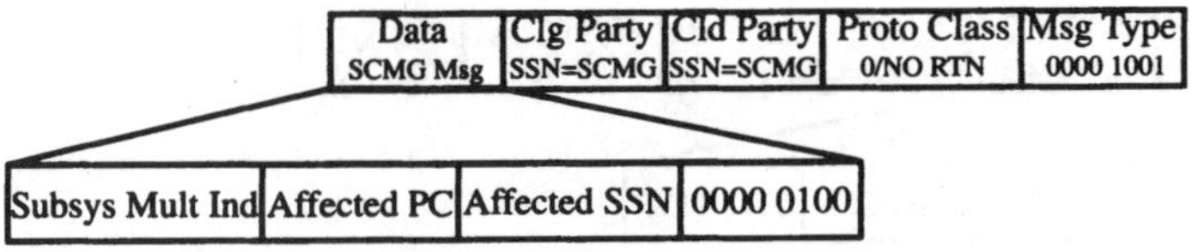

Data SCMG Msg	Clg Party SSN=SCMG	Cld Party SSN=SCMG	Proto Class 0/NO RTN	Msg Type 0000 1001

Subsys Mult Ind	Affected PC	Affected SSN	0000 0100

Subsystem-Out-of-Service-Request (SOR)	**0 0 0 0**	**0 1 0 0**
Affected Subsystem Number	0 0 0 0	0 0 0 1
Affected Point Code	0 0 0 0	0 0 1 0
Subsystem Multiplicity Indicator	0 0 0 0	0 0 1 1

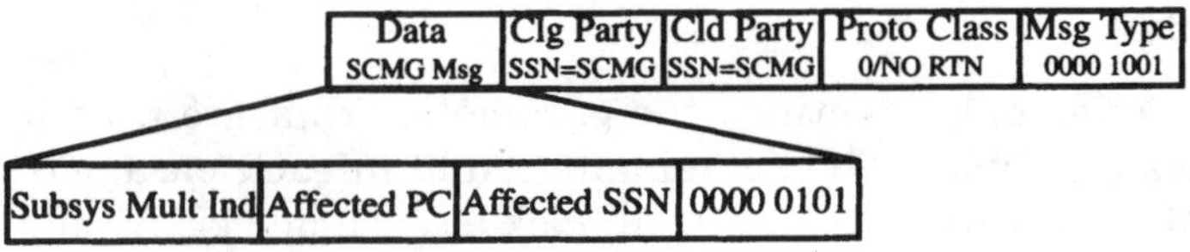 Data SCMG Msg	Clg Party SSN=SCMG	Cld Party SSN=SCMG	Proto Class 0/NO RTN	Msg Type 0000 1001

Subsys Mult Ind	Affected PC	Affected SSN	0000 0101

Subsystem-Out-of-Service-Grant (SOG)	**0 0 0 0**	**0 1 0 1**
Affected Subsystem Number	0 0 0 0	0 0 0 1
Affected Point Code	0 0 0 0	0 0 1 0
Subsystem Multiplicity Indicator	0 0 0 0	0 0 1 1

Data SCMG Msg	Clg Party SSN=SCMG	Cld Party SSN=SCMG	Proto Class 0/NO RTN	Msg Type 0000 1001

Subsys Mult Ind	Affected PC	Affected SSN	1111 1101

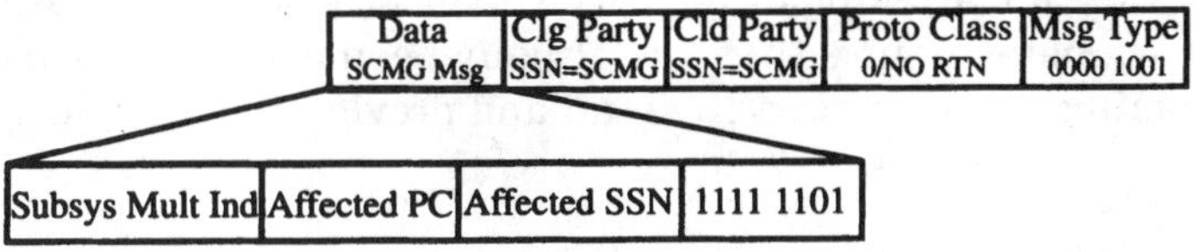

Subsystem-Backup-Routing (SBR)	**1 1 1 1**	**1 1 0 1**
Affected Subsystem Number	0 0 0 0	0 0 0 1
Affected Point Code	0 0 0 0	0 0 1 0
Subsystem Multiplicity Indicator	0 0 0 0	0 0 1 1

Data SCMG Msg	Clg Party SSN=SCMG	Cld Party SSN=SCMG	Proto Class 0/NO RTN	Msg Type 0000 1001

Subsys Mult Ind	Affected PC	Affected SSN	1111 1110

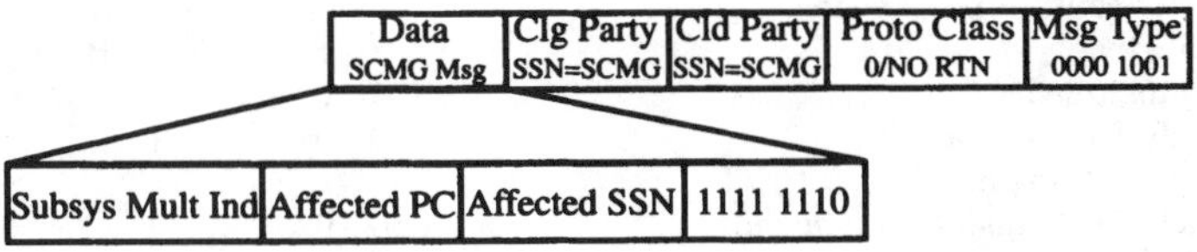

Subsystem-Normal-Routing (SNR)	**1 1 1 1**	**1 1 1 0**
Affected Subsystem Number	0 0 0 0	0 0 0 1
Affected Point Code	0 0 0 0	0 0 1 0
Subsystem Multiplicity Indicator	0 0 0 0	0 0 1 1

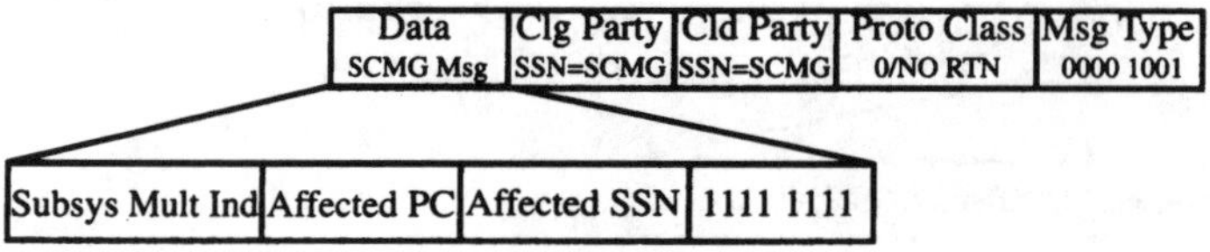

Data SCMG Msg	Clg Party SSN=SCMG	Cld Party SSN=SCMG	Proto Class 0/NO RTN	Msg Type 0000 1001

Subsys Mult Ind	Affected PC	Affected SSN	1111 1111

	HGFE	DCBA
Subsystem-Routing-Status-Test (SRT)	**1 1 1 1**	**1 1 1 1**
Affected Subsystem Number	0 0 0 0	0 0 0 1
Affected Point Code	0 0 0 0	0 0 1 0
Subsystem Multiplicity Indicator	0 0 0 0	0 0 1 1

SCCP Parameters

The previous section only identified the parameters that are used by the various message types. The actual data found in each parameter was not indicated. In this section, we will look at each one of the individual parameters and explain the various data values possible within each particular parameter.

The parameters differ, depending on the message type they are used with. Some parameters are common and can be found in several message types. Others are specific to a particular message type and contain information relevant only to the specific message type.

This section will not attempt to identify which message types use each of these parameters, since that has already been discussed. This section will explain the parameters in detail and provide the bit values of each one.

	H	G	F	E	D	C	B	A
Called/Calling Party Address	**0**	**0**	**0**	**0**	**0**	**0**	**1**	**1**
Octet One								
Address Indicator	**H**	**G**	**F**	**E**	**D**	**C**	**B**	**A**
Subsystem Indicator								
Address contains subsystem number								1
Address does not contain subsystem number								0
Point Code Indicator								
Address contains point code							1	
Address does not contain point code							0	
Global Title Indicator	**H**	**G**	**F**	**E**	**D**	**C**	**B**	**A**
No global title included			0	0	0	0		
Global title includes translation type, numbering plan, and encoding			0	0	0	1		
Global title includes translation type only			0	0	1	0		
Not assigned in U.S. networks			0	0	1	1		
			to					
			1	0	0	0		
Spare			1	0	0	1		
			to					
			1	1	1	0		
Reserved for extension			1	1	1	1		

Routing Indicator	H	G	F	E	D	C	B	A
Route using global title only		0						
Route using point code/subsystem number		1						
National/International Indicator								
Address indicator coded as international	0							
Address indicator coded as national	1							

The subsystem indicator and the point code indicator are used to indicate whether or not one of these two entities is found in the address. In fact, this entire octet is used to indicate what elements of the address are found and which elements are to be used by routing.

The routing indicator is used to instruct the routing function which element of the address to use for routing the message. If the routing indicator is equal to 0, this indicates that global title translation is necessary. The receiving signaling transfer point, if it has global title capability, should then provide global title translation. If the STP does not have this functionality, it should route the message to its destination. Usually, the end STP will provide the global title translation, before routing to an adjacent SCP for routing to the actual database or application.

Octet Two

Subsystem Number	H	G	F	E	D	C	B	A
Subsystem number unknown/not used	0	0	0	0	0	0	0	0
SCCP Management	0	0	0	0	0	0	0	1
Reserved	0	0	0	0	0	0	1	0
ISDN User Part	0	0	0	0	0	0	1	1
Operations, Maintenance, and Administration Part	0	0	0	0	0	1	0	0
*Mobile Application Part (MAP)	0	0	0	0	0	1	0	1
*Home Location Register (HLR)	0	0	0	0	0	1	1	0
*Visited Location Register (VLR)	0	0	0	0	0	1	1	1
*Mobile Switching Center (MSC)	0	0	0	0	1	0	0	0
*Equipment Identification Register (EIR)	0	0	0	0	1	0	0	1
*Authentication Center	0	0	0	0	1	0	1	0
Spare	0	0	0	0	1	0	1	1
				to				
	1	1	1	1	1	1	1	0
Reserved for expansion	1	1	1	1	1	1	1	1

*These are implemented in Bellcore networks for the use of cellular internetworking.

The subsystem number identifies the application to which the message is being addressed, or from where the message was originated. New subsystems have been added recently in the Bellcore recommendations to include the various database functions found in the cellular network. The Mobile Application Part (MAP) is actually an application entity, and uses the services of the TCAP and SCCP protocols to deliver control and signaling information through the network. MAP is used today in conjunction with IS-41 for seamless roaming and hand-off procedures within the cellular network.

The registers are actually databases which store information regarding cellular subscribers. The home location register (HLR) stores information regarding subscribers within the provider's calling area. The visited location register (VLR) provides information regarding those subscribers outside of their home calling area, using roaming numbers. The VLR is constantly updated by the cell sites as the cellular phone broadcasts location signals and, in turn, updates the HLR using SCCP and TCAP.

Octet Three

Point Code	**H G F E**	**D C B A**
Member Number	0 0 0 0	0 0 0 0
	to	
	1 1 1 1	1 1 1 1

Octet Four

Point Code	**H G F E**	**D C B A**
Cluster Number	0 0 0 0	0 0 0 0
	to	
	1 1 1 1	1 1 1 1

Octet Five

Point Code	**H G F E**	**D C B A**
Network Identification	0 0 0 0	0 0 0 0
	to	
	1 1 1 1	1 1 1 1

The point code is represented in the same format as the destination point code and origination point code address found in the routing label. The same rules apply here as to the routing label (regarding the ranges allowed for point codes).

Octet Six

If the global title indicator in the address indicator is equal to 0001, the following format is used for the global title parameter.

Translation Type	**H G F E**	**D C B A**
Reserved	0 0 0 0	0 0 0 0
891 Telecommunications Credit Cards	0 0 0 0	0 0 0 1
14-Digit Calling Card	0 0 0 0	0 0 1 0
Cellular Nationwide Roaming Service	0 0 0 0	0 0 1 1
Global Title = Point Code	0 0 0 0	0 1 0 0
Calling Name Delivery	0 0 0 0	0 1 0 1
Call Management Application	0 0 0 0	0 1 1 0
Message Waiting Application	0 0 0 0	0 1 1 1
Internetwork Applications	0 0 0 0	1 0 0 0
	to	
	0 0 0 1	1 1 1 1

Network Specific Applications	1	1	0	0	0	0	0	0
				to				
	1	1	1	1	1	0	0	0
Message Waiting Application	1	1	1	1	1	0	0	1
Network Specific Applications	1	1	1	1	1	0	1	0
Call Management Application	1	1	1	1	1	0	1	1
14-Digit Calling Card Application	1	1	1	1	1	1	0	1
*800 number LIDB Application	1	1	1	1	1	1	1	0

*This translation type has already been defined in many networks for 800 number translations. However, this number can be used for other network-specific applications. In the event that this translation type is being used for something other than 800 number translations, consideration towards internetworking should be taken.

Translation types help route messages between networks to the proper function within a signaling point. They are optional and network dependent. The Bellcore recommendations provide several predefined codes (shown above), but every network can assign its own translation types. The only rule here is that the translation type name and the translation type number be used consistently in any one signaling point and across the network.

Octet Seven

Encoding Scheme	H	G	F	E	D	C	B	A
Unknown					0	0	0	0
Binary Coded Decimal, odd number of digits					0	0	0	1
Binary Coded Decimal, even number of digits					0	0	1	0
Spare					0	0	1	1
						to		
					1	1	1	1

Numbering Plan	H	G	F	E	D	C	B	A
Unknown	0	0	0	0				
ISDN/Telephony Numbering Plan	0	0	0	1				
Reserved	0	0	1	0				
Data Numbering Plan	0	0	1	1				
Telex Numbering Plan	0	1	0	0				
Maritime Mobile Numbering Plan	0	1	0	1				
Land Mobile Numbering Plan	0	1	1	0				
ISDN/Mobile Numbering Plan	0	1	1	1				

The numbering plan identifies the format used for the global title. For example, in the United States, telephone numbers use the formula stipulated by the North American Numbering Plan. This is classified as the ISDN/Telephony numbering plan. Cellular networks use the land mobile numbering plan.

The encoding scheme identifies the format used for the digits. Digits are always displayed in BCD format (unless value is unknown). This parameter must always be of even length (eight-bit multiples), so an indication of whether the parameter represents an even or an odd number of digits is provided. In the event that an odd number of digits is represented, then the last four bits (bits EFGH) are padded with all zeros.

Octet Eight and beyond

The actual address (which can be dialed digits or any other number) is divided into four-bit segments using BCD numbering. If the address includes an odd number of digits, the last four bits of the octet are set to all zeros and the encoding scheme specifies an odd number of digits.

Address Signal	H/D	G/C	F/B	E/A
Digit 0	0	0	0	0
Digit 1	0	0	0	1
Digit 2	0	0	1	0
Digit 3	0	0	1	1
Digit 4	0	1	0	0
Digit 5	0	1	0	1
Digit 6	0	1	1	0
Digit 7	0	1	1	1
Digit 8	1	0	0	0
Digit 9	1	0	0	1
Spare	1	0	1	0
*Code 11	1	0	1	1
*Code 12	1	1	0	0
Spare	1	1	0	1
Spare	1	1	1	0
ST	1	1	1	1

*Use of these codes is not fully defined to date and is under further study.

If the global title indicator in the address indicator is equal to 0010, the format is the same as above, except for the absence of the numbering plan and encoding scheme parameters.

The address signal provides the first digit in bits ABCD, the second digit in bits EFGH, and the third digit in the ABCD bits of the next octet. This pattern is repeated for every digit.

The called and calling party address provide adequate information for receivers of all Signaling Connection Control Part messages. The calling party address is generated by the originator of a message, providing enough information that a response can be returned to the correct signaling point.

The calling party address should include at least the point code and subsystem number of the originator, or the global title if one exists. The objective is to provide enough information that will identify the originator even after global title translation. If an SCCP message is global title translated, then the origination point code may change to that of the signaling point providing the global title. But if the subsystem number of the originator and the global title of the originator are provided, then the responses can be returned to the true originator of the message.

Credit 0000 1001

This parameter is a one-octet field, not counting the indicator. Following the indicator is a one-octet field indicating the number of messages that may be sent without acknowledgment. This parameter is used only with connection-oriented services to allow for a "sliding window" size (flow control).

When an acknowledgment is sent, the sender of the acknowledgment will set the credit field to a particular value indicating the number of messages which the originator may send. This is based on the current status of the signaling point sending the acknowledgment.

The originator of the connection may then negotiate for a higher window size, if it deems it necessary (based on the number of messages it has to send). If no negotiation is necessary, then the credit parameter is accepted and the originator begins sending data. An acknowledgment is then expected within the given window size.

The purpose of this parameter is to allow for more flexible flow control. In most protocols, the window size is a preset configuration, which may not be changed dynamically according to the current status of the node. With this parameter, the protocol may adjust its window size using this credit parameter when traffic increases and the signaling point needs more control over its resources.

Data 0000 1111

The data that is actually being carried by the Signaling Connection Control Part follows this indicator. The data may be TCAP, or it may be some other protocol. TCAP may be carrying actual application data, as is the case with the MAP or IS-41. This means that there may be another layer of protocol before the actual data.

This should be taken into account when transmitting through the network using the services of SCCP and other transport protocols, such as TCAP, because each additional protocol will require some of the space in the data field. The maximum size for this field is 272 octets.

SCCP management also uses this field to transport management messages. SCCP management will always use the unitdata format.

Destination/Source Local Reference 0000 0001

Octets One, Two, and Three

This three-octet field (four, counting the indicator) consists of the indicator value and the three-octet number assigned by the destination. This is different from the source local reference, which is assigned by the local originator. The purpose of this parameter is to identify a connection within a signaling point.

Each entity in a connection assigns a number for reference. There should be a source (inbound) and destination (outbound) local reference number. This number is then used during connection-oriented calls for establishing, maintaining, and releasing connections.

The numbers are of local significance only; in other words, they have no meaning to other entities other than as the originator of the number. They are included in the message for return messages to reference responses to.

End of Optional Parameters **0 0 0 0 0 0 0 0**

This parameter is found at the end of every SCCP message that has optional parameters. If there are no optional parameters used in the message, then this is not used.

Error Cause	**0 0 0 0**	**1 1 0 1**
Local ref. # (LRN) mismatch—unassigned dest. LRN	0 0 0 0	0 0 0 0
Local ref. # mismatch—inconsistent source LRN	0 0 0 0	0 0 0 1
Point code mismatch	0 0 0 0	0 0 1 0
Service class mismatch	0 0 0 0	0 0 1 1
Unqualified	0 0 0 0	0 1 0 0
Spare	0 0 0 0	0 1 0 1
	to	
	1 1 1 1	1 1 1 1

The error cause parameter is used only in the error message. The error message is used with connection-oriented services. This is returned to the originator of a message that was received in error. The error listed is not caused by transmission problems, but by the originator, and represents a protocol error rather than a data error.

Protocol Class	**0 0 0 0**	**0 1 0 1**
Class Indicator		
Class 0		0 0 0 0
Class 1		0 0 0 1
Class 2		0 0 1 0
Class 3		0 0 1 1
Message Handling (Classes 0 and 1 only)		
Discard message on error	0 0 0 0	
Spare	0 0 0 1	
	to	
	0 1 1 1	
Return message on error	1 0 0 0	
Spare	1 0 0 1	
	to	
	1 1 1 1	

The class indicator identifies the type of services to be provided by the Signaling Connection Control Part. Class 0 is basic connectionless service. In basic connectionless service, messages can be delivered out of sequence. Most messages use class 0 services in today's networks.

Class 1 provides sequenced connectionless services. The sequence numbering is provided through the sequencing/segmenting parameter. The Message Transfer Part has the ultimate responsibility of guaranteeing in-sequence delivery. This is accomplished by using the same route for all sequenced messages, and not using the "bit rotation" scheme for link load sharing. This ensures that all SCCP messages travel the same path, thus guaranteeing in-sequence delivery.

Messages that must be broken down into smaller messages are segmented and sent in multiple SCCP messages. These are then sent using class 1 SCCP. The messages are divided into equal lengths so that all SCCP segments are equal in size, or as close as possible to equal in size.

Class 2 services provide basic connection-oriented delivery of messages. Basic connection-oriented services do not guarantee any specific level of service, other than establishing a connection and delivering data through the established connection. Once the data transmission is complete, the connection is released.

Class 3 services add flow control to the connection-oriented function. Flow control allows the starting and stopping of data flow according to resources and their congestion status (level-four resources, not level-three).

Class 4 adds error recovery as well as flow control. Error recovery consists of retransmission of SCCP messages which are received in error. This is also a connection-oriented service.

United States networks do not use any of the connection-oriented services of the SCCP. In fact, this author does not know of any networks currently using connection-oriented services of the Signaling Connection Control Part.

Receive Sequence Number	**0 0 0 0 0 1 1 1**
Spare	0
Sequence Number	0 0 0 0 0 0 0
	to
	1 1 1 1 1 1 1

This parameter indicates the next expected sequence to be received. It is sent in the backwards direction to indicate an acknowledgment of received messages. Unlike the Message Transfer Part, which uses the backward sequence number as an acknowledgment, this indicates the next sequence number that the SCCP *expects* to receive. MTP indicates the last *received* sequence.

Refusal Cause	**0 0 0 0 1 1 1 0**
End user originated	0 0 0 0 0 0 0 0
End user congestion	0 0 0 0 0 0 0 1
End user failure	0 0 0 0 0 0 1 0
SCCP user originated	0 0 0 0 0 0 1 1
Destination address unknown	0 0 0 0 0 1 0 0
Destination inaccessible	0 0 0 0 0 1 0 1
Network resource—QoS not available/permanent	0 0 0 0 0 1 1 0
Network resource—QoS not available/transient	0 0 0 0 0 1 1 1
Access failure	0 0 0 0 1 0 0 0
Access congestion	0 0 0 0 1 0 0 1
Subsystem failure	0 0 0 0 1 0 1 0
Subsystem congestion	0 0 0 0 1 0 1 1
Expiration of the connection establishment timer	0 0 0 0 1 1 0 0
Inconsistent user data	0 0 0 0 1 1 0 1
Not obtainable	0 0 0 0 1 1 1 0
Unqualified	0 0 0 0 1 1 1 1
Spare	0 0 0 1 0 0 0 0 to 1 1 1 1 1 1 1 1

The refusal cause parameter is found in the connection refused message. It indicates the reason that a connection request has been denied. This is used only with the connection-oriented class of messages.

The value "network resource—Quality of Service (QoS) not available" indicates that the requested quality of service could not be provided either on a permanent or a temporary condition.

Release Cause	**0 0 0 0 1 0 1 0**
End user originated	0 0 0 0 0 0 0 0
End user busy	0 0 0 0 0 0 0 1
End user failure	0 0 0 0 0 0 1 0
SCCP user originated	0 0 0 0 0 0 1 1
Remote procedure error	0 0 0 0 0 1 0 0
Inconsistent connection data	0 0 0 0 0 1 0 1
Access failure	0 0 0 0 0 1 1 0
Access congestion	0 0 0 0 0 1 1 1
Subsystem failure	0 0 0 0 1 0 0 0
Subsystem congestion	0 0 0 0 1 0 0 1
Network failure	0 0 0 0 1 0 1 0
Network congestion	0 0 0 0 1 0 1 1
Expiration of reset timer	0 0 0 0 1 1 0 0
Expiration of receive inactivity timer	0 0 0 0 1 1 0 1
Not obtainable	0 0 0 0 1 1 1 0
Unqualified	0 0 0 0 1 1 1 1
Spare	0 0 0 1 0 0 0 0 to 1 1 1 1 1 1 1 1

The release cause parameter is found in the released message and is used to indicate the reason for releasing a specific connection. Keep in mind that these are logical connections and have nothing to do with any voice circuits or any other physical connections within the Public Switched Telephone Network.

The release cause is found only when using connection-oriented services and is only used in the released message. All other release or return messages use other specific parameters. Although there may be a lot of similarities between these various parameters, they serve a significantly different purpose.

The user indicated in the first few "cause codes" refers to the application entity, which may be the Transaction Capabilities Application Part, or a higher level user, such as the Mobile Application Part.

Reset Cause	**0 0 0 0**	**1 1 0 0**
End user originated	0 0 0 0	0 0 0 0
SCCP user originated	0 0 0 0	0 0 0 1
Message out of order—incorrect P(s)	0 0 0 0	0 0 1 0
Message out of order—incorrect P(r)	0 0 0 0	0 0 1 1
Remote procedure error—message out of window	0 0 0 0	0 1 0 0
Remote procedure error—incorrect P(s) after reinit.	0 0 0 0	0 1 0 1
Remote procedure error—general	0 0 0 0	0 1 1 0
Remote end user operational	0 0 0 0	0 1 1 1
Network operational	0 0 0 0	1 0 0 0
Access operational	0 0 0 0	1 0 0 1
Network congestion	0 0 0 0	1 0 1 0
Not obtainable	0 0 0 0	1 0 1 1
Unqualified	0 0 0 0	1 1 0 0
Spare	0 0 0 0	1 1 0 1
	to	
	1 1 1 1	1 1 1 1

This parameter is found only in the reset request message, which is used for connection-oriented services. The parameter provides the reason for requesting the reset of a virtual connection. The end user is the application entity, such as the Mobile Application Part which is using the services of the Signaling Connection Control Part. If the request is successful, the connection is reset and communications are reestablished.

Return Cause	**0 0 0 0**	**1 0 1 1**
No translation for an address of such nature	0 0 0 0	0 0 0 0
No translation for this specific address	0 0 0 0	0 0 0 1
Subsystem congestion	0 0 0 0	0 0 1 0
Subsystem failure	0 0 0 0	0 0 1 1
Unequipped user	0 0 0 0	0 1 0 0
Network failure	0 0 0 0	0 1 0 1
Network congestion	0 0 0 0	0 1 1 0
Unqualified	0 0 0 0	0 1 1 1
SCCP hop counter violation	0 0 0 0	1 0 0 0
*Error in message transport	0 0 0 0	1 0 0 1

*Error in local processing	0 0 0 0	1 0 1 0
*Destination cannot perform reassembly	0 0 0 0	1 0 1 1
Spare	0 0 0 0	1 1 0 0
	to	
	1 1 1 1	1 0 0 0
*Invalid ISNI routing request	1 1 1 1	1 0 0 1
Unauthorized message	1 1 1 1	1 0 1 0
Message incompatibility	1 1 1 1	1 0 1 1
*Cannot perform ISNI constrained routing	1 1 1 1	1 1 0 0
*Redundant ISNI constrained routing information	1 1 1 1	1 1 0 1
*Unable to perform ISNI identification	1 1 1 1	1 1 1 0
Reserved for extension	1 1 1 1	1 1 1 1

*Applies only to XUDTS message.

This parameter is found only in class 0 and class 1 messages, where connectionless services are provided. The purpose is to identify the reason a particular SCCP message was returned to the originator.

When an SCCP message is received in error, or any one of the above events occurs, the original SCCP message is returned with its data to the originator. The Unitdata Service or Extended Unitdata Service message structure is used to return the data and the return cause to the message originator. It is then up to the originator to determine a plan of action, either retransmit or abandon the transaction.

Segmentation	**0 0 0 1**	**0 0 0 0**
Octet One		
Remaining Segments	**H G F E**	**D C B A**
Last segment		0 0 0 0
One segment left		0 0 0 1
Two segments left		0 0 1 0
Three segments left		0 0 1 1
Four segments left		0 1 0 0
Five segments left		0 1 0 1
Six segments left		0 1 1 0
Seven segments left		0 1 1 1
Eight segments left		1 0 0 0
Nine segments left		1 0 0 1
Ten segments left		1 0 1 0
Eleven segments left		1 0 1 1
Twelve segments left		1 1 0 0
Thirteen segments left		1 1 0 1
Fourteen segments left		1 1 1 0
Fifteen segments left		1 1 1 1
Spare	0 0	
In-Sequence Delivery Option (ISDO)	**H G F E**	**D C B A**
In-sequence delivery	1	
Not in-sequence delivery	0	
"First" Bit		
First segment	1	
All other segments	0	

The remaining segments parameter identifies how many segments are to follow that are associated with this message. This is used whenever the data to be sent by SCCP is larger than the 255 octets, and the data is broken into smaller segments for transmission. The segments are as close to equal length as possible.

In addition to the remaining segments parameter, an indicator is provided that instructs the originator and any intermediate signaling points whether or not in-sequence delivery is to be used. The first segment is always sent with the "first bit" parameter set to a 1.

In-sequence delivery is used with both connection-oriented and connectionless services. This parameter is used only with class 1 messages, and is part of the extended unitdata and the unitdata SCCP messages.

Segmenting/Reassembling	**0 0 0 0 0 1 1 0**
No more data	0
More data	1
Spare	0 0 0 0 0 0 0

This parameter only shows if additional data, or segments, are to follow. No other information is provided. It is found in the data form 1 (DT1) message only, which is used in connection-oriented data transfer. The remaining seven bits in this parameter are currently reserved for future implementation.

Sequencing/Segmenting	**0 0 0 0 1 0 0 0**
Octet One	
Spare	0
Sending Sequence Number—P(s)	0 0 0 0 0 0 0
	to
	1 1 1 1 1 1 1
Octet Two	
More Data Indicator	H G F E D C B A
No more data	0
More data	1
Received Sequence Number—P(r)	0 0 0 0 0 0 0
	to
	1 1 1 1 1 1 1

The first octet of this parameter is used to indicate the sending sequence number. This is not the same as what we discussed with the sequence number parameter earlier. This parameter is not used with unitdata messages (connectionless). This parameter is used with connection-oriented classes, data form 1 (DT1), and data form 2 (DT2) messages.

The second octet provides the received sequence number, which is the same as an acknowledgment. If there is to be another SCCP message carrying additional data associated with this particular segment, then the *more data* indicator is used. This is only needed when the data has been divided among several SCCP segments.

Chapter

8

Overview of TCAP

The Transaction Capabilities Application Part (TCAP) is designed for non-circuit-related messages. These messages are destined for database entities as well as actual end office switches. The TCAP protocol provides a means for reliable transfer of information from one application at a switch location to another application within another network entity. To understand this, it is probably best to look at some of the actual applications and problems that we face in the telephone network today.

The first usage of the TCAP protocol was 800 number translation. An 800 number cannot be routed through the telephone network, because the area code "800" does not specify any particular exchange. To overcome this problem, the number must be converted into a routable number. This requires a database.

The database for 800 numbers provides a routing number which the local office can then use to route the call through the Public Switched Telephone Network (PSTN). This database is usually centrally located within the service provider's network. It does not make good sense to place this database in too many multiple locations, since that would make maintenance of the database more difficult.

The problem with centralized databases is providing access to them. All switches in the network must be able to access the database and retrieve the routing number for the 800 numbers used in their network. To compound the problem, in today's network, all 800 numbers must be routable by all carriers. This means that no matter which telephone company "owns" the 800 number, all other telephone companies must be able to access the proper database and retrieve the routing number for that 800 number. This is known as transportable 800 numbers.

Making 800 numbers transportable also allows subscribers to keep the 800 numbers they have had provided by another carrier, even when they change carriers. Previous to the transportability ruling, if subscribers changed carriers, they had to surrender their 800 number and obtain a new number from the new carrier. This is no longer the case.

The TCAP protocol provides the parameters and the services to maintain a dialog with a database. The protocol contains message types which are used by the Service Control Point (SCP) to query a database for specific information. This information is then carried back to the requesting central office switch using the same TCAP protocol. In short, TCAP messages are designed for accessing either a database or another switch and either retrieving information or invoking features using its parameters and message types.

Cellular networks have the same types of needs, since they are very dependent on databases and remote control of switch features. The cellular networks have previously used proprietary networks, prohibiting the ability to access remote databases in other networks. It is for this reason that the cellular industry has begun deploying SS7.

TCAP provides the mechanism for transferring information from one switch to another switch, even if they are a substantial distance apart. The information is not related to any one circuit (such as in an ISDN User Part message), and the information must be transferred through the network using end-to-end signaling.

ISUP messages do not use end-to-end signaling, and must follow the same path used to establish the circuit connection. This means the message must be passed along from one exchange to another, with intermediate Signaling Transfer Points through-switching the ISUP messages to the next exchange. This is one of the fundamental differences between TCAP and ISUP.

Another difference between the two protocols is the transport used. ISUP uses the Message Transfer Part (MTP) for routing the message from one exchange to another. The MTP protocol does not support end-to-end signaling. So TCAP must use an additional protocol as a transport. The Signaling Connection Control Part (SCCP) protocol is used with MTP to route messages end to end.

The SCCP protocol provides the additional controls needed when passing messages from end to end. However, MTP must also be used to provide the routing functions from one node to the next node. The MTP also provides the basic error detection and correction needed for reliable message transfer.

Now that we have identified the mechanism used for through-switching informational messages from one exchange to another, we can begin looking at specific applications. There are many applications within the telephone company network. Many of these applications have not even been developed yet. We will talk about both present and future applications.

We have already discussed the use of TCAP for accessing a database (using the 800 number scenario). There are many other databases used in the telephone network besides the 800 database. Every telephone number has records associated with it. These records identify who the subscriber is and what types of services they have subscribed to.

These Line Information Databases (LIDBs) belong to the individual telephone companies that provide services to their subscribers. For this discussion, we will use the feature *call forwarding* as an example. Call forwarding allows subscribers to forward their phone calls to other subscriber telephone numbers, until the call forwarding is canceled. All calls to the subscriber telephone number are then redirected to the forwarded number.

This feature requires the use of the Line Information Databases to verify that the subscriber is allowed to use this feature. When a subscriber invokes the forwarding feature, the information about where the call is forwarded to is stored in the end switch. The database is used when the feature is accessed (by dialing some sort of feature code). The database verifies that the subscriber is allowed to use this feature.

This is probably the simplest of applications. Let's look at a more sophisticated function of TCAP. The TCAP protocol also provides the mechanism to access other remote switches and activate features within that switch. The switch must have the feature capability already; TCAP only invokes the feature remotely.

Later, we talk about a scenario using another feature called *automatic callback.* In this feature, when a subscriber dials a number that is busy, the subscriber can enter in a feature code and hang up. When the dialed number becomes available, the local exchange notifies the caller's local switch by sending a TCAP message. This TCAP message allows the local switch to ring the phone of the caller. The distant switch has reserved the called party's line so no other phone calls can be routed to it.

When the calling party answers the phone, normal call setup procedures are used to establish the connection between the two exchanges. TCAP, in this case, serves as an alerting mechanism, sending an informational message (not circuit related) to another entity within the network. This can be extended to many other types of applications where remote invocation is an option.

In the cellular network, TCAP has become the solution to roaming. Prior to the deployment of SS7 in the cellular network, when cellular subscribers took their phones to other areas serviced by other cellular providers, they would have to call ahead and obtain roaming numbers. The roaming number was only good for a specific geographic area (regional service area, or RSA).

When the roaming number was dialed, the cellular network immediately knew how to route the call, because the roaming number could be routed like any other Plain Old Telephone Service (POTS) number. The problem was that roaming was not seamless and required user intervention.

Some cellular subscribers found themselves having to have two and three numbers for their cellular phone, depending on where they were. This defeats the purpose of having a mobile telephone. This was the reason for the seamless roaming.

The missing element was the ability to update the network databases with the current location of the cellular telephone subscriber. This information is updated every few minutes by the cell site sending a message to the mobile switching center identifying the mobile subscriber and the cell site reporting the subscriber's presence.

An application entity called IS-41 provides procedures for updating databases on the status of cellular subscribers. Every cellular subscriber has a home database, called the Home Location Register (HLR), which keeps a record on where the cellular subscriber is located. This record is what gets updated every few minutes.

TCAP is used to carry these update messages from one database (the Visitor Location Register, or VLR) to the subscriber's Home Location Register. When a call comes in for the subscriber, the call is routed to the home regional service area, which must then look at the HLR to determine how to connect the call to the subscriber. The HLR then provides the information (location and status) so the Public Switched Telephone Network knows how to connect the call.

When the subscriber moves to another area, the TCAP protocol is used once again to update the Home Location Register on the new location and cancel the subscriber's registration in the previous regional service area. This is completely transparent to the subscriber and allows cellular subscribers the freedom to move around the network without ever having to register with other service providers. The network keeps track of their locations the entire time the cellular phone is activated.

In the Intelligent Network, TCAP is the protocol that will be used to invoke features in remote switches. As we discussed earlier with the automatic callback feature, TCAP allows features to be activated and deactivated remotely. In the Intelligent Network, services will be activated and deactivated the same as features.

Services include high-speed data transmission or video circuit connections. TCAP will provide access to the database, which will activate these services for the subscriber. As we discussed in the first chapter, subscribers will activate and deactivate these services through a terminal located on their premises and connected to the SS7 network via a data link.

TCAP Functionality

The Transaction Capabilities Application Part provides a way for end users in the SS7 network to access other end users on a peer-to-peer level. In the SS7 network, end users are seen as applications within the network entities.

The SS7 protocol provides a means for database access between signaling points as well as access to remote operations. The latter function is somewhat newer to the SS7 protocol stack, and is slowly finding applications in the network. The Advanced Intelligent Network (AIN) will depend heavily on the ability for a signaling point to access another signaling point remotely and to invoke an operation or feature within the remote signaling point.

The cellular network also depends today on remote access to other signaling nodes. For example, when a cellular switching center needs to hand over control of a call to another switching center, signaling information regarding that call must be transferred to the remote switching center. SS7 provides the means to accomplish this task.

In the Intelligent Network, features such as automatic callback require the ability for an end office switch to send a message to another end office switch concerning the status of a previously called number. SS7 provides this ability.

The protocol used for these types of transactions is the Transaction Capabilities Application Part. The TCAP protocol originally provided database access, yet the intention has always been there to provide the facilities within the protocol to invoke remote features and access remote applications within the signaling network.

Description of TCAP

An application uses an application process to communicate with the other entities in the network. The application process then communicates with the TCAP and other protocols to transfer the information across the network.

The function that allows applications to communicate with one another is a communications function called the *application service element* (ASE). The Transaction Capabilities Application Part and the Mobile Application Part are both examples of ASEs (Fig. 8.1).

An application service element provides the communications services for applications within any signaling point. These communications are peer-level communications. An application process acts as the coordinator of network services such as ISDN call setup and mobile services. The application process can be found below the function of applications, but above the ASE.

An application can use more than one application service element for any one transaction. This is necessary to provide flexibility in the communications function, allowing the addressing of multiple entities within one message.

Another method of reaching several entities within one transaction is the use of subsystem numbers for addressing. The subsystem numbers found in the called and calling party addresses in SCCP are the addresses of applications. By using a subsystem number rather than a point code,

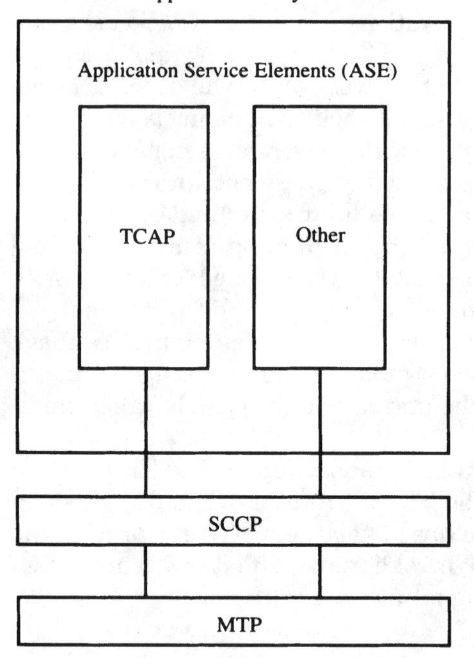

Figure 8.1 The application service element (ASE) provides a communication interface to application entities (AEs) such the Mobile Application Part (MAP) and the Operations, Maintenance, and Administration Part (OMAP).

the transaction message can be sent to multiple signaling points or can even be rerouted by network management in the event of a failure.

One of the newer protocols to join TCAP is the Mobile Application Part. This protocol is found in IS-41 cellular networks, and is used for transporting roaming information and other signaling from one cellular network to another.

The TCAP and the MAP protocols are capable of transferring information and invoking operations within other entities. This is probably the most important aspect to understanding that these protocols not only transfer information from one to another, but also invoke operations (or tasks) within a remote entity. An operation may be a particular function within the switching equipment or a database located at an end node.

The TCAP protocol is capable of both invoking an operation and returning the results of that operation, using the services of the Signaling Connection Control Part protocol. SCCP provides the network services, while the Message Transfer Part provides the physical layer and data link layer functionality.

An application process is considered a function above layer seven of the OSI model. An example of an application may be the Line Information Database (LIDB) or Operations, Maintenance, and Administration Part (OMAP). These applications are found in various signaling points, depending on the location of the signaling point in the network and its function within the network (i.e., Signal Transfer Point, gateway STP, or translation STP).

In order for an application to send information and communicate to another application at a remote signaling point, a communications protocol must be used. This protocol should be able to address the application and invoke an operation or task within the application. For example, if a node needs to invoke a CLASS feature at a distant node for a call in progress, a message would need to be sent to the distant signaling point informing it of the operation needed. The receiving signaling point should be given enough information about the call to be able to invoke the specified CLASS feature.

The address of the message sent would be of the application providing the operation. This address is not an address used within the rest of the network, but references only the various applications supported within a network. The addressing scheme used for these applications is the subsystem number. The subsystem number does not have to be known at every node within the network. Global title can be used to route a message to the adjacent node of the application, at which point global title translation must provide the subsystem number and the point code of the application.

Today, only connectionless services (datagram) are supported. In some cases, a connection with a peer application may be necessary, in which case connection-oriented services must be employed to establish a connection with the application and maintain the connection until the originating signaling point has completed its transactions.

In the United States, support of connection-oriented services is not yet endorsed. Both ANSI and Bellcore standards mention connection-oriented services and even define their functions. Yet, no application requiring connection-oriented services has emerged. Connectionless services are supported and are what U.S. networks use for communicating between application processes.

Because datagrams are used for information transfer between applications, some sort of reference must be used to associate multiple messages to a single transaction and multiple components to a single operation. Two forms of referencing are used: the transaction ID and the invoke ID.

A transaction ID is used as a reference within a dialog so the receiver can associate the received message with a transaction in progress. This transaction ID is significant only to the local receiver, but is sent to the remote as a reference to be used in any responses.

An invoke ID is used to identify invoke components which are used to invoke an operation within the entity. One transaction may consist of several invokes. The receiver of an invoke is typically expected to respond with a return result, in which case a correlation ID is used to identify which invoke component the result is referencing. The correlation ID is a mirrored image of the invoke ID.

ASP Services

The Application Service Part (ASP) consists of the layers above the Signaling Connection Control Part and below the Transaction Capabilities Application Part. It provides the functions of layers four through six of the OSI model. These functions are not presently required in the SS7 network, and are under further study; however, the ITU-TS and ANSI standards do reference these as viable functions.

The lack of connection-oriented services in today's network is why the Application Services Part is not currently needed. However, as the network matures and new technologies emerge, connection-oriented services will become a necessity for certain applications. This will force the need for the functions of these middle layers.

TCAP Message Structure

The message units within TCAP are partitioned into three portions: the *transaction portion*, the *component portio,* and the *dialog portion*. The transaction portion provides the information necessary for the signaling point to route the component information to its destination. Included in the transaction portion is the *transaction ID,* which is used as a reference for tracking all TCAP messages.

The component portion gets its information from the *operation protocol data unit* (OPDU), received from the application. The OPDU contains the primitives and parameters necessary to invoke an operation or request services from another entity (such as database query).

The dialog portion is used to identify the version of transaction being used. It also provides security information if encryption is used on the transaction. The dialog portion is an optional field within TCAP.

To fully understand the structure of these primitives, review the section in the chapter for SCCP, titled "Primitives." This section discusses the use of primitives and interfaces between SCCP and MTP. The structure of all primitives is the same throughout the layers of SS7. The interface between TCAP and the lower layers of the stack looks like:

P-UNITDATA Generic Name Specific Name Parameter

The **P-** indicates the location of the interface. These primitives consist of the same generic names described in the SCCP chapter (Chap. 7). The specific names used are described in the following paragraphs.

The operation protocol data unit consists of several types of data units. Each type indicates the type of service or operation to be invoked by the receiver. These OPDUs are then carried through the SS7 network in the form of a TCAP message. The component portion of the TCAP message is used to transport these OPDUs to their destinations. The following OPDUs are found in TCAP:

Invoke

Return result

Return error

Reject

The *Invoke* OPDU is used to request an operation from another application. For example, for the Mobile Application Part to be able to notify a remote Mobile Switching Center (MSC) to take control of a cellular call in progress, an *Invoke* OPDU must be created from the originating MSC. The invoke OPDU is then used to form a TCAP message which, in turn, is transferred to the destination MSC through the SS7 network.

Upon receipt of the *Invoke* message, the remote Mobile Switching Center then initiates the operations required using the parameters provided in the component portion of the TCAP message. The *Invoke* may require a return message to indicate successful completion of an operation. The *Return Result* OPDU is used by the destination application to indicate successful completion of an operation. This, in turn, creates the appropriate TCAP message, with the component portion carrying the *Return Result* and its parameters.

If the operation was unsuccessful, a *Return Error* or *Reject* OPDU is created at the destination Mobile Switching Center and used to create the TCAP message to be returned to the originator of the invoke message.

TCAP messages contain information elements that are used to convey the information being transported to the remote application. These information elements are always structured the same, regardless of their contents. The first field of every information element is the *tag* (referred to as an *identifier* in ANSI).

The tag (or identifier) indicates the type of information element being sent and, in doing so, allows the receiver to determine how the contents will be interpreted. The tag is one octet and is coded to describe the handling of the contents (see Fig. 8.2).

Following the tag is the length field. The length field indicates the number of octets to be found in the contents field. The contents field is the location of the information being conveyed. This is not a fixed field, but a variable, depending on the type of components used.

The contents field consists of one or more components. In the case where more than one information element may exist in the contents

Tag Class	F	Tag Code
2 bits	1 bit	5 bits

Figure 8.2 Tag class from ITU Q.733.

field, the same structure as previously described is used (that is, tag, length, and contents) for each information element. However, the length field indicates the length of the individual information element rather than the whole information element and all its components.

The tag is coded with a class, form, and tag code. The tag code may exceed the one-octet length in some cases, but is usually maintained within the first octet. The structure of a tag is shown in Fig. 8.2.

Tag class

The tag class is used to indicate whether this particular information element uses a common structure or if the contents are proprietary. This information is used by the receiver to determine how the message is to be interpreted and handled. As seen in Fig. 8.2, the tag class uses bits HG. There are four values in this field:

Universal	0 0
Application-wide	0 1
Context-specific	1 0
Private use	1 1

The universal tag is one that is compliant with ITU-TS Recommendation X.208 and is used for all types of application entities. Universal tags are compatible with other recommendations and can be used with X.400 MHS, as well as with other ITU-TS standards.

Application-wide indicates that the tag can be used with all applications within the SS7 standard. These tags do not conform to any other recommendations. Application-wide tags in the United States refer to the international standardized TCAP.

Context-specific indicates that the context of the received information is determined by the preceding component (information element). When more than one component is necessary, the first component received indicates how subsequent components are to be interpreted (if they carry a tag class of context-specific). The same component in a different message may be interpreted differently, depending on its usage within that message. Context-specific is used with components that are part of a series of components. These are referred to as *constructors,* meaning that each component builds off another.

The private use class indicates that the component is specific to a national standard (such as ANSI) or a private proprietary standard.

ITU-TS recommendations indicate that this field is used for national or private use and does not define any of the components that use this tag class. The ANSI standards do define specific national TCAP components not found in ITU-TS that are coded as private class.

Form

An information element may require subsequent information elements or may be a single value. The F bit is used to code each information element as either a primitive (single value) or constructor (multiple information elements). The values are shown as follows:

Primitive	0
Constructor	1

Tag code

The code indicates the type of information element being sent. This field is expandable beyond the five-bit structure by using extension bits. In the event that an extension is needed, the first five bits (EDCBA) of the first octet in the identifier are set to 11111. This indicates that the tag code is found in the second and subsequent octets. If the second octet is extended, bit H is set to 1, indicating an extension. In each subsequent octet, bit H indicates whether there is an extension (1) or no extension (0).

In the following discussions, each information element and its contents will be described. The coding (national, private, primitive, etc.) will be provided within the description.

The TCAP message (Fig. 8.3) consists of three parts, as mentioned previously. The first part is labeled the transaction portion, and provides the information necessary to identify the nature of the transaction. The transaction portion is a mandatory field for all TCAP messages.

The second part is labeled the dialog portion, and is used to identify the version of the transaction as well as security information (in the event the transaction is encrypted). This part was recently added to TCAP.

The third part is labeled the component portion and is the part that contains the contents of the primitives sent down from the various applications. The component portion also contains the parameters that identify the specific details of the TCAP message. A dialog is maintained by sending a series of components in one or more TCAP messages and correlating those which are associated with a specific transaction and operation.

The protocol provides various parameters within the component portion that allows it to emulate a logical connection (hence, fulfilling the need for connection-oriented services without the need for connection-oriented SCCP).

Following is a description of all the fields within the transaction portion.

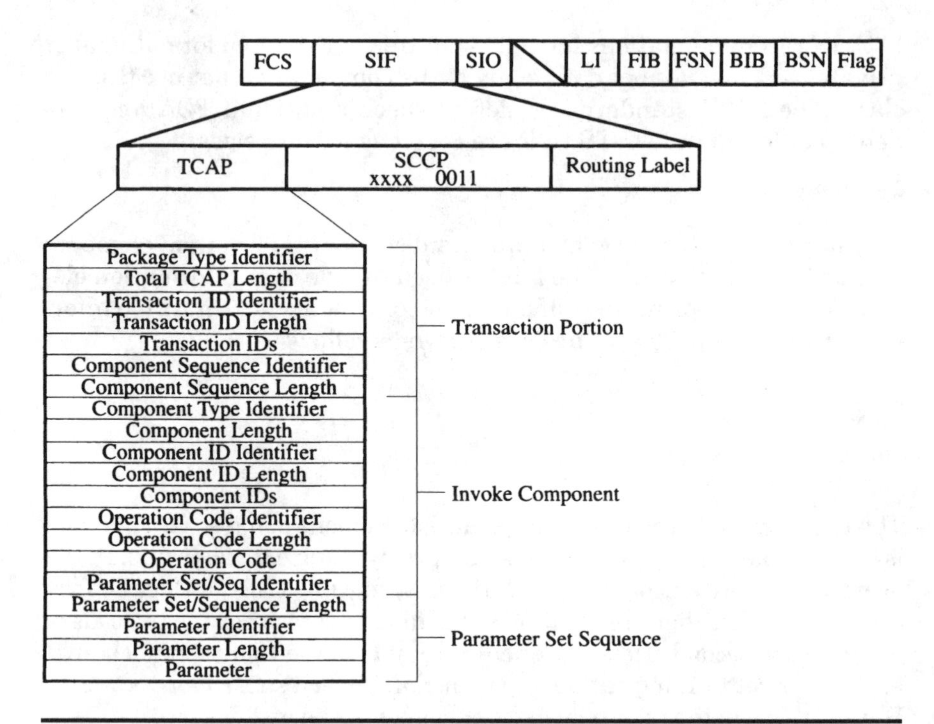

Figure 8.3 This figure shows the components of a TCAP message.

Package type identifiers

The package type identifier describes the type of transaction being sent (Fig. 8.4). A transaction may be a one-way transmittal, or it may require a two-way dialog. These requirements are determined by the package type. The descriptions provided below describe the function of each of these package types in a connectionless environment. Connection-oriented services are discussed separately, since these services are not yet implemented in U.S. networks.

Following is a description of each package type identifier and its mandatory values. An asterisk in front of a parameter indicates that the field is mandatory but may be empty (set to all zeros).

Unidirectional

This type of TCAP is sent in one direction only and does not require a return message (or response). No transaction identifier is required for *unidirectional* messages. Unidirectional messages are used for sending information to an application when a transaction does not need to be established. No correlation between multiple components is necessary.

The *unidirectional* TCAP consists of the following fields (an M indicates a mandatory field, while O indicates an optional field):

Package Type	HGFE	DCBA
Unidirectional	1 1 1 0	0 0 0 1
Query w/Permission	1 1 1 0	0 0 1 0
Query w/out Permission	1 1 1 0	0 0 1 1
Response	1 1 1 0	0 1 0 0
Conversation w/Permission	1 1 1 0	0 1 0 1
Conversation w/out Permission	1 1 1 0	0 1 1 0
Abort	1 1 0 1	0 1 1 0

Figure 8.4 These are the package types used in TCAP.

Package type identifier (M)

Total TCAP message length (M)

*Transaction identifier (null) (M)

*Transaction ID length (null) (M)

Dialog portion (O)

Component sequence identifier (M)

Component sequence length (M)

Query With Permission

A *Query* is used to access information stored within a database. It is also used to initiate the transaction, and triggers the assignment of a transaction identifier. If a dialog is to take place, a *Query* will be followed by a *Conversation.* The dialog allows for information to be exchanged between the two applications without impediment.

The receiving signaling point of a *Query With Permission* is granted permission to end a transaction if it deems it necessary. Upon receipt of a *Query With Permission,* the receiving application must decide whether it wishes to establish a transaction and maintain a dialog (using *Conversation* components). In the event that it does not wish to establish a transaction, the *Response* component is sent. This releases the application from any transaction. No transaction ID is established.

If the receiving application does wish to establish a transaction, the *Conversation* component is sent (either with or without permission), and a dialog can be maintained between the two applications.

When an application needs to enter into a transaction with a remote application, and it does not anticipate sending additional components for the transaction, a *Query With Permission* is generated.

The *Query With Permission* consists of the following fields (an M indicates a mandatory field, while O indicates an optional field):

Package type identifier (M)

Total TCAP message length (M)

Transaction ID identifier (M)

Transaction ID length (M)

Originating transaction ID (M)

Dialog portion (O)

Component sequence identifier (M)

Comment sequence length (M)

Component sequence (M)

Query Without Permission

The *Query Without Permission* is identical to the *Query With Permission* for the exception of the permission granted. The receiver of this message is not granted permission to end the transaction. This, of course, does not include termination for network management reasons. As mentioned in the preceding paragraph, ending a transaction is accomplished by releasing the transaction ID.

When an application needs to enter into a transaction with a remote application, and the originator of the transaction anticipates additional components related to the transaction to be sent, a *Query With Permission* is generated. This prevents the remote application from ending the transaction before all the components have been received.

The *Query Without Permission* consists of the following mandatory parameters (an M indicates a mandatory field, while O indicates an optional field):

Package type identifier (M)

Total TCAP message length (M)

Transaction ID identifier (M)

Transaction ID length (M)

Originating transaction ID (M)

Dialog portion (O)

Component sequence length (M)

Component sequence identifier (M)

Component sequence (M)

Response

The *Response* is used to end a TCAP transaction. In the case of a *Query*, the *Response* is used to return the requested data. In the case of a dialog between two applications, the *Response* is the last transmission sent.

The *Response* consists of the following parameters (an M indicates a mandatory field, while O indicates an optional field):

Package type identifier (M)
Total TCAP message length (M)
Transaction ID identifier (M)
Transaction ID length (M)
Responding transaction ID (M)
Dialog portion (O)
Component sequence identifier (M)
Component sequence length (M)
Component sequence (M)

Conversation With Permission

This TCAP message is sent after a *Query,* and is used to carry out a dialog between two applications. The package type *Conversation* continues between the two entities until the transaction is complete, at which point a *Response* is sent by either party.

When the *Conversation* component is sent, the transaction ID received from the originating node is duplicated and placed in the originating transaction ID field. Besides the originating transaction ID, the receiving node also creates a transaction ID corresponding to the transaction established locally. This transaction ID is significant only to the remote application and is placed in the responding transaction ID field.

Once the *Conversation* component is sent, all subsequent messages must contain one of the *Conversation* components (with or without permission). This component is then used to maintain a dialog between the two applications until the transaction is finished and one of the applications sends a *Response* component.

The *Permission* indicator grants the receiver of this message the ability to end the transaction (by releasing the transaction ID). Either party can end the transaction.

Permission is granted when an application has responded to all received components and does not anticipate the need to send further components related to a transaction. In essence, the application uses the permission indicator to ensure that the transaction is maintained until all the components related to a specific transaction can be sent. This prevents the remote application from ending the transaction prematurely.

If an application during a dialog determines the need to gain control over the release of a transaction (even if it previously relinquished control), then it may send a *Conversation Without Permission* even when it had previously relinquished control.

A transaction is ended by an application sending a *Response* component. The transaction can be ended even when permission is not granted in special circumstances (resource management intervention, for example).

The *Conversation With Permission* consists of the following parameters (an M indicates a mandatory field, while O indicates an optional field):

Package type identifier (M)
Total TCAP message length (M)
Transaction ID identifier (M)
Transaction ID length (M)
Originating transaction ID (M)
Responding transaction ID (M)
Dialog portion (O)
Component sequence identifier (M)
Component sequence length (M)
Component sequence (M)

Conversation Without Permission

This message is the same as the *Conversation With Permission* for the exception of being able to end a transaction. As previously described, this component is sent to prevent the remote application from ending a transaction before the originating application has completed transmission of all its components. It consists of the following parameters (an M indicates a mandatory field, while O indicates an optional field):

Package type identifier (M)
Total TCAP message length (M)
Transaction ID identifier (M)
Transaction ID length (M)
Originating transaction ID (M)
Responding transaction ID (M)
Dialog portion (O)
Component sequence identifier (M)
Component sequence length (M)
Component sequence (M)

P-Abort

A *P-Abort* is used when the originating entity must end a transaction. It is important to remember that when a transaction is aborted or ended under normal conditions, the application is responsible for the action. A transaction can be ended by the protocol or lower layers of the stack in a number of ways. At this level, the TCAP identifies why the abort was necessary (as determined by the application) and forwards the reason or causes to the remote entity.

The *P-Abort* consists of the following parameters (an M indicates a mandatory field, while O indicates an optional field):

Package type identifier (M)

Total TCAP message length (M)

Transaction ID identifier (M)

Transaction ID length (M)

Responding transaction ID (M)

Dialog portion (O)

P-Abort cause identifier (M)

P-Abort cause length (M)

P-Abort cause (M)

U-Abort

The User abort is used when the User of TCAP aborts a transaction. It is used in the same way as the P-Abort, but contains slightly different information. Tbe *U-Abort* consists of the following parameters (an M indicates a mandatory field, while O indicates an optional field);

Package type identifier (M)

Total TCAP message length (M)

Transaction ID identifier (M)

Transaction ID length (M)

Responding transaction ID (M)

Dialog portion (O)

U-Abort information identifier (M)

U-Abort information length (M)

U-Abort information (M)

Each of the package types identified here contains a number of additional parameters (as noted in the preceding descriptions). These parameters contain the specific information regarding the transaction taking place.

These package types are identified in the transaction portion. Following the transaction portion is the dialog portion. The dialog portion contains information regarding the version of the transaction and security information if any of the transaction is encrypted. The dialog portion is an optional part of TCAP. Following is a listing of each of the dialog portion fields:

Protocol version identifier

Protocol version length (O)

Application context identifier (O)

Application context length (O)

Application context name (O)

User information identifier (O)

User information length (O)

User information (O)

Security context identifier (O)

Security context length (O)

Security context (O)

Confidentiality identifier (O)

Confidentiality length (O)

Confidentiality information (O)

The application context fields are only used in unidirectional, query, user abort, and the first backward conversation or response TCAP messages.

Following the dialog portion is the component portion, which consists of one or more components. The component structures were described earlier as "information elements." Each information element may be a *primitive* (single value) or a *constructor* (multiple components).

The component types listed consist of the information elements described earlier. An identifier is used as a tag for an information element. The parameters for each of these fields are described later in the chapter. They are listed here as an introduction to the structure of a TCAP message. Following is a listing of each of the component types and the fields that are sent with each component:

Invoke Component

Component type identifier

Component length

*Component ID identifier

*Component ID length

*Component IDs

Operation code identifier

Operation code length

Operation code
*Parameter set/sequence identifier
*Parameter set/sequence length
*Parameter set/sequence

Return Result Component

Component type identifier
Component length
*Component ID identifier
*Component ID length
*Component IDs
*Parameter set/sequence identifier
*Parameter set/sequence length
*Parameter set/sequence

Return Error Component

Component type identifier
Component length
*Component ID identifier
*Component ID length
*Component IDs
Error code identifier
Error code length
Error code
*Parameter set/sequence identifier
*Parameter set/sequence length
*Parameter set/sequence

Reject Component

Component type identifier
Component length
*Component ID identifier
*Component ID length
*Component IDs
Problem code identifier
Problem code length

Problem code

*Parameter set/sequence identifier

*Parameter set/sequence length

*Parameter set/sequence

All component fields are mandatory. Those with an asterisk indicate mandatory fields that can be empty (all bits set to zero, or null). The length indicators provide the number of octets for the fields immediately following, not including the length field itself. This and all other fields are explained further later in this chapter.

Again, the two portions of the TCAP message are used in different ways. The transaction portion is used by the receiver of this message to determine which transaction this message is associated with (if one is in progress) or to begin a new transaction. The component portion is used by the receiver to invoke an operation at a remote application. The component portion may consist of one or more information elements providing a sequence of operations to be performed. Each of these information elements can be correlated through a correlation ID, which is used to associate multiple components to an operation.

Connectionless TCAP Functionality

The connectionless services deployed in today's network provide for some connection-oriented emulation. However, there have been procedures defined for connection-oriented services in both the ANSI standards and Bellcore standards. This indicates that there may come a day when TCAP and SCCP will need to support full connection-oriented services.

There are three levels of identification provided for transactions and their operations. The first and highest level of reference is the *transaction ID*. The transaction ID is used when multiple transactions are sent to an application to correlate the received transaction with other transactions already in progress. But within a transaction may be components related to multiple operations.

The *correlation ID* is used to correlate multiple components to a component already received. The receiver of a TCAP transaction associates the transaction ID to an earlier transaction, and then correlates the various components to operations already in progress based on the correlation ID. It is also the correlation ID that keeps multiple components associated with one another.

When an *Invoke* component is sent, it may or may not require a response. The response sent must reference the *Invoke* it is associated with. This is done through the *invoke ID*. The invoke ID is then mirrored in the *Return Result* component's correlation ID. When a component is responding to an *Invoke* component, it must also be determined whether this is the last component or if additional components will be sent as a response. This is determined by the TCAP function using the *Invoke* (last/not last) and *Return Result* (last/not last) components.

You cannot have multiple operations using the same invoke ID. The invoke ID cannot be reassigned until all the components expected have been received and all *Return Results* have been sent. The application will then inform the application process to end the transaction, which releases all transaction IDs, correlation IDs, and invoke IDs associated with the transaction.

The application process provides the necessary data to TCAP for invoking an operation. When an invoke is requested, the components and their parameters are provided to TCAP for inclusion in an *Invoke* message. There may be more than one transaction at one time, and each transaction may have multiple operations running concurrently. TCAP must be able to address these operations individually and as a group.

TCAP can send multiple components and address multiple operations in one TCAP transaction. This means that a single TCAP message may contain information for several operations that are not associated. The use of the correlation ID and transaction ID keeps the various parameters straight within a TCAP transaction.

Handover procedures

An application can request the application process to send a component or multiple components to another remote application entity for processing. When this occurs, a *handover* must be generated by TCAP. The handover may be temporary or permanent.

The purpose of the handover is to send the information required of the remote application to perform the operations being requested of it. A single component may be all that is required, or several components may be required. In the case of a temporary handover, any responses by the new application are sent to the application that initiated the handover.

When a permanent handover occurs, the new application can be directed to send all responses directly to the originating application. In essence, the application is directed to assume control of the transactions and operations being requested. The application does not have any knowledge of the application that requested the handover. All transactions are treated as if they were addressed directly to the application.

The information required of the application is provided in an *Invoke* component, which possesses the handover operation parameter. For example, if a dialog is in progress between applications A and B, with B being the remote application and A being the originating application, B would send an *Invoke* component to application C with the temporary handover operation. Besides the temporary handover operation, the transaction ID of B, the SCCP calling party address of B, and the package type that application C should use in its messages to application A are all provided as parameters of the *Invoke* component.

Recovery procedures

In the event that an error occurs, TCAP invokes one of three levels of recovery procedures. The three levels of recovery correspond directly with the three types of errors that could occur. Errors can be classified as protocol errors, application errors, or end-user errors. Application errors are detected first, then application errors, and then end-user errors.

TCAP defines only the procedures used to recover from errors between TCAP and the application process. Any error procedures within the application process or the application are implementation dependent and beyond the scope of the SS7 standards. TCAP is responsible for reporting the errors to the application process, which, in turn, will report it to the application, which determines the correct recovery procedure if the error is at the application level.

Protocol errors

Protocol errors involve incorrect TCAP messages. Incorrect means the TCAP message contained package types or components that are invalid (or unrecognized), called for an operation that the application process did not recognize, or referenced a transaction ID or correlation ID that was not in progress.

Protocol errors can be detected by TCAP or the application process. It is reported to the remote application process using the *Reject* component. The *Reject* component must also provide the type of error (cause). This is sent to the remote application process that is then responsible for recovering from the error.

These types of errors are different from other protocol errors in the way recovery is invoked. In most protocols, when an error is detected by a remote entity, the remote entity discards the erred packet and requests a retransmission. In TCAP, the erred packet is still discarded, but a *Return Result* is sent back to the originator to inform them why the packet was rejected. It is then up to the originator of the erred message to determine if retransmission is necessary. If the application process decides to retransmit, the message is re-sent as if for the first time.

Application errors

Application errors are errors involving the application process, and indicate a violation of the application process procedures. These errors can also indicate common resources (such as recordings) that are unavailable.

An unexpected sequence of components and an unexpected data value are also types of errors that occur at the application process level. These errors mean that the application process was expecting components in an order different from how they were received (in comparison to a script or some other program). An unexpected data value means that data was received which does not match what should have been received for the received component.

A missing customer record and overdue reply are also listed as possible causes of application errors. These errors are reported using the *Return Error* component.

End-user abnormalities

These errors are found last by the application process, and indicate an error caused by the end user of the application. Two causes cited in the ANSI standard are caller abandonment and improper call response.

Caller abandonment indicates that a caller hung up before the transaction could be completed. This does not necessarily mean that an error occurred, but that the transaction could not be completed as normal.

Improper caller response indicates that a caller did not dial the proper information during a call where callers are asked by a recording to input some form of information (such as the calling card number when billing a call to a calling card).

Reject component

The *Reject* component reports many of these errors. In the *Reject* component, the type of problem is identified and the problem is identified. The type of problem is divided into categories that correspond to the various portions of the TCAP message structure. Errors are identified as transaction portion, general, *Invoke* component, *Return Result* component, or *Return Error* component.

Errors in the transaction portion are identity errors occurring within the transaction portion of the TCAP message. These include the package type and transaction ID. The errors are reported using the *Reject* component, and they report errors in the transaction portion of the message.

The type of errors that can occur in the transaction portion include an unrecognized package type, a badly structured transaction portion, an unrecognized transaction ID, a permission release problem, or an unavailable resource.

Unknown package types indicate that a package type was indicated but is not defined as received in the signaling network. In the case of an international network, this would indicate that a package type is not defined in the ITU-TS standards. The package type is the parameter that identifies what type of TCAP message is being sent and is used by the receiver to determine how to handle the received message.

A badly structured transaction portion indicates a problem with the encoding of the message. For example, the length may not be as indicated, or may be in conflict with the expected length for the indicated package type. The *total TCAP message length* is used to indicate the length of the entire TCAP message.

An unrecognized transaction ID indicates that the transaction ID indicated in the message is not currently in progress. Either the transaction has been ended and the transaction ID released or the message has an incorrect transaction ID. The receiver of this message is responding to a previous *Invoke* or *Conversation* component that provided the transaction ID of the originator. This message could also indicate that the originating *Invoke* component may have had an incorrect originating transaction ID, which is mirrored and returned with any responses.

A *permission to release* problem is an error currently under study and is not presently defined. The only mention of this error is in the ANSI T1.114 standard. It does not appear in the Bellcore TR-NWT-000246 publication.

When applications require other resources, such as recordings, there may come an instance when those resources are not available. When this happens, the operation being requested is rejected. The originating application can then determine whether to try again later.

General problems are related to problems recognizing the component portion, or indicate some problem with the component portion. This includes an unrecognized component type, an incorrect component portion, or a badly structured component portion. In all these cases, the component portion cannot be recognized and is rejected. These errors are not related to problems with the various components inside the component portion. These errors indicate a problem with the entire component portion itself.

Along with problems with the component portion, there may occur errors within each component. The protocol is capable of reporting errors related to specific component types. Error reports for *Invoke* components, *Return Result* components, and *Return Error* components can be generated when a problem is detected with the component itself. A problem with one component does not reflect a problem with the entire transaction—only with the specified component. Therefore, the rest of the components that may exist within the component portion may be processed without error, unless they are all associated with the same operation, in which case the whole transaction is affected.

Invoke component errors include the receipt of duplicate invoke IDs, unrecognized operation codes, incorrect parameters, or unrecognized correlation IDs. Duplicate invoke IDs and correlation IDs indicate that these numbers have already been assigned for an operation or transaction already in progress and cannot be reused until they are released. An unrecognized operation code indicates that the received operation code is not presently defined, and an incorrect parameter indicates that a parameter other than what was expected was received.

There are three types of errors related to the *Return Result* component. When a component is received with a correlation ID that is not recognized (because it does not match any transactions in progress), an error is returned as *unrecognized correlation ID*. An *Invoke* component that was not successful may result in the return of a result that is not expected (other than a success indication). In this event, an *unexpected return result* error is generated. If a parameter is undefined or unexpected (does not match what should have been received for the type of component), an error of *incorrect parameter* is returned.

Errors related to the *Return Error* component include unrecognized correlation ID, unexpected return error, unrecognized error, unexpected error, and incorrect parameter. An unrecognized correlation ID indicates that the correlation ID received does not match any operation presently in progress.

An unexpected return error occurs when a return error component is received that does not report failure of the invoked operation. A return error may also contain an unrecognized error, which is one not defined by the application process. If the returned error is not applicable to the invoked operation, then it is marked as unexpected. An incorrect parameter or unexpected parameter will also generate an error.

Return Error component

This component is used to report a failure of an operation. It is also possible for this component to report the success of an operation and the failure of an operation. At any rate, the operation fails.

Included in this component is the error that occurred and the application process that was in error. If an invoke ID or correlation ID was provided in either the *Invoke* component or the *Return Result* component, then it also is reflected in the *Return Error* component.

Return Result component

This component reports the success of an operation. It also reports any end-user errors that may have occurred. The successful operation is identified, as well as parameters that identify the end-user error.

An end-user error does not necessarily cause an operation to fail. The invoked operation may be successful, but may have caused a problem within the application entity. This would cause the *Return Result* to report the end-user failure.

Problems relating to the transaction portion of the TCAP message follow a different procedure. Since the transaction portion is what the receiver uses to determine the handling of a received TCAP, any errors result in the application process not knowing how to handle the message. The usual procedure in this case is to discard the message. There are instances, however, when this is not favorable. Whenever enough information can be obtained from the message, some sort of error report should be returned to the user, indicating the type of error. The *Abort* component is used to report such errors.

The *Abort* component is used whenever any type of component is received where either the originating and/or responding transaction ID can be derived from the message. In all other instances, the TCAP is discarded, with no report to the user. Without the transaction ID, any report to the user would be fruitless.

Definition of TCAP Parameters

The preceding section describes the various functions of the TCAP message and its components. This section identifies the values of these components and their parameters. The values given are derived from the Bellcore TR-NWT-000246 publication, Issue 2, Revision 2, December 1992.

Transaction portion

The transaction portion (Fig. 8.5) identifies whether or not the component portion consists of a single transaction or multiple transactions, and alerts the receiving application as to how to handle the message. Important to the receiver is the type of structure used within TCAP. This is determined by the package type. Also significant is the length of the TCAP message in its entirety.

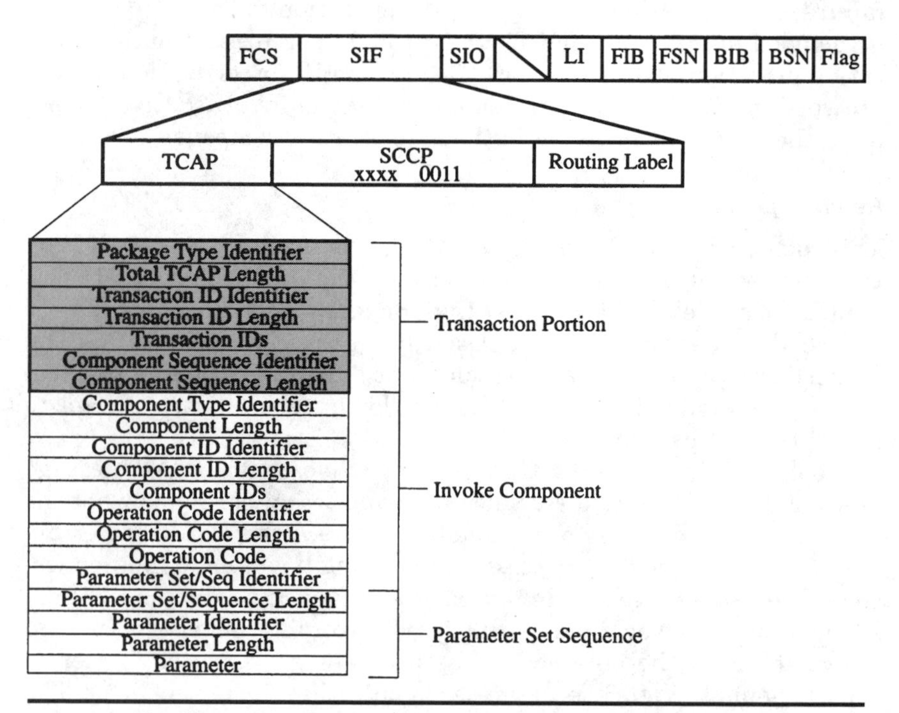

Figure 8.5 This figure identifies the components of the transaction portion.

The transaction portion provides the necessary data for the receiver to be able to determine the nature of the message and its relation to any existing transactions in progress. Following are the package types and their bit values used in the transaction portion to identify the type of TCAP message being presented.

Package Type Identifiers	H G F E	D C B A
Unidirectional	1 1 1 0	0 0 0 1
Query With Permission	1 1 1 0	0 0 1 0
Query Without Permission	1 1 1 0	0 0 1 1
Response	1 1 1 0	0 1 0 0
Conversation With Permission	1 1 1 0	0 1 0 1
Conversation Without Permission	1 1 1 0	0 1 1 0
Abort	1 1 0 1	0 1 1 0

Total TCAP message length. This indicates the total length of the TCAP message (including all components and parameters). Within each component in the component message is a length indicator to indicate the length of that individual component. All variable parameters also include length indicators.

Transaction ID identifier 1100 0111. This indicates that a transaction ID is present and will follow this field, it is coded as national, primitive (see discussion in previous section describing the encoding used for national and international messages). As mentioned in the discussion at the beginning of this chapter, the transaction ID is of local significance only. This identifier only suggests the presence of the transaction ID and does not represent the ID itself.

Transaction ID length. This indicates the length of the transaction ID, including both the originating and responding transaction ID, if applicable. If the package type is unidirectional, the length will be equal to zero. Only the unidirectional package type will have a length of zero, because a transaction ID is not assigned to this package type.

If the package type is one which contains only the originating transaction ID, the length will be equal to four octets. The originating ID is sent to the remote application to be used as a reference in any responses.

If the remote application is sending a component which will require a response of some nature, then the originating transaction ID and a response ID will be sent. The originating transaction ID is sent to associate the response to the transaction of the originator. The response transaction ID is sent to be used by the originator when it sends a response back to the remote application. These dual transaction IDs occur when there is a dialog between two applications, and *Invoke* components are sent from one entity to another.

Transaction IDs. The number of transaction IDs given depends on the package type. The following indicates when multiple transaction IDs are provided.

Package Type	Originating ID	Responding ID
Unidirectional	No	No
Query With Permission	Yes	No
Query Without Permission	Yes	No
Response	No	Yes
Conversation With Permission	Yes	Yes
Conversation Without Permission	Yes	Yes
Abort	No	Yes

Originating transaction ID. This identifies the transaction ID assigned by the originator of the message. The length is four octets, and it is the first field when multiple transaction IDs are presented. As previously described, this is sent by the originator to the remote application to be used as a reference in any responses.

Responding transaction ID. This identifies the transaction ID generated by the responding application. This is independent of the originating transaction ID, and is of significance to the responding application only. This field is used when the responding application expects a response from the originator of an *Invoke* operation component. This is a four-octet field and follows the originating transaction ID.

P-Abort cause identifier 1101 0111. This identifier indicates that the *Abort* cause value follows. This identifier is coded as national, primitive. An *Abort* cause is sent when a transaction is aborted before it had time to get completed. An *Abort,* as described earlier, indicates a problem within the application entity rather than a protocol error.

P-Abort cause length. The length does not include this and the identifier field—it includes only the cause field. The cause field is always one octet in length; thus, this field will always carry a value of one octet.

P-Abort Cause	H G F E	D C B A
Unrecognized package type	0 0 0 0	0 0 0 0
Incorrect transaction portion	0 0 0 0	0 0 0 1
Badly structured transaction portion	0 0 0 0	0 0 1 0
Unrecognized transaction ID	0 0 0 0	0 0 1 1
Permission to release problem	0 0 0 0	0 1 0 0
Resource unavailable	0 0 0 0	0 1 0 1

The cause code informs the receiver of an *Abort* and what caused the application process at the remote end to suddenly abort a transaction in progress. An unrecognized package type indicates that the package type received is not currently defined at that application entity.

An incorrect transaction portion indicates that an identifier in the transaction portion was unexpected (not what should have been received in comparison with the rest of the transaction portion and the transaction in progress).

A badly structured transaction portion indicates that there is a problem with the encoding of the transaction portion. This could mean that the transaction portion is not of the correct length for the indicated identifiers, or that there are missing fields.

An unrecognized transaction ID occurs when a TCAP is received with a transaction ID that does not reflect a transaction currently in progress. The receiving application does not know to which transaction and operations to associate the message; therefore, it will abort the transaction and return the cause to the sender.

The cause, *permission to release problem,* is not currently defined and is for further study. The last cause code, *resource unavailable,* is returned when a transaction is aborted because resources necessary to perform the operations specified (such as recordings or databases) are not available.

User abort information identifier 1101 1000. The identifier indicates that the user abort information field is to follow. This identifier is coded as national, primitive. A user can initiate an *Abort* when an unexpected event occurs, such as when the caller hangs up before the transaction can be completed. The user in this message is not the calling or called party, but an application entity.

User abort information length. The length does not include this and the identifier field—it includes only the user abort information field. The information field is a variable field.

User abort information. This variable field is used by TCAP to provide information regarding a user abort. The user abort indicates the reason that a user (application) aborted a transaction before it could be completed. The abort codes are not defined in ANSI or Bellcore, because these are implementation-specific codes defined by the various manufacturers.

Component sequence identifier 1110 1000. This field is used to notify the receiver that a sequence of components will follow this field. This identifier is coded as national, constructor. The component sequence identifier does not identify how many components are to follow, only that they will follow. This is really used as a header to the component portion, and all TCAP messages will contain this identifier.

Component sequence length. The length of the entire component portion is indicated in this field. The length does not include the length field itself or the identifier before it.

Dialog portion

The dialog portion is an optional part of TCAP. It contains information regarding security, encryption, and protocol version of data contained in the user part of TCAP. Following are the parameters of the dialog portion.

Dialog Portion	**H**	**G**	**F**	**E**	**D**	**C**	**B**	**A**
Dialog portion	1	1	1	1	1	0	0	1
Protocol version	1	1	0	1	1	0	1	0
Integer application context	1	1	0	1	1	0	1	1
Object identifier application context	1	1	0	1	1	1	0	0
User information	1	1	1	0	1	0	0	0
Integer security context	1	0	0	0	0	0	0	0
Object idfentifier security context	1	0	0	0	0	0	0	1
Confidentiality	1	0	1	0	0	0	1	0
Integer confidentiality algorithm	1	0	0	0	0	0	0	0
Object identifier confidentiality algorithm	1	0	0	0	1	0	0	1

The *dialog portion* identifier indicates that a dialog portion is included in the TCAP message. This is an optional field, used when the user data contained in the TCAP message is secure data. Following the dialog portion identifier is the dialog portion length field. This provides the length (in octets) of the entire dialog portion of the TCAP message, not including this length field or the dialog portion identifier.

The *protocol version identifier* indicates the protocol version follows. The protocol version length indicates the length of just the protocol version part of the dialog portion. The protocol version field indicates which version of the T1.114 protocol is being sent in the TCAP message. This allows nodes to determine which TCAP version of the T1.114 protocol is being sent in the TCAP message. This allows nodes to determine which TCAP version (for example, ANSI T1.114, 1997) is being used to encode the SS7 message.

The *application context identifier* indicates that an application context name follows. This is used by the applications at the receiving nodes. It is accompanied by user data in the user information fields of the dialog portion. The identifier is followed by the application context length, which provides the length of the application context name only.

The *user information identifier* indicates that user information is provided in the dialog portion. The user information length follows, indicating the length of the user information portion only, not counting the user information identifier or the length fields

The *security context identifier* indicates that security information is contained in the dialog portion. It is followed by the security context length field. The security context provides information regarding the encryption of user information when encryption is used. This allows for secure transmission of user information when TCAP is being used to access databases which may have sensitive data.

The last part of the dialog portion is the confidentiality fields. The *confidentiality identifier* indicates that a confidentiality algorithm follows, after the *confidentiality length* field.

Component portion

The component portion contains one or more components and their associated parameters. These components are used to invoke operations and return results to the invoking application.

A component may consist of the component itself, which indicates how the receiver will respond (if any response is necessary), the operation to be performed (optional), and any required parameters to be used when invoking the specified operation.

Components will always have an identifier, length field, and contents, which specify what the component is asking of the receiver. Their values and descriptions are shown as follows.

Component Type Identifier	**H**	**G**	**F**	**E**	**D**	**C**	**B**	**A**
Invoke (last)	1	1	1	0	1	0	0	1
Return Result (last)	1	1	1	0	1	0	1	0
Return Error	1	1	1	0	1	0	1	1
Reject	1	1	1	0	1	1	0	0
Invoke (not last)	1	1	1	0	1	1	0	1
Return Result (not last)	1	1	1	0	1	1	1	0

An *Invoke* (last) is used to invoke an operation at a remote application entity. For example, if a signaling point wanted to invoke some feature (operation) at another remote signaling point, a TCAP with an *Invoke* component would be sent to the remote signaling point, which would then act upon the *Invoke* request accordingly. The indication of "last" means there are no more *Invoke* components for this transaction.

The *Invoke* (not last) is the same as just described, with the exception of the "not last" indicator, which alerts the receiver that additional *Invokes* are to be sent in relation to this transaction.

The *Return Result* (last) is used to return the results of an invoked operation. Not all operations require the *Return Result* to be sent. The indicator of "last" means there will be additional results sent in response to the specified *Invoke* and transaction. These are usually accompanied by a correlation ID, which is used to associate the *Return Result* with an earlier *Invoke* component.

Return Error components are used to return an error code in the event of an operational error. An operational error is one which occurs at the application level, and does not necessarily indicate a protocol problem.

The *Reject* component indicates a problem with protocol and means that the component was rejected because of an incorrect component portion in an earlier *Invoke*.

Component length. The length field identifies how many octets are left in the component. This field and the identifier are not included in the length. There may be multiple components in one component portion; each component includes a component length field. This is not the length of the entire component portion (as seen in the transaction portion). This is only the length for this component (in which it is carried).

Component ID identifier 1100 1111. This field indicates that a component has an invoke ID and, possibly, a correlation ID. As mentioned earlier, these IDs are of local significance only, and are sent so that *Return Results* can be correlated with an earlier *Invoke* request. The invoke ID is assigned by the originator of an *Invoke,* and requires that any *Return Result* will require a correlation ID. The correlation ID is the same ID as the invoke ID, and is returned within the *Return Result* so the originator of an *Invoke* can correlate returns with their respective *Invokes.*

Component ID length. The length field indicates the total length of the component ID field only. The ID length may be zero (unidirectional only), four (invoke ID only), or eight octets in length (both invoke ID and correlation ID provided). A *Return Result* can have only a correlation ID; therefore, the length for a *Return Result* will always be four octets. The same is true for *Return Error* and *Reject.*

As seen in the following table, only the *Invoke* can carry both an invoke ID and a correlation ID.

Component IDs. The component may have an invoke ID and a correlation ID, based on the following criteria. These IDs are used to associate multiple components with operations and transactions already in progress.

Component Type	Invoke ID	Correlation ID
Invoke	Optional	Yes
Return Result	No	Yes
Return Error	No	Yes
Reject	No	Yes

Invoke ID. The invoke ID is a one-octet identifier assigned to a component that is invoking an operation. This is an optional field, and is of local significance only.

Correlation ID. The correlation ID is used whenever a component is responding to another component. If responding to a component that contained an invoke ID, the correlation ID is mandatory. The correlation ID is a mirror image of an invoke ID, and is used as a reference for associated invoke IDs with previous invokes.

Operation Code Identifier	H	G	F	E	D	C	B	A
National TCAP	1	1	0	1	0	0	0	0
Private TCAP	1	1	0	1	0	0	0	1

National TCAP operation codes are defined in both ANSI and Bellcore standards. These are TCAP messages which must be common in all systems interworking with Bell Operating Company networks and other ANSI networks.

Private TCAP operation codes are of significance only to private networks and are not compatible with any networks that interwork. Private networks cannot be connected to the Public Switched Telephone Network because the message types will conflict with national standards.

Operation code length. The length field identifies the length of the operation code field only and does not include itself or the identifier field. If the operation code is national TCAP, the length will be two octets. If the operation code is private TCAP, there is no limitation on the length.

Operation codes

The operation codes are implementation specific, and are used by the application entities as instructions on how to carry out the component action. TCAP as a protocol does not interact with the operations; it only delivers them to the appropriate application process.

Operation codes may vary from network to network, because they are very network dependent. This section identifies those operation codes used within the Bellcore networks as identified in the Bellcore publication, TR-NWT-000246, T1.114.5, Issue 2, Revision 3, December 1992. The EIA/TIA has also specified operation codes for usage in the cellular network. The operation codes used in cellular provide a means for mobile switching centers to pass information from one to another, as well as to invoke operations at remote MSCs.

The operation code is a two-octet field divided into the operation family and the operation specifier. The operation family is a seven-bit field (bits A–G), and identifies the group or category related to this operation. Bit H indicates whether or not a response is expected.

The second octet of the operation code consists of the operation specifier, which is used to identify the operation being requested. The specifier is the specific instruction being asked of the application.

Operation Families	G	F	E	D	C	B	A
Parameter	0	0	0	0	0	0	1
Charging	0	0	0	0	0	1	0
Provide Instructions	0	0	0	0	0	1	1
Connection Control	0	0	0	0	1	0	0
Caller Interaction	0	0	0	0	1	0	1

Send Notification	0 0 0 0	1 1 0
Network Management	0 0 0 0	1 1 1
Procedural	0 0 0 1	0 0 0
Operation Control	0 0 0 1	0 0 1
Report Event	0 0 0 1	0 1 0
Miscellaneous	1 1 1 1	1 1 0

The operation family does not provide enough information for the application to process the operation. This only serves as a category for the operation code. As you will notice in the rest of this section, the operation codes themselves are not all unique codes. They each use an ascending order beginning with the binary value of 0000 0001. The operation family field then becomes the delimiter between the various operation codes.

Parameter	**H G F E**	**D C B A**
Provide value	0 0 0 0	0 0 0 1
Set value	0 0 0 0	0 0 1 0

The parameter family provides instructions on how to use the parameters accompanying this parameter. There are two options: *provide value* and *set value*. The *provide value* indicator instructs the receiver to provide the requested value for the given parameter. The parameter portion of the TCAP message contains the actual parameter to which this operation refers.

The option *set value* instructs the receiver to set the value for the given parameter. This is for further study and has not yet been implemented. One example of how this operation code may be used is in the case where a signaling point has control of a call, but does not have the necessary resources available to continue processing (such as recordings). A temporary handover procedure is invoked, and the *provide value* indicator shown here could be included in the parameters sent to the remote application to indicate the need for resources.

Charging	**H G F E**	**D C B A**
Bill call	0 0 0 0	0 0 0 1

Presently, there is only one option for this operation code. The *bill call* option is used to notify the receiving application that a billing record is to be created for the calling party indicated in the parameters.

Provide Instructions	**H G F E**	**D C B A**
Start	0 0 0 0	0 0 0 1
Assist	0 0 0 0	0 0 1 0

This operation code is used to request instructions during an assist procedure. The *start* option is for further study, while the *assist* option indicates that the assist procedure has been requested and the receiver of an assist is asking the sender for instructions.

Connection Control	H G F E	D C B A
Connect	0 0 0 0	0 0 0 1
Temporary connect	0 0 0 0	0 0 1 0
Disconnect	0 0 0 0	0 0 1 1
Forward disconnect	0 0 0 0	0 1 0 0

The four codes used in this operation are grouped into associated pairs. The *connect* and *disconnect* are related, and will require further study. When a *connect* is issued, the *disconnect* is used to terminate the connection.

The *temporary connect* is used when an error is encountered and a connection to another database for completion of processing is required. The *temporary connect* operation must also be accompanied by parameters providing the subsystem number of the database, the routing number of the exchange to which a connection is being requested, and a reference number for correlation of transactions between the database and the requesting exchange.

When the *temporary connect* is received, the receiving entity knows that the *forward disconnect* will follow. The *forward disconnect* is used to terminate a temporary connection.

Caller Interaction	H G F E	D C B A
Play announcement	0 0 0 0	0 0 0 1
Play announcement and collect digits	0 0 0 0	0 0 1 0
Indicate information waiting	0 0 0 0	0 0 1 1
Indicate information provided	0 0 0 0	0 1 0 0

The caller interaction family of operations allows announcements to be specified and provides a mechanism for application processes to communicate with one another regarding the state of expected information. The first two options, *play announcement* and *play announcement and collect digits,* are identical, with the exception of the latter, which waits and collects dialed digits from the user. The dialed digits can then be routed to a voice response unit for processing.

The indicators for information waiting allow one application process to inform another application process that there is information waiting and, when the information has been transferred, that the information has been provided.

Send Notification	H G F E	D C B A
When party free	0 0 0 0	0 0 0 1

The operation in this operation family is used for certain CLASS features such as automatic callback, where a caller reaches a busy signal and requests notification from the network when the party becomes available. When the called party becomes available by placing the receiver back on-hook, the remote exchange is notified via TCAP that the called party is available. ISUP call setup is then used to set the call up as if it were a normal call.

Network Management	**H G F E**	**D C B A**
Automatic code gap	0 0 0 0	0 0 0 1

The *automatic code gap* is used by network management to temporarily inhibit specified codes for the specified time. Additional parameters indicate the time the codes are to be inhibited and the duration between operations.

Procedural	**H G F E**	**D C B A**
Temporary handover	0 0 0 0	0 0 0 1
Report assist termination	0 0 0 0	0 0 1 0
Security	0 0 0 0	0 0 1 1

This operation family is used to control procedural operations. The *temporary handover* operation indicates that a temporary handover is presently in progress. When the temporary handover procedure has been completed, the receiver of this message will then release all resources dedicated to this operation.

The *report assist termination* is used to end an assist procedure.

Operation Control	**H G F E**	**D C B A**
Cancel	0 0 0 0	0 0 0 1

The *cancel* operation is used with the send notification operation to cancel a *when party free* operation. This is presently the only operation that can be canceled. The *service key* parameter accompanies this operation and provides the called party number.

Report Event	**H G F E**	**D C B A**
Voice message available	0 0 0 0	0 0 0 1
Voice message retrieved	0 0 0 0	0 0 1 0

This operation family is used with the Voice Message Storage Retrieval (VMSR) systems. When a subscriber uses such a service, and the VMSR system is located at another exchange (other than the subscriber's), the *voice message available* option is used to alert the subscriber's exchange of a message. The calling number of the party that left the message can be included, as well as the time the message was left. A *Return Result* is sent when the operation has been successful.

The *voice message retrieved* operation is used to remove the message available indicator from a subscriber's Voice Message Storage Retrieval. Both the subscriber's number and the identification of the VMSR system used by the subscriber are provided with this operation code.

Miscellaneous	H G F E	D C B A
Queue call	0 0 0 0	0 0 0 1
Dequeue call	0 0 0 0	0 0 1 0

Error codes

These error codes are used to indicate the reason for an unsuccessful completion of an operation. The error code is sent in a *Return Error* component to the originator of the operation request. The error codes are coded as either national or private. The error codes identified as follows are national error codes as defined in Bellcore publication TR-NWT-000246, Issue 2, Revision 2, December 1992.

Error Code	H G F E	D C B A
Unexpected component sequence	0 0 0 0	0 0 0 1
Unexpected data value	0 0 0 0	0 0 1 0
Unavailable resource	0 0 0 0	0 0 1 1
Missing customer record	0 0 0 0	0 1 0 0
Spare	0 0 0 0	0 1 0 1
Data unavailable	0 0 0 0	0 1 1 0
Task refused	0 0 0 0	0 1 1 1
Queue full	0 0 0 0	1 0 0 0
No queue	0 0 0 0	1 0 0 1
Timer expired	0 0 0 0	1 0 1 0
Data already exists	0 0 0 0	1 0 1 1
Unauthorized request	0 0 0 0	1 1 0 0
Not queued	0 0 0 0	1 1 0 1
Unassigned DN	0 0 0 0	1 1 1 0
Spare	0 0 0 0	1 1 1 1
Notification unavailable to destination DN	0 0 0 1	0 0 0 0
VMSR system ID did not match user profile	0 0 0 1	0 0 0 1

An *unexpected component sequence* indicates that one or more components were received which did not match what should have been received (was expected), considering the previous components received from the same originator. Likewise, an *unexpected data value* indicates that data received within a component was not what should have been sent, considering the type of component.

When resources required to carry out an *Invoke* are not available, the receiving application will return an *error of unavailable resources.* The originator must then determine if it will petition another application at another signaling point, or possibly even within the same signaling point. Resources include things like recordings and databases.

When accessing a database to add information to a customer database or when retrieving information from a customer record, the identity of the called party will be used to identify the record. If the record does not exist, an *error of missing customer record* will be sent to the originating application. Customer records are used to determine the billing options for a call as well as the type of features a subscriber can use.

With certain CLASS features, the customer record will also provide specific instructions on how to handle specific services on a per-call basis.

Data unavailable could indicate that the database containing requested information is not available. This could be the case if there was a failure within the subsystem or if the application process failed and was unable to reach the database.

Task refused is returned when the application entity has been requested to perform some sort of task, but the entity chosen cannot perform the task. No reason is given, only the *Reject.*

With custom calling features and CLASS, certain features require a queue to temporarily store numbers. For example, automatic callback and automatic recall require the last dialed number or the last number that called to be stored in a queue until the feature is invoked. The feature (application entity) then accesses this queue to complete the processing of the feature. There are three states associated with these queues: *queue full, no queue,* and *not queued. Not queued* is used when a number is to be removed from a called party queue.

When a parameter is sent with data that has already been received, the reject cause will be *data already exists.* Only a parameter change operation can change data that has already been received. Several reject causes are related to directory numbers (DNs). These reject causes are sent for a variety of reasons, but are usually related to the lack of resources or *unauthorized to access services and/or databases* being requested.

A Voice Message Storage Retrieval (VMSR) system is now being offered in many areas. This allows voice-mail equipment to be installed at the central office, rather than the subscriber purchasing a voice-mail system. The subscriber must purchase the service, and the service must be an entry in the customer record before access is allowed. If access is attempted and the directory number is not a subscriber to the VMSR system, a reject cause of *VMSR system identification did not match user profile* is sent.

Parameters

Parameters are associated with individual components and are the last items in the component portion of the TCAP message. There are three elements in any parameter: the parameter identifier, length, and contents.

The parameter identifier is used to identify the individual parameters, and consists of a one-octet field. All parameter identifiers are listed in the following table. The length field indicates the length of the contents field, which is a variable field. The contents can be one-octet specifiers or implementation-dependent codes.

The following parameters are found in the Bellcore publication, TR-NWT-000246, Issue 2, Revision 2, December 1992.

Parameter Name	H	G	F	E	D	C	B	A
Timestamp	0	0	0	1	0	1	1	1
ACG indicators	1	0	0	0	0	0	0	1
Standard announcement	1	0	0	0	0	0	1	0
Customized announcement	1	0	0	0	0	0	1	1
Digits	1	0	0	0	0	1	0	0
Standard user error code	1	0	0	0	0	1	0	1
Problem data	1	0	0	0	0	1	1	0
SCCP calling party address	1	0	0	0	0	1	1	1
Transaction ID	1	0	0	0	1	0	0	0
Package type	1	0	0	0	1	0	0	1
Service key	1	0	1	0	1	0	1	0
Busy/idle status	1	0	0	0	1	0	1	1
Call forwarding status	1	0	0	0	1	1	0	0
Originating restrictions	1	0	0	0	1	1	0	1
Terminating restrictions	1	0	0	0	1	1	1	0
DN to line service type mapping	1	0	0	0	1	1	1	1
Duration	1	0	0	1	0	0	0	0
Returned data	1	0	1	1	0	0	0	1
Bearer capability requested	1	0	0	1	0	0	1	0
Bearer capability supported	1	0	0	1	0	0	1	1
Reference ID	1	0	0	1	0	1	0	0
Business group	1	0	0	1	0	1	0	1
Signaling network identifier	1	0	1	1	0	1	1	0
Generic name	1	0	0	1	0	1	1	1
Message waiting indicator type	1	0	0	1	1	0	0	0
Look ahead for busy	1	0	0	1	1	0	0	1
Circuit identification code	1	0	0	1	1	0	1	0
Precedence identifier	1	0	0	1	1	0	1	1
Call reference identifier	1	0	0	1	1	1	0	0

Parameter values

Following are all the parameters just listed and the values of their contents. Parameters complement the components already described and their operation codes. Operation codes may choose any number of the following parameters as a set. These are all national parameters as described, and are defined in the Bellcore publication TR-NWT-000246, Issue 2, Revision 2, December 1992.

Timestamp	H G F E D C B A
Octets 1–2	Year in binary (e.g., 93)
Octets 3–4	Month in binary (e.g., 07)
Octets 5–6	Day in binary (e.g., 28)
Octets 7–8	Hour in binary (e.g., 17)
Octets 9–10	Minutes in binary (e.g., 30)
Octet 11	+ or − (ahead or behind GMT)
Octets 12–13	Hours (see above description)
Octets 14–15	Minutes (see above description)

The timestamp provides the time and date an event occurred. Both local time and the difference between local time and Greenwich Mean Time (GMT) are provided. The first hour and minutes are those of local time and the second hour and minutes fields reflect the difference between local time and GMT. For example, if the local time is 1730, and the local time is in Atlanta (which is five hours behind Greenwich time), the second hour and minutes field would reflect a difference of −0500.

Automatic Code Gap	**H G F E**	**D C B A**
Control Cause Indication		
Vacant code	0 0 0 0	0 0 0 1
Out-of-band	0 0 0 0	0 0 1 0
Database overload	0 0 0 0	0 0 1 1
Destination mass calling	0 0 0 0	0 1 0 0
Operation Support System (OSS) initiated	0 0 0 0	0 1 0 1

The Automatic Code Gap (ACG) is a network management function that allows the network to throttle traffic for a specified period of time. ACG can be initiated manually or automatically. The preceding codes indicate the cause for invoking ACG.

Vacant code indicates that calls are being received for an unassigned code. Out-of-band is related to calls for a band that a subscriber does not subscribe to. Database overload indicates a database that is overloaded, while destination mass calling indicates that an excessive number of calls are being received for a destination.

When the Automatic Code Gap is initiated manually, it is initiated by an Operations Support System (OSS). When this is the case, the cause code will indicate that the OSS initiated the ACG.

Duration (in seconds)	**H G F E**	**D C B A**
Not used	0 0 0 0	0 0 0 0
1	0 0 0 0	0 0 0 1
2	0 0 0 0	0 0 1 0
4	0 0 0 0	0 0 1 1
8	0 0 0 0	0 1 0 0
16	0 0 0 0	0 1 0 1
32	0 0 0 0	0 1 1 0
64	0 0 0 0	0 1 1 1
128	0 0 0 0	1 0 0 0
256	0 0 0 0	1 0 0 1
512	0 0 0 0	1 0 1 0
1024	0 0 0 0	1 0 1 1
2048	0 0 0 0	1 1 0 0

One-octet field indicating the time duration in seconds that an ACG should be applied.

Gap (in seconds)	**H G F E**	**D C B A**
Remove gap control	0 0 0 0	0 0 0 0
0.00	0 0 0 0	0 0 0 1

0.10	0 0 0 0	0 0 1 0
0.25	0 0 0 0	0 0 1 1
0.50	0 0 0 0	0 1 0 0
1.00	0 0 0 0	0 1 0 1
2.00	0 0 0 0	0 1 1 0
5.00	0 0 0 0	1 0 0 0
15.00	0 0 0 0	1 0 0 1
30.00	0 0 0 0	1 0 1 0
60.00	0 0 0 0	1 0 1 1
120.00	0 0 0 0	1 1 0 0
300.00	0 0 0 0	1 1 0 1
600.00	0 0 0 0	1 1 1 0
Stop all calls	0 0 0 0	1 1 1 1

One-octet field indicates the interval between applications of the ACG control. Time is measured in seconds.

Standard Announcement	**H G F E**	**D C B A**
Not used	0 0 0 0	0 0 0 0
Out-of-band	0 0 0 0	0 0 0 1
Vacant code	0 0 0 0	0 0 1 0
Disconnected number	0 0 0 0	0 0 1 1
Reorder (120 pulses per minute) tone	0 0 0 0	0 1 0 0
Busy (60 pulses per minute)	0 0 0 0	0 1 0 1
No circuit available	0 0 0 0	0 1 1 0
Reorder recording	0 0 0 0	0 1 1 1
Audible ringing	0 0 0 0	1 0 0 0

When an announcement is to be applied to a particular call, either a standard announcement or a customized announcement may get requested. The standard announcements for a Bellcore network are depicted in the preceding table.

Customized announcement. These are implementation dependent and allow networks to address their own network-specific announcements. The parameter consists of two elements: the announcement set and the individual announcement. Length is variable.

Customized announcements are unique within the network in which they reside. Independent companies and private networks may implement their own announcements for usage within their own networks. Both standard and customized announcements can be used within the same network. These two conventions allow either one to be requested for a specific call.

Digits	**H G F E D C B A**
	Type of digits
	Nature of number
	Number plan | Encoding
	Number of digits
	Digits

Type of digits		
Not used	0 0 0 0	0 0 0 0
Called party number	0 0 0 0	0 0 0 1
Calling party number	0 0 0 0	0 0 1 0
Caller interaction	0 0 0 0	0 0 1 1
Routing number	0 0 0 0	0 1 0 0
Billing number	0 0 0 0	0 1 0 1
Destination number	0 0 0 0	0 1 1 0
Local Access and Transport Area (LATA)	0 0 0 0	0 1 1 1
Carrier	0 0 0 0	1 0 0 0
Last calling party	0 0 0 0	1 0 1 0
Calling directory number	0 0 0 0	1 0 1 1
VMSR identified	0 0 0 0	1 1 0 0
Original called number	0 0 0 0	1 1 0 1
Redirecting number	0 0 0 0	1 1 1 0
Connected number	0 0 0 0	1 1 1 1

When digits are being received in TCAP for invoking features, the type of digits (source) and the coding of the digits (BCD) must be specified so that the receiving entity knows how to decode the digits. The preceding types indicate the source type of the digits and allow the receiver to determine how to handle the received digits.

In some cases (CLASS features, that is), the subscriber may be requested to dial digits. When this is the case, the digits type is *caller interaction.* An example of this would be when a caller inputs a calling card number when requested by an announcement.

Dialed numbers can also be used for network routing (routing number) or billing information (billing number). When information is being requested regarding a particular line, the *destination number* may be provided as a reference to the subscriber who owns that line.

Digits can also specify the Local Access Transport Area a particular caller is calling from, to be used in routing and accessing line information. All *carriers* are numbered as well. When callers wish to dial their long distance carrier's operator direct, they dial a 10xxx number, the last three digits being the carrier number. These digits can be carried through TCAP, as well, to access a billing record and record usage of a carrier or when a customer line record is accessed to specify the long distance carrier for that subscriber.

Last calling party digits identify the last directory number to dial a certain number. This is used with CLASS features such as automatic callback to identify the directory number of the last party to call a number. *Last party called* identifies the last party that was called (directory number).

A Voice Mail Send/Receive (VMSR) is also identified by digits, and is referred to in messages accessing and/or invoking a VMSR system for a call.

Redirecting number identifies the number of the party who last invoked a forwarding to a specific number. For example, with follow-me forwarding, a number can be reforwarded from any phone whenever needed. The redirecting number indicates the last directory number to forward the specified number.

Nature of number	**H**	**G**	**F**	**E**	**D**	**C**	**B**	**A**
National								0
International								1
No presentation restriction							0	
Presentation restriction							1	

Encoding	**H**	**G**	**F**	**E**	**D**	**C**	**B**	**A**
Not used					0	0	0	0
Binary Coded Decimal (BCD)					0	0	0	1
IA5					0	0	1	0

Numbering plan	**H**	**G**	**F**	**E**	**D**	**C**	**B**	**A**
Unknown or not applicable	0	0	0	0				
ISDN numbering	0	0	0	1				
Telephony numbering	0	0	1	0				
Data numbering	0	0	1	1				
Telex numbering	0	1	0	0				
Maritime mobile numbering	0	1	0	1				
Land mobile numbering	0	1	1	0				
Private numbering plan	0	1	1	1				

Number of digits	**H G F E**	**D C B A**
For BCD encoding:	2nd digit	1st digit
	nth digit	$(n-1)$th digit

Digits are coded as follows:				
Digit 0 or filler	0	0	0	0
Digit 1	0	0	0	1
Digit 2	0	0	1	0
Digit 3	0	0	1	1
Digit 4	0	1	0	0
Digit 5	0	1	0	1
Digit 6	0	1	1	0
Digit 7	0	1	1	1
Digit 8	1	0	0	0
Digit 9	1	0	0	1
Spare	1	0	1	0
Code 11	1	0	1	1
Code 12	1	1	0	0
*	1	1	0	1
#	1	1	1	0
ST	1	1	1	1

All of the preceding identify the type of digits as well as the digits themselves. These are used when presenting digits into the TCAP message so the receiver can understand the origin of the digits and know how they are to be decoded. Digits appear in a variety of TCAP transactions, and are being used more and more as Advanced Intelligent Network (AIN) features begin working their way into the network.

Standard User Error Code	H G F E	D C B A
Not used	0 0 0 0	0 0 0 0
Caller abandon	0 0 0 0	0 0 0 1
Improper caller response	0 0 0 0	0 0 1 0

User error codes define the reason for a user-induced failure. When a transaction is interrupted and aborted because of subscriber actions, these cause codes are used to describe the reason. Presently, only two causes are defined.

Caller abandon indicates that a subscriber (calling party) hung up before the transaction could be completed. The transaction could have been a calling card call to another number, which involves TCAP to access the calling card database for verification and billing. If any type of call forwarding or access to a voice-mail system (VMSR) is used, TCAP is required for completing the informational transactions.

Improper caller response indicates that the caller was queried to enter digits and did not enter the correct digits, or entered in the wrong number of digits. A good example of this would be entering in a calling card number when prompted. Another example is with VMSR systems where a caller may be prompted by an announcement to enter in a callback number, but enters in an incorrect number of digits.

Problem data. The problem data cites the specific reason for the error. This field is implementation dependent and coded as contextual, primitive. The format is the same as for other parameters: a one-octet parameter identifier, the length of the parameter contents, and the parameter contents (variable-length field).

This is provided in the protocol as an optional source of information when a problem occurs. Systems can elect to send additional information and codes regarding a specific set of problems encountered.

SCCP calling party address. This is the address field used by the receiver of a handover procedure to determine the calling party address. The address structure is the same as discussed in Chap. 7. The address can consist of global title digits, a point code, or a subsystem number. This information is part of the temporary handover parameter.

Transaction ID. This is the same format used in the transaction portion. The difference is that this is used during the temporary handover procedure by the receiver of a temporary handover message. This information is part of the temporary handover parameter.

Package type. The package type is used here to notify the receiver of the temporary handover parameter what package type to respond with. The package type indicated here will be used to return some sort of response to the calling party address (if included in the transaction).

Service key. The service key is used to specify which parameters should be used to access a record. This is used in database queries, which are used to access customer database records and service records. The service key is coded as contextual, constructor.

Busy/Idle Status	**H**	**G**	**F**	**E**	**D**	**C**	**B**	**A**
Not used	0	0	0	0	0	0	0	0
Busy	0	0	0	0	0	0	0	1
Idle	0	0	0	0	0	0	1	0

Busy and *idle* statuses are used to provide information regarding the status of a subscriber line when particular CLASS features and custom calling features are deployed.

Call Forwarding Status	**H**	**G**	**F**	**E**	**D**	**C**	**B**	**A**
Call Forwarding Variable								
Service not supported	0	0						
Active	0	1						
Not active	1	0						
Spare	1	1						
Call Forwarding on Busy	**H**	**G**	**F**	**E**	**D**	**C**	**B**	**A**
Service not supported			0	0				
Active			0	1				
Not active			1	0				
Spare			1	1				
Call Forwarding Don't Answer	**H**	**G**	**F**	**E**	**D**	**C**	**B**	**A**
Service not supported					0	0		
Active					0	1		
Not active					1	0		
Spare					1	1		
Selective Forwarding	**H**	**G**	**F**	**E**	**D**	**C**	**B**	**A**
Service not supported							0	0
Active							0	1
Not active							1	0
Spare							1	1

These call forwarding parameters present the status of a line using call forwarding features. These are used when custom calling or

CLASS is being offered in a calling area. *Call forwarding variable* is used to forward a call immediately, before ringing the called party.

Call forwarding on busy will forward a call only if the called party is off-hook. *Call forwarding don't answer* forwards a call when the called party does not answer within the predetermined number of rings (defined by the subscriber). Selective forwarding allows only select calls to be forwarded, based upon criteria defined by the user (such as calling party number).

Originating Restrictions	**H**	**G**	**F**	**E**	**D**	**C**	**B**	**A**
Denied origination	0	0	0	0	0	0	0	0
Fully restricted origination	0	0	0	0	0	0	0	1
Semirestricted origination	0	0	0	0	0	0	1	0
Unrestricted origination	0	0	0	0	0	0	1	1

Restrictions are assigned to business groups (such as Centrex service) to define the type of outside calling a station is permitted to make. Denied origination indicates that calls are not allowed to be originated from the specified line. Business groups are used to define Centrex lines and other similar services.

Fully restricted origination allows a line to originate calls to lines within the business group, but not to the attendant (local business group operator, usually at the reception desk of a business) and not to lines outside of the business group.

Semirestricted origination allows a line to call outside the business group, but not by direct dial. The line can be forwarded, conferenced, or transferred via an attendant but cannot direct dial an outside line.

Unrestricted origination allows a line to call any number within the business group or outside the business group. This allows full unrestricted access to any outside line.

Terminating Restrictions	**H**	**G**	**F**	**E**	**D**	**C**	**B**	**A**
Denied termination	0	0	0	0	0	0	0	0
Fully restricted termination	0	0	0	0	0	0	0	1
Semirestricted termination	0	0	0	0	0	0	1	0
Unrestricted termination	0	0	0	0	0	0	1	1
Call rejection applies	0	0	0	0	0	1	0	0

Terminating restrictions are like the originating restrictions, but apply only to incoming calls. *Denied termination* prevents any calls from being terminated to the line. The line may be allowed to dial within the group, or may even be allowed to dial outside the group (depending on the originating restriction applied).

Fully restricted termination prevents all calls from outside the business group being terminated to this line. Calls cannot be transferred from any other line or the attendant or forwarded.

Semirestricted lines are allowed to receive calls from within the business group. Outside calls must be transferred or forwarded to the line. Lines outside the business group cannot direct dial this line.

Unrestricted termination allows full access to the line from within the business group and outside the business group. Call rejection allows a line to request rejection of an incoming call. This is communicated using electronic Centrex phones with displays. The display shows the calling party number (ANI), allowing the called party to determine whether to accept the call or reject the call.

Directory Number to Line Service Type Mapping

	H	G	F	E	D	C	B	A
Match Status	**H**	**G**	**F**	**E**	**D**	**C**	**B**	**A**
Spare	0	0						
No match	0	1						
Match	1	0						
Spare	1	1						
Line Service Type	**H**	**G**	**F**	**E**	**D**	**C**	**B**	**A**
Individual			0	0	0	0	0	0
Coin			0	0	0	0	0	1
Multiline hunt			0	0	0	0	1	0
PBX			0	0	0	0	1	1
Choke			0	0	0	1	0	0
Series completion			0	0	0	1	0	1
Unassigned DN			0	0	0	1	1	0
Multiparty			0	0	0	1	1	1
Nonspecific			0	0	1	0	0	0
Temporarily out of service			0	0	1	0	0	1

This identifies the type of line service for a given subscriber line. This information must be retrieved from a database line record for a given subscriber line. Numbers which are not in service or have been disconnected are also indicated with these codes.

Duration	**H G F E**	**D C B A**
	Hours	Hours
	Minutes	Minutes
	Seconds	Seconds

Duration is used for features such as automatic callback, where the called party must be monitored until it is free. The duration parameter allows a duration to be set for monitoring the called party. After the duration period, if the called party is still busy, the feature is restarted for another duration.

Bearer capability requested. This information is related to the type of bearer capability a subscriber is allowed. The information is stored in the line information database, and retrieved via TCAP when queried by an end office.

Octet 1

Extension indicator	**H**	**G**	**F**	**E**	**D**	**C**	**B**	**A**
Octet extended to next octet	1							
Octet not extended to next octet	0							
Coding standard	**H**	**G**	**F**	**E**	**D**	**C**	**B**	**A**
ITU-TS standardized		0	0					
Reserved for other international standards		0	1					
National standard		1	0					
Reserved		1	1					
Information transfer capability	**H**	**G**	**F**	**E**	**D**	**C**	**B**	**A**
Speech				0	0	0	0	0
Unrestricted digital information				0	1	0	0	0
Restricted digital information				0	1	0	0	1
3.1-kHz audio				1	0	0	0	0
7-kHz audio				1	1	0	0	1
15-kHz audio				1	0	0	1	0
Video				1	1	0	0	0

Octet 2

Extension indicator	**H**	**G**	**F**	**E**	**D**	**C**	**B**	**A**
Octet extended to next octet	1							
Octet not extended to next octet	0							
Transfer mode	**H**	**G**	**F**	**E**	**D**	**C**	**B**	**A**
Circuit mode		0	0					
Packet mode		1	0					
Information transfer rate	**H**	**G**	**F**	**E**	**D**	**C**	**B**	**A**
Channel size				0	0	0	0	0
64 kbps				1	0	0	0	0
384 kbps				1	0	0	1	1
1536 kbps				1	0	1	0	1
1920 kbps				1	0	1	1	1

The information transfer rate can be used to indicate the transfer rate in both directions, through the use of octet 2b. When octet 2b is not included, the bidirectional transfer rate is symmetrical with the rate indicated in octet 2. When octet 2b is included, then the transfer rate in the direction origination to destination is that of the rate indicated in octet 2b. This allows for setting different transfer rates in each direction or the same transfer rate in both directions.

Octet 2a

Extension indicator	**H**	**G**	**F**	**E**	**D**	**C**	**B**	**A**
Octet extended to next octet	1							
Octet not extended to next octet	0							

Structure	H	G	F	E	D	C	B	A
Default (see note for default values)		0	0	0				
8-kHz integrity		0	0	1				
Service data unit integrity		1	0	0				
Unstructured		1	1	1				

Note: The default values assigned (if field = 000, or octet 2a is omitted) are as follows:

Transfer mode	*Transfer capability*	*Structure*
Circuit	Speech	8-kHz integrity
Circuit	Unrestricted digital	8-kHz integrity
Circuit	Restricted digital	8-kHz integrity
Circuit	Audio	8-kHz integrity
Circuit	Video	8-kHz integrity
Packet	Unrestricted digital	Service data unit integrity

Configuration	H	G	F	E	D	C	B	A
Point-to-point					0	0		
Multipoint					1	0		

Establishment	H	G	F	E	D	C	B	A
Demand							0	0

Octet 2b

This octet can be omitted, unless it is desirable to indicate a different transfer rate in one direction other than what is specified in octet 2. When this octet is used, the transfer rate indicated is applicable to the direction origination to destination.

Extension indicator	H	G	F	E	D	C	B	A
Octet extended to next octet	1							
Octet not extended to next octet	0							

Symmetry	H	G	F	E	D	C	B	A
Bidirectional symmetric		0	0					
Bidirectional asymmetric		0	1					
Unidirectional (origination to destination)		1	0					
Unidirectional (destination to origination)		1	1					

Information transfer rate	H	G	F	E	D	C	B	A
Channel size				0	0	0	0	0
64 kbps				1	0	0	0	0
384 kbps				1	0	0	1	1
1536 kbps				1	0	1	0	1
1920 kbps				1	0	1	1	1

Octet 3

This is an optional octet which can be omitted or repeated. When there is a need to identify more than one protocol at higher layers, this field can be repeated to reflect each of the other protocols.

Extension indicator	H	G	F	E	D	C	B	A
Octet extended to next octet	1							
Octet not extended to next octet	0							

Multiplier or layer identification	H	G	F	E	D	C	B	A
Bearer capability multiplier		0	0					

User information layer 1 protocol		0	1					
User information layer 2 protocol		1	0					
User information layer 3 protocol		1	1					
User information layer 1 protocol identification	**H**	**G**	**F**	**E**	**D**	**C**	**B**	**A**
ITU-TS (CCITT) Rec. I.412		0	0	0			0	0
Rate Adaption (see note)				0	0	0	0	1
ITU-TS (CCITT) Rec. G.711 u-law		0	0	0			1	0
ITU-TS (CCITT) Rec. G.711 A-law		0	0	0			1	1
ITU-TS (CCITT) Rec. G.721 32 kbps ADPCM				0	0	1	0	0
ITU-TS (CCITT) Rec. G.722, G.725, 7-kHz audio		0	0	1			0	1
User information layer 2 protocol identification	**H**	**G**	**F**	**E**	**D**	**C**	**B**	**A**
Undefined				0	0	0	0	0
ITU-TS (CCITT) Rec. Q.921 (I.441)				0	0	0	1	0
ITU-TS (CCITT) Rec. Q.710				0	0	0	1	1
ITU-TS (CCITT) Rec. X.25 link level				0	0	1	1	0
User information layer 3 protocol identification	**H**	**G**	**F**	**E**	**D**	**C**	**B**	**A**
Undefined				0	0	0	0	0
ITU-TS (CCITT) Rec. Q.931 (I.451)				0	0	0	1	0
ITU-TS (CCITT) Rec. X.25 packet level				0	0	1	1	0

Note: When the multiplier or layer identification field indicates the user information layer 1–3, the user information protocol identification fields are indicated in the same octet. Each of these protocol identifiers is directly related to the layer specification in bits GF. For example, if bits GF indicate layer-1 protocol, then the bits EDCBA equal that of the appropriate layer-1 protocol as shown in the preceding table under *User information layer 1 protocol identification.*

Bearer Capability Supported	**H**	**G**	**F**	**E**	**D**	**C**	**B**	**A**
Not used	0	0	0	0	0	0	0	0
Bearer capability is supported	0	0	0	0	0	0	0	1
Bearer capability is not supported	0	0	0	0	0	0	1	0
Bearer capability not authorized	0	0	0	0	0	0	1	1
Bearer capability not presently available	0	0	0	0	0	1	0	0
Bearer capability not implemented	0	0	0	0	0	1	0	1

Reference ID. This field is used to identify the transaction between the database and the exchange during the assist service. The format consists of an identifier, a length indicator, and the variable contents field. The contents field carries the identification number (four octets) used by this service to correlate the transaction with the database access. The identification number is assigned by the database.

Business group identifier

Octet 2

Length of parameter

The length indicates the entire length of the parameter, not counting this field.

Octet 3

	H	G	F	E	D	C	B	A
Attendant status	**H**	**G**	**F**	**E**	**D**	**C**	**B**	**A**
No indication		0						
Attendant line		1						
Business group identifier	**H**	**G**	**F**	**E**	**D**	**C**	**B**	**A**
Multilocation business group			0					
Interworking private number			1					
Line privileges information indicator	**H**	**G**	**F**	**E**	**D**	**C**	**B**	**A**
Fixed line privileges				0				
Customer-defined line privileges				1				
Party selection	**H**	**G**	**F**	**E**	**D**	**C**	**B**	**A**
No indication					0	0	0	0
Calling party number					0	0	0	1
Called party number					0	0	1	0
Connected party number					0	0	1	1
Redirecting number					0	1	0	0
Original called number					0	1	0	1

Octet 4 and 5

Subgroup ID	**H**	**G**	**F**	**E**	**D**	**C**	**B**	**A**
No indication	0	0	0	0	0	0	0	0
Customer assigned subgroup codes	0	0	0	0	0	0	0	1
				to				
	1	1	1	1	1	1	1	1

Octet 6

This field is associated with the *line privileges* field in the preceding table. If the line privileges field indicates customer-defined line privileges, then this one-octet field is used to indicate those customer-defined line privilege codes. If the line privileges field indicates fixed line privileges, this field is divided into two subfields. The bits HGFE indicate the terminating restrictions, while the bits DCBA indicate the originating restrictions.

Line privileges	**H/D**	**G/C**	**F/B**	**E/A**
Unrestricted	0	0	0	0
Semirestricted	0	0	0	1
Fully restricted	0	0	1	0
Fully restricted intraswitch	0	0	1	1
Denied	0	1	0	0

Business groups are used to offer PBX-type services to business customers who do not wish to purchase a PBX. Probably the most common type of service offered under this category is Centrex. These parameters are used to identify what class of Centrex is being used with the specified line.

In addition to defining the business group, these parameters also allow the definition of the various members within a business group, and allow lines to be placed into subgroups. Subgroups may be located within the same area or may be in another calling area. When this is the case, this information must be shared with the remote offices.

Signaling networks identifier. The signaling network identifier is used to indicate to the receiver which networks the sender anticipates going through to reach the destination. Each network ID requires two octets. The signaling networks identifier is a variable parameter.

Generic Name

	H G F E	D C B A
Type of Name	H G F E	D C B A
Spare	0 0 0	
Calling name	0 0 1	
Original called name	0 1 0	
Redirecting name	0 1 1	
Connected name	1 0 0	
Spare	1 0 1	
	to	
	1 1 1	
Availability	H G F E	D C B A
Name available/unknown	0	
Name not available	1	
Presentation	H G F E	D C B A
Presentation allowed		0 0
Presentation restricted		0 1
Blocking toggle		1 0
No indication		1 1

Message Waiting Indicator Type

H G F E	D C B A
2nd digit	1st digit

The *message waiting indicator* is a two-octet parameter predefined by the customer. This parameter and its contents notify the customer about the type of message waiting in a telephone-company-provided voice-mail system. The message waiting type must be defined at service deployment.

Look Ahead For Busy Response

	H G F E	D C B A
Acknowledgment type	H G F E	D C B A
Path reservation denied	0 0	
Negative acknowledgment	0 1	
Positive acknowledgment	1 0	
Spare	1 1	
Location	H G F E	D C B A
User		0 0 0 0

	H G F E	D C B A
Private network serving the local user		0 0 0 1
Public network serving the local user		0 0 1 0
Transit network		0 0 1 1
Public network serving the remote user		0 1 0 0
Private network serving the remote user		0 1 0 1
Local interface controlled by this signaling link		0 1 1 0
International network		0 1 1 1
Network beyond interworking point		1 0 0 0
Acknowledgment type	H G F E	D C B A
Path reservation denied	0 0	
Negative acknowledgment	0 1	
Positive acknowledgment	1 0	
Spare	1 1	

Circuit identification code. The circuit identification code (CIC) identifies the trunk circuit used to send the voice or data to the called party. The CIC is also indicated in any ISUP messages, which are used to set up the connection between end offices. The CIC is a two-octet field.

Precedence. Used in military systems to determine the type of call precedence used. Call precedence allows calls in progress to be terminated and the trunk to be released so that those of a higher military rank may use the trunk for an outgoing call. This feature allows the military to maintain a low number of trunks at all military installations, while still allowing high-ranking officials access to lines when they need them. All lines within the military installation are coded, and all users are prompted to input an identification number, which identifies their rank and precedence level.

Octet 1

Precedence level	H G F E	D C B A
Flash override		0 0 0 0
Flash		0 0 0 1
Immediate		0 0 1 0
Priority		0 0 1 1
Routine		0 1 0 0
Service		

These fields are used to indicate the service code as assigned by the code administrator that applies to this call. The service codes are used to identify the type of service subscribed to in this particular network.

Call reference. Call references are only used with military installation calls. These codes allow tracking of calls independent of the circuit number. The call reference pertains to the identification input when callers dial their IDs and telephone numbers (called party). This allows tracking of both the line number and the individual code used to place the call.

This is a six-octet field, divided into three-octet sections. The first three octets identify the identification number assigned to this call. This identification number is used to identify a particular call, and is separate from the circuit identification code.

The next three octets are used to indicate the point code in which the identification code has been assigned. The identification code is only significant to the signaling point indicated in the last three octets.

Summary

As new services are defined, these parameters and operation codes will grow. The TCAP protocol itself is becoming an important aspect of the SS7 network and will become more important as the Intelligent Network grows.

New switches become more and more sophisticated, and offer new and improved services and features. These features will require the support of TCAP for remote activation. There is no doubt that TCAP will be an important part of the communications network well into the next decade.

The current traffic mix in SS7 networks today tends to be primarily ISUP messages with some TCAP traffic. This is quickly changing as TCAP becomes the predominant traffic generator and as the SS7 network expands and becomes more sophisticated.

Local Number Portability (LNP) has increased the amount of TCAP traffic across the SS7 network in giant proportions. As subscribers begin switching to competitive local access provders, the amount of TCAP traffic will grow exponentially.

As cellular services and Personal Communications Services become more and more popular, they will demand more and more of the network's resources. TCAP will prove to be the most valuable of all the SS7 protocols.

Chapter

9

Overview of ISUP

The ISDN User Part (ISUP) has been used in U.S. networks for many years now, as an alternative to the European equivalent, the Telephone User Part (TUP). In early implementations of SS7, TUP was found to be far too limited for the scope of North American networks and was modified to align with the future services of ISDN and many other network features still under development. Today, many of those features are under implementation, and the SS7 network is being utilized more and more. However, much of its potential is still untapped.

ISUP has been a good protocol for circuit-related messages, but is already under modification to support new broadband services soon to be offered by major telephone companies. The new broadband services being offered for tomorrow's networks will require a new version of ISUP called B-ISUP.

The ISDN User Part is used to set up and tear down all circuits used for data or voice calls in the Public Switched Telephone Network (PSTN). In addition to its usage in the PSTN, ISUP can also be found in wireless networks for establishing trunk connections between switching centers.

ISUP is not widely used throughout the world, in fact, the United States was the first to adopt ISUP for usage in its networks. The ITU is currently developing an international version of ISUP, which will be used in the international plane. Otherwise, other countries use ISUP's predecessor, the Telephone User Part.

The Telephone User Part does not offer the same services and capabilities as ISUP, which was designed with the Integrated Services Digital Network (ISDN) in mind, and is fully compatible with the signaling in ISDN. It was for this reason that ISUP quickly replaced TUP in U.S. networks.

In fact, the Telephone User Part is considered as almost obsolete for those wishing to offer more control over their circuits. TUP is good for physical circuit connections, but is not capable of handling virtual circuits, permanent in digital networks.

Another shortcoming of the Telephone User Part is its inability to support "bearer" circuits. In a digital network, there is both a physical circuit and logical circuits which are dependent on the amount of data being sent by the "user." This bearer traffic determines how many virtual circuits will be needed to accommodate the data. The ISDN User Part provides the mechanisms for supporting bearer traffic, but does not fully support broadband signaling, which uses a different scheme altogether.

Additional work is currently under way to accommodate the new broadband services to be offered by the telephone companies. Asynchronous Transfer Mode (ATM) and broadband ISDN (BISDN) are making their way into the PSTN, replacing the existing DS3 and DS1 facilities used for so many years between exchanges. With these new facilities will come new configuration parameters and choices to be made by the protocol.

The support of broadband ISDN (which will become the subscriber interface to the broadband network) through the SS7 network is accommodated by a new signaling protocol, broadband ISUP. There are many similarities between ISUP and B-ISUP; in fact, the same procedures and message types are used in both. The exception in B-ISUP lies in additional message types and changes in how circuits will be assigned to a call connection.

Another fundamental difference being introduced with broadband signaling is the advent of fully associated signaling, rather than quasi-associated signaling. Fully associated signaling is accomplished by using the same path as the voice circuit, as would be the case when a channel from a DS3 circuit is used for signaling, and the other channels are used for voice and data. Once ATM has been deployed in the telephone networks, SS7 will be sent through the ATM network along with the voice and data.

This will work just fine, and accomplishes the same task as quasi-associated signaling, which relies on signal transfer points (STPs) to relay the messages from the originating exchange to the destination exchange. ATM will not eliminate the need for SS7 networks, but it will change the protocols and add additional functions. Signaling ATM adaptation layer (SAAL) will eliminate MTP L2, for example, on ATM links.

The signal transfer point (STP) will not disappear, but its role may change somewhat. The STP is still needed as a gateway into networks, or even a gateway into certain regions within a network. The STP will continue to provide global title translation services, as well as database access. Additional features and functions will likely be placed on STPs to justify their existence.

There is no problem with sending all of the B-ISUP traffic through the ATM network, and leaving the SS7 network for database access and other control functions. In fact, as the Advanced Intelligent Network (AIN) is deployed and implemented in existing networks, the traffic mix within the SS7 network will become predominantly TCAP and SCCP anyway.

Broadband ISUP has been included in this chapter because of the similarities to normal ISUP. The new protocol is explained in less detail than normal ISUP, as the standards are still being defined.

ISUP Services

There are two types of ISUP services, basic and supplementary. Basic service provides the support for establishing connections for circuits within the network. These circuits can be audio circuits for voice transmission or data circuits for any digital information, voice or data. Supplementary services are all other circuit-related services, which typically encompass message transport after a call path is established.

In addition to the two types of services, ISUP uses two methods for end-to-end signaling. End-to-end signaling is the process of sending circuit-related information from one exchange to a distant exchange. These two exchanges may be adjacent to each other or across LATAs.

The method currently used for passing signaling information to the distant exchange is called the "pass-along" method. With the pass-along method, the signaling information moves from one exchange to the next. All subsequent information related to the same circuit is then passed using the same path that was used to send the initial call setup information. This of course means that information must follow the same ISUP hops as the setup messages, which is not the most efficient method of routing.

The alternative method is called the "SCCP" method, and uses the services of the SCCP protocol to route the message through the network. When using the SCCP protocol, the information does not have to follow the same path as the call setup information. In fact, it can follow any path, provided the final destination is the same.

The SCCP method uses true network routing, and is probably more favorable for services that require information to be shared between exchanges when a call is in progress. However, today this method is not used.

The ISUP message provides important data regarding the service being requested of the remote exchange. These services are related to the circuit specified in the Initial Address Message (IAM), which is the initial setup message used in this protocol. The receiver of an IAM then must determine if it has the resources necessary to provide the type of service being requested.

The IAM provides the distant exchange with the calling and called party numbers, as well as information regarding the availability of SS7 signaling, whether or not the ISUP protocol is required end-to-end, and the type of network signaling available (if SS7 is not used throughout the network). The IAM also indicates whether or not further information will be available using subsequent messages.

The ISUP protocol uses the services of the Message Transfer Part (MTP) to send signaling messages from one signaling point to another. The ANSI standard and the ITU standards do allow for usage of SCCP services as well, although there are currently no applications in U.S. networks for this. The concept of using SCCP with ISUP is to allow end-to-end signaling without having to send messages to each intermediate exchange.

An example of how ISUP messages travel from one exchange to another is found in the diagram in Fig. 9-1. This diagram shows that an ISUP setup message or IAM is used to connect both ends of the voice trunk between the originating exchange and the next exchange, or tandem exchange. Once the connection is established, another connection must be set up between exchange B and exchange C by sending another ISUP setup message (IAM) from exchange B to exchange C.

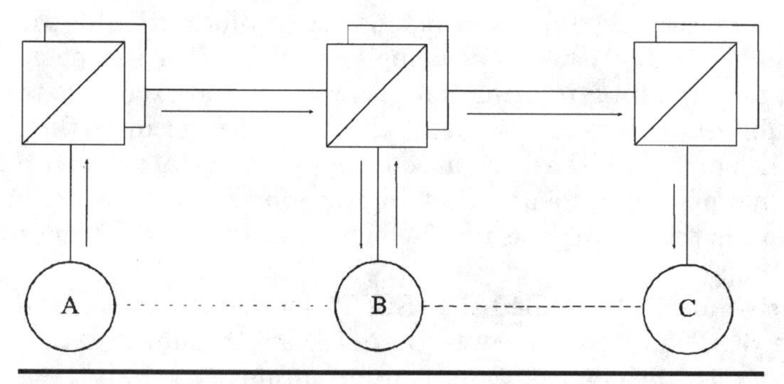

Figure 9.1 This figure depicts the path of ISUP setup messages for a typical telephone call. The first setup takes place from A to B. Exchange B must then initiate a circuit connection toward exchange C.

This continues through the network until there are voice circuits connected from end to end. End-to-end signaling using ISUP therefore requires many ISUP hops. If any message needs to be sent from one exchange to another exchange during the duration of the call, the same path is used.

The reason for this is due to the limited routing capabilities of the MTP. The MTP routing function only has knowledge of the adjacent signaling points, and only provides node-to-node routing. The SCCP routing function knows the final destination of the message, and is capable of providing routing based on the final destination, or endpoint, without having to know all the intermediate signaling points.

This would allow ISUP messages to travel through the network with a minimal number of hops and still be able to maintain association with a call in progress, without using the same path of the call setup messages. At this time it is unknown if this feature will ever be implemented, although it is very likely that the Intelligent Network will find this a useful feature.

Another fundamental change in signaling with ISUP is the handling of the service tones used by the local exchange. Before ISUP and SS7, when a local call was placed to another exchange the service tones (busy, ringback, etc.) were set by the distant exchange through the voice circuit to the calling party. With SS7, this is no longer necessary. In fact, the voice circuit does not need to get connected until the called party answers.

The service tones can be sent by the originating exchange. This is accomplished by the distant exchange sending ISUP messages indicating the status of the call (status information is implied within many ISUP messages, such as Address Complete). For example, when the distant exchange receives an IAM, it will send an Address Complete Message (ACM) in return. The ACM is used as an acknowledgment, and it also implies that ringing is being sent to the called party.

In most networks the service tones are sent by the destination exchange. The trunk circuits that have been reserved along the call path are not yet cut-through in both directions, but they are cut-through in one direction; from the destination exchange back to the originating exchange. This allows service tones to be sent in the backward direction to the originating exchange. If there is no answer, or the call is disconnected for any other reason, the trunks can be quickly released in the one direction. (See Fig. 9.2.)

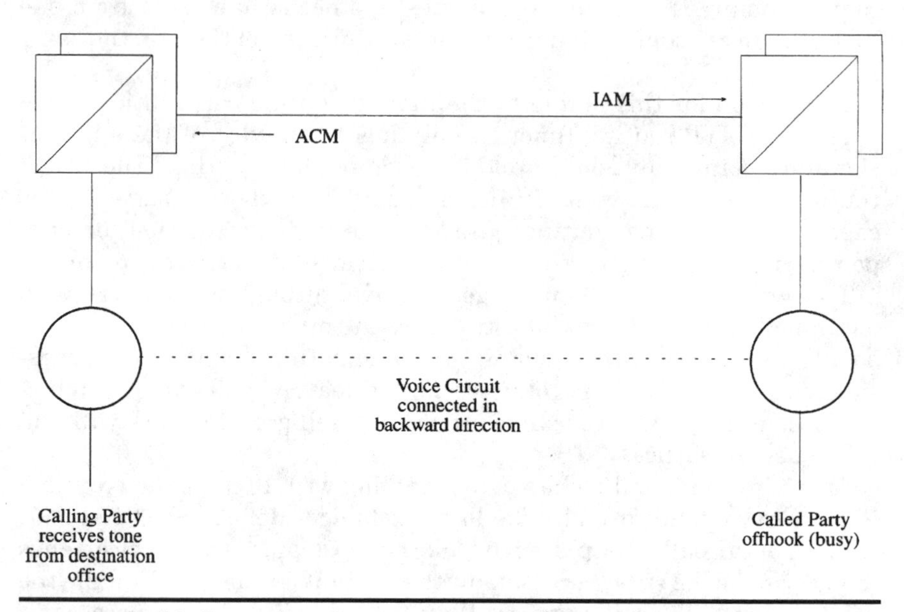

Figure 9.2 In this figure, the voice circuit is reserved but not cut-through. Service tones are sent by the local office instead of by the remote exchange.

If the calling party or the called party is using an ISDN interface, then the call setup information and call status information can be much more complex than with service tones. In an ISDN call, the ISUP protocol is used to carry setup information as well as call status information through the PSTN and to the exchange. This is especially advantageous when the called or calling party is terminated at a Public Branch Exchange (PBX).

When a PBX is used, the PBX can share status and setup information with the distant PBX. This is a feature never before possible with conventional signaling, because conventional signaling was all analog and unable to support such a broad base of information.

Even information about the station users within a PBX, such as class of service information and dialing privileges, can be shared through the PSTN to the distant PBX through the use of the ISUP protocol. Large corporations with multiple PBXs can enjoy the benefits of a large network without leasing private lines between PBXs. (See Fig. 9.3.)

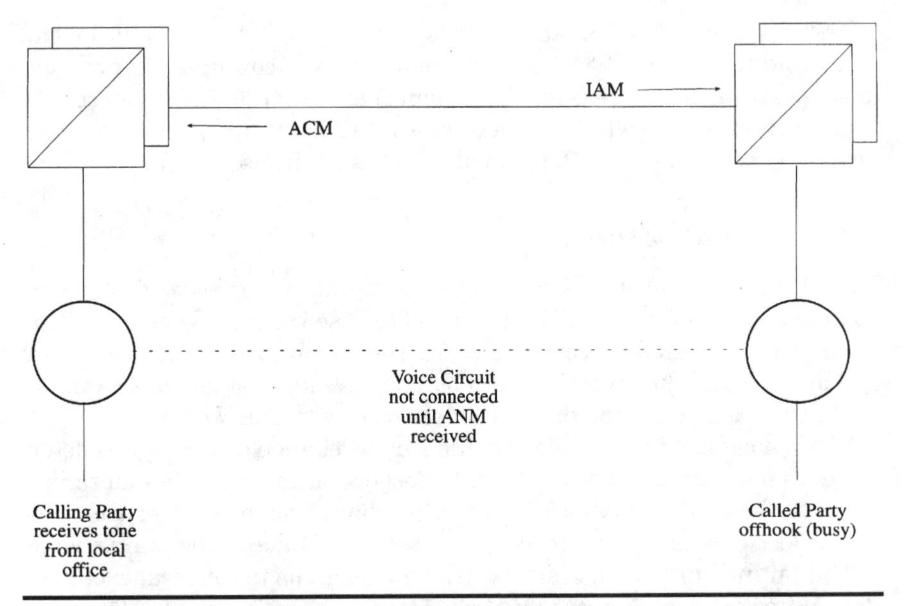

Figure 9.3 The ANM message is used to trigger cut-through on the voice trunk when the called party is within the United States. If an international call, the trunk is cut-through upon receipt of the ACM message.

The voice circuit in an ISUP call setup is not actually cut-through (connected) in both directions until the called party goes off-hook and answers the call. The voice circuit does get reserved for the call, and can even be tested before the setup begins. But the connection is not made in both directions until the called party answers the call.

The most commonly asked question, then, is how does SS7 save time in call setup. The answer lies in the speed of digital call setup and tear-down, versus analog signaling.

Digital messages travel at high speeds, and allow calls to be setup much quicker than analog signaling. If you don't believe this concept, pick up a phone, and dial a number, then time how long it takes for the call to connect. Now compare the setup time with a call five years ago, before SS7 was deployed throughout the network. Many times, ring-back is heard as soon as the final digit is dialed.

When a called party is not available because of a busy condition, the reserved voice circuit is released and is immediately available for another call. In the meantime, a digital message is sent back to the originating exchange notifying the originating exchange of the busy condition. The originating exchange then sends a busy tone to the calling party.

While the originating exchange is sending a busy tone to the calling party, the voice circuit has been released (between the exchanges), and only the circuit between the calling party and the local exchange is maintained.

There are many other advantages of using ISUP and TUP for call setup and teardown. If one understands the basic concepts of how a call is set up and then released, the advantages become much clearer. In the next section, we will talk about how a call is set up in greater detail, providing examples of different call setup situations.

Call Setup and Teardown

To understand how ISUP works, and its advantages, we need to first understand the basics of how a call is set up and released in the SS7 network. The ISUP protocol is used to accomplish this. These examples will assume there are analog lines being used between both parties and the telephone company. We will later examine the procedures used with ISDN circuits.

When a caller lifts the receiver, the local exchange (exchange A) determines that the caller is off-hook by the presence of current on the subscriber line interface (DC signaling). The local exchange acknowledges the presence of loop current by sending a service tone (dial tone) to the calling party.

The calling party then dials digits, which signals to the local exchange the address (telephone number) of the distant called party. The local exchange must wait until all digits have been dialed, then examine the first three digits dialed to determine if the calling party dialed an area code or a prefix (determined by the North American Numbering Plan). (See Fig. 9.4.)

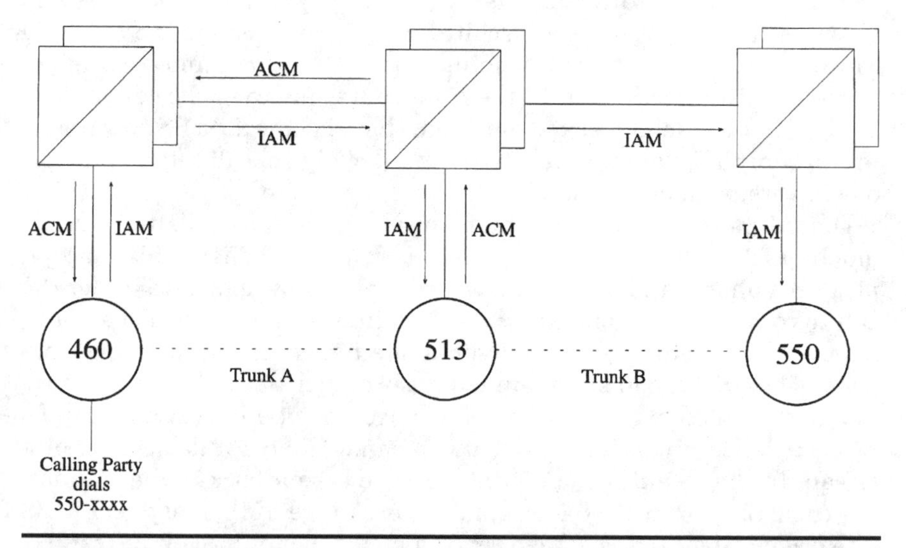

Figure 9.4 When the calling party completes dialing digits, the local exchange determines which trunk needs to be reserved for the call. An Initial Address Message (IAM) is then generated toward the first exchange. When the Address Complete Message (ACM) has been received, the local exchange begins delivering ring-back tone, even before the called party's telephone begins ringing, in some cases.

If the exchange determines that the number dialed was a long-distance number, the call is routed to a long-distance carrier through a Point-of-Presence (POP) in the Local Access Transport Area (LATA) of the calling party. The prefix and the subscriber number (last four digits of the telephone number) are then routed to the distant exchange.

The local exchange determines how it will connect this call based on information in its trunk routing tables. These routing tables identify which voice circuits to use to establish an end-to-end circuit with the least number of hops. When it determines which voice circuits to use, a call setup message is created and sent to the exchange which will provide the first voice connection (exchange B).

This exchange may not be the final destination for this call. In fact, it may be a tandem, used as an intermediate switch to reach the final destination. Intermediate tandem switches are used to prevent all exchanges having to have voice circuits to all other exchanges within any given LATA.

The call setup is sent using the ISUP protocol through the SS7 network. The Signal Transfer Point (STP) serves as a network router for these messages, simply routing SS7 messages to their proper destination, and does not play any real significant role in setting up the voice circuits. In general, the STP has no real knowledge of the ISUP message; it only delivers it to the proper exchange.

The Initial Address Message is created by the local exchange (exchange A) and sent to the intermediate tandem exchange (exchange B). In this IAM can be found all the information necessary for the tandem exchange to establish a connection. The IAM message does not contain information for the final destination, as it is not establishing a connection from originating exchange A to destination exchange C.

The tandem exchange will acknowledge receipt of the IAM by sending an Address Complete Message to the originating exchange A when it has received an ACM from the destination exchange. This indicates that the circuit designated as reserved in the IAM has been reserved at exchange B, and a connection can be made when ready.

The tandem exchange can begin setting up the next circuit between itself and the destination exchange C. This is accomplished by generating another IAM, including the called and calling party address information provided by the originating exchange A, and sending the IAM to the destination exchange C.

The IAM will also specify the signaling method to be used for this call. If the IAM specifies that the ISUP protocol is to be used end-to-end (required), then the call must be set up using the ISUP protocol. If the tandem exchange (B) cannot set up the call using ISUP (in the event the exchange does not support ISUP to the destination, or there are no facilities available which use ISUP), the call is rejected, and a message indicating the reason for rejection is sent back to the originator.

If the IAM indicates that ISUP is preferred, the receiving exchange will check for available resources to determine if the call can be set up using ISUP. If not, the call is still set up, but using a different method, such as MF signaling or Telephone User Part (TUP) protocol.

The IAM can also indicate that ISUP is not required "all the way," in which case the call is set up using whatever method is available. These methods could be ISUP, TUP, or a non-SS7 signaling method such as MF (discussed in Chapter 1).

An intermediate exchange (such as the tandem in our example) can change some of the information in the IAM. The first six digits of the called party number can be modified, information regarding the connection (nature of connection), and the end-to-end method indicator can be modified. All other fields are passed transparently to the distant exchange.

As is sometimes the case, C does not have to be the final destination. There can be more exchanges or tandem exchanges required establishing this end-to-end voice circuit, but for simplicity we will only discuss three connection points.

Upon receipt of the IAM, the distant exchange must examine the message to determine if there will be any further information in subsequent messages. If not, the called party number is examined and the exchange determines if the called party is available or busy. If the called party is busy, a release message (REL) is sent to the originator and the circuit is immediately released for another call.

The distant exchange may find that the called party number is not included in the IAM. When this occurs, the distant exchange (exchange C) must request the called party number using the protocol services specified in the IAM. Two methods can be used, end-to-end (SCCP services) or link-to-link (pass-through using MTP). Currently, only the pass-through method is used in ANSI networks.

If the called party is not busy and the call can be accepted, an Address Complete Message is sent to exchange B. Exchange B then sends an ACM to the originating exchange. The distant exchange (exchange C) then signals the called party that there is a call by sending ringing to the called party on its subscriber line (DC signaling).

No message is returned until the called party answers the phone. When exchange C determines the called party has lifted their receiver by detecting loop current (DC signaling) on the subscriber interface, an Answer Message (ANM) is generated and returned to the tandem exchange (exchange B). The same path used to send the IAM from exchange B to exchange C is used for the ANM. This means the same links and the same STPs are used for all associated ISUP messages. The voice circuit between exchange C and exchange B is immediately cut-through when exchange B receives the ANM.

When the tandem exchange receives the ANM, it sends an ANM to the originating exchange A using the same path the IAM was received on. The originating exchange can then begin cut-through on the voice circuit between itself (exchange A) and the tandem exchange (exchange C).

Once the voice circuit is connected, conversation can begin, and no messages are necessary through the SS7 network until either party goes on-hook. There are some features associated with CLASS and some other calling features that may require exchanges to share information with each other during the duration of the call. Presently, the communications is handled using the same path as the setup messages. However, the standards do allow the use of SCCP to carry such information from one end to the distant end without following the call setup path. (See Fig. 9.5.)

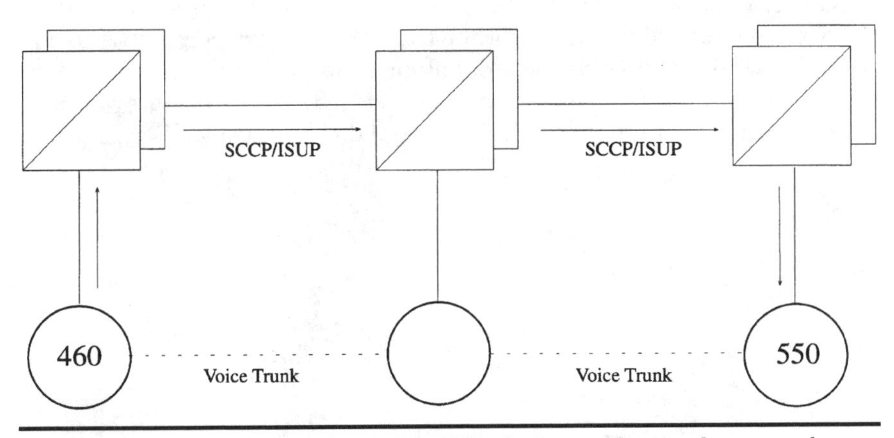

Figure 9.5 If SCCP services are used for ISUP, the intermediate exchange can be bypassed and ISUP messages sent directly to the remote exchange. This applies only to messages between two exchanges after a connection is established and allows the switches at both ends of a call to exchange status information.

The above discussion applies to normal routing procedures. Local Number Portability (LNP) has changed the way calls are routed. The ISUP protocol has been modified to provide additional information needed to route calls to numbers that have been "ported." Numbers are considered ported when the subscriber changes from one local service provider to another, keeping their telephone number. In the past, you had to give up your telephone number and obtain a new telephone number from the new service provider.

With LNP, calls are no longer routed based on the digits dialed. Each call made to an NPA-NXX that has a number ported in it requires a database transaction to obtain additional routing information. The database will identify whether or not the dialed number has been ported, and if so where to route the call.

For routing purposes, each end office (and tandem) is assigned a 10-digit location routing number (LRN). The LRN is usually the same NPA-NXX currently assigned to the end office switch, with all zeroes for the last four digits. For example, the end office serving (919) 460-xxxx will be assigned the LRN of (919) 460-0000.

The called party number field in the ISUP IAM will now contain the LRN if the dialed number has been ported to another carrier's network. The dialed digits will be placed in the generic access parameter (GAP). Before a call can be routed, the end office must first access the LNP database within its region, and obtain the LRN for the dialed number. For more information about LNP, see Chapter 10.

If the digits dialed are for an international number, the call setup sequence is the same. The only difference is the usage of the international network to route the call. In the international plane, there is no knowledge of the dialed digits other than the country code and the city code. The STPs at the international plane route messages based on their international country code and the city code is used by the gateway STP to determine how to route the call within its own network. (See Fig. 9.6.)

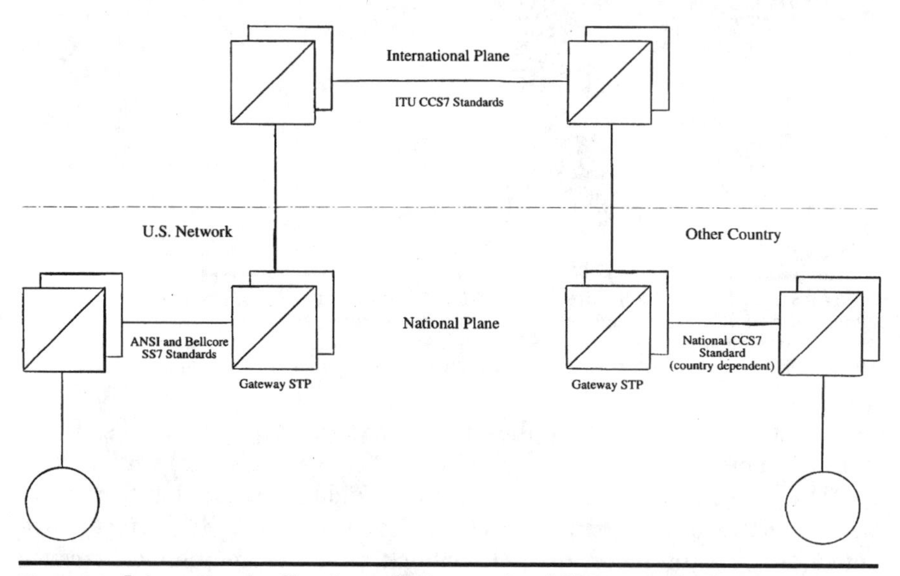

Figure 9.6 International calls rely on the international SS7 network. This network uses ITU-TSS SS7 standards, which are somewhat different from the national standards. The national standard used in the United States is ANSI, with Bellcore standards used within the Bell Operating Company networks.

The addressing of SS7 entities is virtually the same format. The primary differences between national and international addressing are the point code structure of each SS7 signaling point. The ITU-TS standard calls for a 14-bit point code structure; divided as 3-bit zone identification, 8-bit area/network identification, and 3-bit signaling point identification.

National point codes follow a simple 14-bit point code, and are usually found to have no divisions. The exception to this rule is in the case of the ANSI point code structure, which uses a 24-bit point code, with a network identification, cluster identification, and member identification (refer to the chapter on level three MTP).

The call setup messages must be routed according to the dialed digits (or the LRN as in the case of LNP), which must then be translated at some point to a point code. The point code translation within the national plane is usually that of a gateway STP, which then translates the point code into an international point code (14 bit) and routes the message to the proper gateway STP according to the country code dialed.

When a gateway is used to route a call setup message into another network, the exit message is used to indicate that a call connection has been completed in the other network. The exit message may include the outgoing trunk group number used to connect the voice circuit to the other exchange, although this information is optional.

The voice circuit in the case of an international connection can be connected as soon as the ACM is received. This depends on the network and may or may not be true. In theory, the trunk does not have to be cut-through until after the called party answers the call, but this does not imply actual practice.

When taking a call down and releasing all circuits associated with that call, several steps must take place. When either caller goes on-hook, the subscriber line is released immediately (DC signaling). The exchange then sends a suspend message (SUS) to its tandem, or whichever exchange it is directly adjacent to (exchange B in Fig. 9.7).

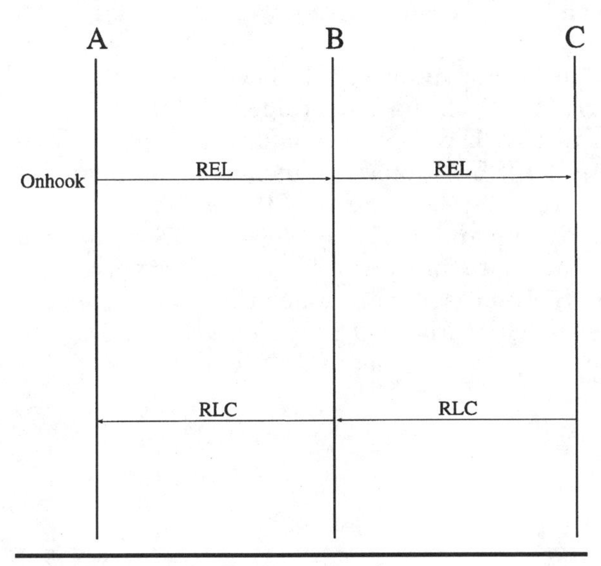

Figure 9.7 Normal call release procedure.

When the adjacent exchange (exchange B) receives the SUS, it sends a suspend message to its adjacent exchange. This continues through the network until the SUS message has reached the originating exchange. As soon as the calling party returns to an on-hook condition, the release (REL) message is sent toward the distant exchange.

When an exchange receives a release (REL) message, it returns a release complete (RLC) message as an acknowledgment. The release complete indicates that the circuit has been returned to an idle condition. The tandem exchange in the diagram must then generate a release message to its adjacent exchange (exchange A in Fig. 9.7), and follow the same procedure to release its circuit. Exchange A in this example would then generate an RLC. Upon receipt of the RLC, exchange B would then release its circuit.

While all this is going on, the circuit between exchange B and exchange C has been released and could be set up for another connection associated with a different call. With conventional analog signaling, this would not be possible; the circuit would remain seized until the distant exchange was ready to release. This often would result in "hung" circuits, especially when the calling party did not hang up their phone.

Call Setup and Teardown of ISDN Circuits

The ISUP protocol was intended for use with digital subscriber interfaces such as ISDN. The intent was to provide a protocol with more versatility, allowing the exchange of status information and other forms of circuit data to be exchanged from local ISDN network to distant ISDN network.

Where distance and other factors prevent ISDN networks from being connected to one another, SS7 can be used to bridge the gap. The ISDN protocol is transparent to SS7, because there is direct mapping from the ISDN parameters to the ISUP protocol. This makes interworking very easy, while still maintaining the security of the network.

The procedures for setting up and tearing down an ISDN connection are somewhat simpler than those for conventional signaling or analog subscriber circuits. The ISDN messages are compatible, but somewhat different than those used in ISUP. Figure 9.8 illustrates how ISDN and ISUP protocol messages map to one another, showing a typical call setup and teardown.

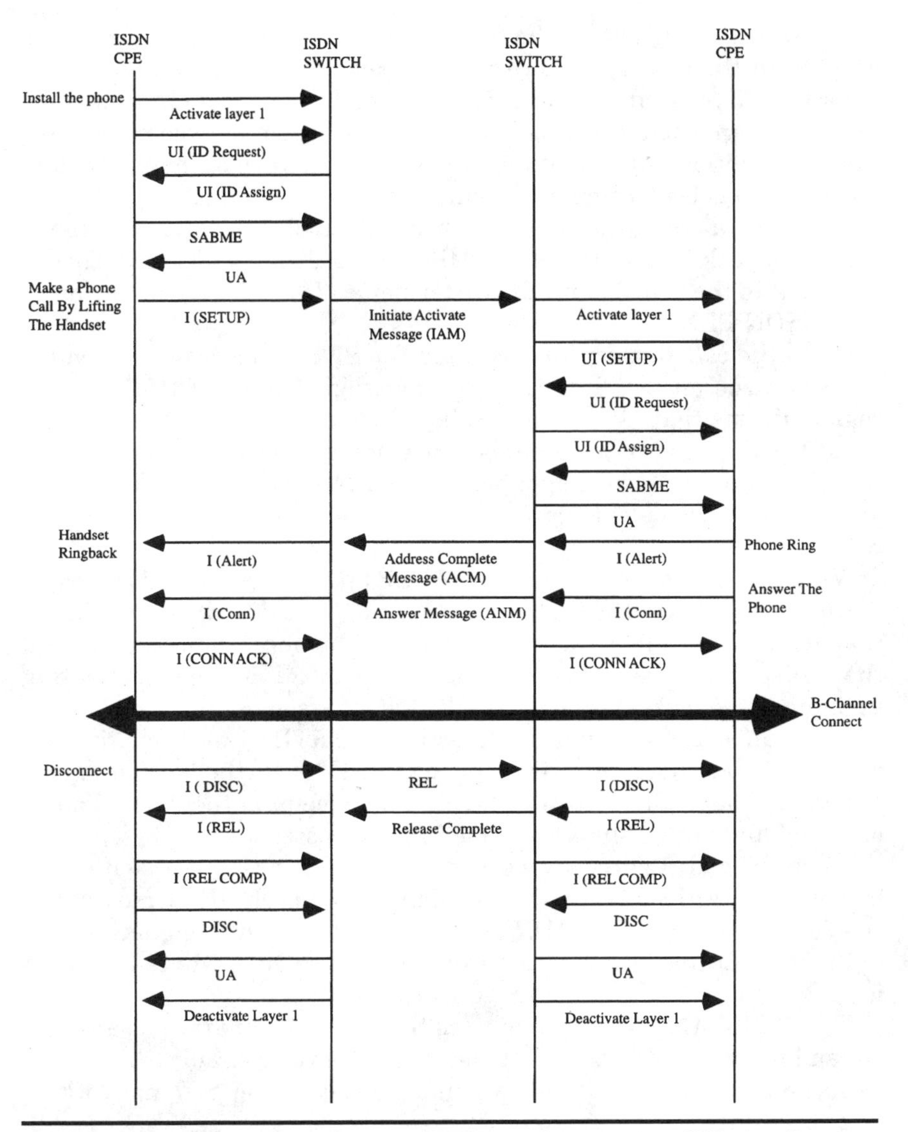

Figure 9.8 This figure illustrates how ISDN signaling maps to SS7 signaling. Backward and forward setup messages are used to exchange additional information regarding a call between two exchanges.

In the ISDN circuit, a SETUP message is sent from the originating (calling) party to the local exchange. The local exchange telephone switch is capable of interpreting ISDN messages at the line circuit level, and converts this into an SS7 IAM message.

The IAM contains the same information as the SETUP message. In fact, there is direct mapping from the ISDN protocol to the SS7 protocol, making them compatible interworking protocols.

The IAM is then sent in the forward direction to the next exchange. As seen in the drawing, there may be several exchanges involved in order for the voice circuit to be connected end-to-end. The IAM must be sent between each exchange and acknowledged between each exchange. Additional information may be sent between exchanges before the circuit has been completed using end-to-end signaling.

When the distant exchange has been reached, the IAM message is converted back into an ISDN SETUP message. The SETUP message is then sent to the called party device (either an ISDN Network Terminal or an ISDN PBX).

When the called party has accepted the SETUP message, the called party returns an ISDN ALERTING message. This ALERTING message indicates that all the addressing signals have been received, and the ISDN terminal (telephone) is now ringing. No ringing generator is sent from the local exchange, because ISDN is digital. The ringing is generated at the telephone device itself, based on the receipt of the SETUP message.

When the exchange receives the ALERTING message, it is converted into the SS7 Address Complete Message (ACM), which indicates that the SS7 exchange has received all of the addressing signals and the called party is being signaled. The ACM at the originating exchange is converted into an ALERTING message.

When an ISDN terminal receives an ALERTING message, it indicates that the called party is being signaled. No further action is necessary, although charging may begin after receipt of the ACM. This is optional, and not widely acceptable in most cases.

When the called party answers, a CONNECT message is sent in the backward direction. The SS7 equivalent to the CONNECT message is the Answer Message (ANM). The ANM is passed through each of the exchanges until the originating exchange is reached, where the ANM is converted back into a CONNECT message.

The CONNECT message indicates that the called party has answered, and it usually triggers cut-through on the voice circuit (this is network dependent). The ANM is usually used in the SS7 network to trigger billing.

When the call has been completed, either party may terminate the call by hanging up. When either party hangs up, the ISDN network creates a DISCONNECT message. The local exchange receives this DISC message and converts it to a Release (REL) message. The REL message is sent to the next exchange, which acknowledges the REL message with a Release Complete (RLC). The RLC triggers the actual release of the voice circuit.

Note that this continues on through the network between each of the exchanges until all of the circuits have been released. As soon as a circuit has been released, it is immediately available for another call. It is quite possible that the local loop can be released and already connected to another call before all the circuits from a previous call have been released end to end, although timers are usually used to prevent this from occurring and creating troubles.

Call Setup and Teardown of Broadband ISDN Circuits

Broadband ISDN brings on many challenges for the ISUP protocol. There are different addressing schemes used in BISDN, with support of virtual paths as well as virtual connections within a virtual path. The ISUP protocol does not support this addressing, and has been modified as broadband ISUP.

The standards for B-ISUP are still evolving, although the first Bellcore publications have been released and included in the GR-246-CORE. In this publication, several new sections have been added describing the services and procedures for B-ISUP.

The ANSI standards for BISDN signaling are the same as those formulated by the ITU-TS, which is good news for all those in international networks. This means simpler provisioning and implementation. This section will define the message types and procedures used to set up and tear down broadband circuits.

Before a circuit may be assigned for a call, it must be determined which signaling point will be responsible for the assignment of bandwidth and virtual path/virtual channels for a given circuit. One of the inherent problems in broadband is the possibility for "glare," or dual seizure. This occurs when two signaling points assign calls to the same virtual path/virtual connection combination. For this reason, end nodes will share the responsibility of assigning these characteristics 50/50.

All odd-numbered Virtual Path Connection Identifiers (VPCIs) are the responsibility of one signaling point, while even-numbered VPCIs are the responsibility of the other signaling point. This prevents the possibility of dual seizure, or glare. The exchange with the highest point code will be responsible for even-numbered VPCIs.

In the event there are no available VPCIs controlled by an exchange, a setup may be issued for a VPCI not controlled by the exchange. In this event, the virtual path and virtual connection ID are not provided (these parameters are found in the connection element identifier parameter, which is normally included in the setup message).

If there is an incoming call on a circuit controlled by the receiving exchange, the connection element identifier is not included. It is a request for service, and the controlling exchange must assign the virtual path and virtual connection. If this information has not been provided, then it must be assumed that the originating exchange does not have any available virtual paths/virtual connections within its control, and needs assignment from the other half.

When assigning a circuit, the exchange must determine what bandwidth is necessary for the call. This is determined by reading the parameters in the setup message. Once it is determined, the exchange assigns the appropriate virtual path connection within its control.

As mentioned above, each exchange is responsible for the assignment of virtual path connections and bandwidth for one-half of the available virtual path connections. If there is not enough bandwidth available at an originating exchange, it sends a request to the remote exchange (for which it wishes to establish a connection) without any virtual path connection information.

The receiving exchange then determines if there is ample bandwidth available for the connection and if so, provides the virtual path connection information to the requester. If there is not enough bandwidth available, then the request is denied using the REL message, with the cause "cell rate not available."

Once an exchange has received all of the information necessary to establish a connection, the receiving exchange must determine if the call is to be routed to another exchange (intermediate exchange) or if the call is to be terminated within itself. If the call is to be routed through another exchange, then normal ISUP routing is duplicated.

Normal ISUP procedures call for the same setup procedures used to establish a connection to the first exchange which must be repeated for all subsequent exchanges and used to reach the final destination. This is also true in the case of broadband ISDN.

Routing information may be stored within the exchange itself or in a central database accessible by all exchanges. The latter is becoming true in cellular networks, where the Home Location Register (HLR) used to store location information as well as subscriber information is moving to a central location, accessible by all other cellular providers.

The Transaction Capabilities Application Part (TCAP) is then used to access the centralized database for additional routing instructions, before call set-up procedures can begin. This allows for fewer resources within the exchange, and optimizes the individual service providers' network.

There are several parameters used to determine the best route for a connection. The called party address is the most useful, but the broadband bearer capability and the Asynchronous Transfer Mode (ATM) cell rate must also be considered. The ATM cell rate determines which interoffice facility will need to be used to get the connection through the Public Switched Telephone Network (PSTN).

So far, we have only discussed procedures for assigning virtual path connections between two exchanges. The message structures for these procedures are virtually the same as for normal ISDN signaling. The exception is the addition of new parameters that specifically support BISDN and some new message types that augment existing message types.

When service is requested, the Initial Address Message (IAM) is sent to the remote exchange. This IAM is almost identical to the one we talked about in the previous sections, but has additional parameters supporting broadband ISDN and ATM (see the section on message structure for broadband ISUP parameters).

Preceding the B-ISUP IAM parameter is the routing label, which may or may not specify the actual ATM circuit information. According to the rules discussed above about controlling exchanges, the IAM may act as a request. The receiving exchange may have to provide the circuit identification.

If this is the case, then an IAM acknowledgment will be sent to the originating exchange providing the circuit identification. This is a new message type for ISUP, and is used only in B-ISUP protocol.

Once the circuit identification has been provided, an address complete can be returned by the originating exchange, which serves as an acknowledgment that the addressing information has been received.

Other than additional or different parameters in the various message types, the sequence of messages is almost identical to normal ISUP. The major changes have to do with the assignment of the actual circuit, which is split between the two exchanges. Also keep in mind that these are logical connections, not physical connections as in Plain Old Telephone Service (POTS). This complicates the procedures somewhat, and expands the message sizes exponentially.

Message types and parameters for broadband ISDN support are found at the end of this chapter, and are labeled as such to differentiate them from normal ISUP messages and parameters. These are still evolving, and what is published in this book represents the first version as published by ANSI and Bellcore.

Interworking with Non-SS7 Networks

While most all of the Public Switched Telephone Networks (PSTN) now use SS7 throughout the network, there are still some segments that are using conventional signaling. Conventional signaling in most of these cases consists of MF signaling.

SS7 must be able to function and interwork with networks using conventional signaling. One of the primary issues is the reserving of a voice circuit. Unlike conventional signaling, the voice circuit does not get connected until the distant party answers, or at least until both exchanges have sent and received all the addressing information required to connect the call.

In MF signaling, the circuit is connected when the calling party completes dialing the digits, and is used to signal the distant exchange.

Therefore, some method must be established for allowing voice circuits to be "reserved" and tested in conjunction with the SS7 protocol.

To accomplish this, a circuit reservation procedure is used. Prior to the sending of an Initial Address Message (IAM), which is used to send addressing information necessary to establish the circuit connection, a Circuit Reservation Message (CRM) is sent to the non-SS7 exchange. This is then converted to MF signaling by that exchange, and from that point on MF signaling may be used. The exchange acknowledges receipt of the circuit reservation message by sending in the backward direction (back to the originating exchange) a Circuit Reservation Acknowledgment (CRA).

The originating exchange can then specify a continuity test to be invoked by the exchange using MF signaling by sending a Continuity Check Request (CCR) message in the forward direction. This message invokes a loopback test at the remote exchange on the voice circuit. The results of the continuity test are then returned to the originating exchange using the Continuity Test (COT) message.

After receipt of the COT message, if the continuity test proved successful, an IAM message is sent by the originating exchange to begin the normal call setup procedure. The voice circuit is reserved even though MF signaling is used, and is ready for cut-through end-to-end when the calling party answers or the address information has been successfully passed from the originating exchange to the destination exchange, depending upon network deployment.

Interworking with MF networks is no longer efficient with Local Number Portability (LNP) becoming commonplace nationwide (and in the future, worldwide). When the MF network is reached, any contents of the ISUP IAM message are lost, with the exception of the dialed digits. The location routing number (LRN) which is found in the called party address is lost, and the dialed digits found in the GAP parameter are sent via MF to the distant or adjacent network.

If the adjacent network is equipped with SS7, it must then perform a database query to regain the lost routing information. This of course is redundant, and adds unnecessary delay to the call setup. It also places additional burden on the network resources. This can be of major concern when you consider the number of database queries the adjacent SS7 network will be responsible for making.

Consider this scenario; the adjacent SS7 network will be receiving scores of calls from the MF network. The SS7 network will have to perform database queries for each of those calls, even though other networks have already done this. The burden lies on the SS7 network to provide this service, when the originating network should have been the responsible network.

Circuit Testing

The SS7 network provides several mechanisms for testing circuits and switches remotely. This testing is usually performed from the Operations Support Systems (OSSs) located regionally within the network. The tests can also be performed locally by testboard technicians or a maintenance center technician.

The OSS is an operations and maintenance center that allows complete network monitoring and testing. These are fairly new (within the last 10 years) and have been deployed within regional areas of the network.

Before SS7 and automation in the network, testing was conducted at every exchange, using testboard positions. These test positions were capable of connecting to every circuit entering and leaving the central office, and allowed technicians to test the continuity, capacitance, and other properties of the circuit.

Today, these testboard positions have disappeared, and all the testing has been moved to remote locations where many exchanges can be tested by one maintenance center. The SS7 network is used for passing those maintenance messages and test messages through the network to the remote switches.

The ISUP protocol also provides a means for testing circuits as well as translations in various nodes. A translation is a routing instruction, which translates dialed digits into a routable address, such as a signaling point code. This section describes two of those tests, the continuity test, and the circuit validation test.

Continuity testing

Because SS7 uses a separate facility than the voice circuit for sending information to another exchange, there is no knowledge of the operational status of the voice circuit. Most voice switching systems today provide some level of circuit testing and fault isolation, providing alarms when a circuit fails. This alarm information allows diagnostics software to "busy out" the trunk, preventing calls from being routed to the failed circuit.

In many networks, however, digital circuits are used to carry the voice and data. These digital facilities are usually DS1 or DS3 facilities, which require a series of multiplexers. For example, connecting to the voice switch is a DS0 circuit (64 kbps). This DS0 is sent through a multiplexer, which aggregates 24 DS0s and groups them into one DS1. The DS1 signal is then sent with other DS1s to another multiplexer, which aggregates 28 DS1s into one DS3. The DS3 is then used to reach another exchange, where these signals must be demultiplexed back down to their original DS0s.

In order for these circuits to reach the proper switch, they are cross-connected through digital cross-connect systems which allow incoming circuits to be electronically routed to the appropriate switch in the central exchange. This only adds to the problem of circuit testing and diagnostics, because the alarms and circuit status information gets lost in the multiplexing and cross-connecting and never makes its way through to the rest of the network.

To counter this issue, SS7 provides the ability to test the voice circuit before connecting a call to it. This takes place even before the IAM message is sent. The test is known as the continuity test, and uses the COT message. There are two instances when the continuity test procedure may be used. The continuity test may be used within the same network or it may be used when connecting to networks which use Exchange Access Signaling (EAS).

Exchange Access Signaling uses a series of tones to indicate signaling. This method has been replaced in most instances here in the U.S. by the SS7 standard of signaling, but may still exist in some rural areas. Because EAS does not support the digital messages of SS7, a conversion must be made converting the SS7 message into the analog format of EAS signaling.

The exchange that will perform this conversion will use both EAS and DC signaling (a method called the "wink," where the polarity of the trunk is temporarily reversed and then returned to its original polarity as an acknowledgment).

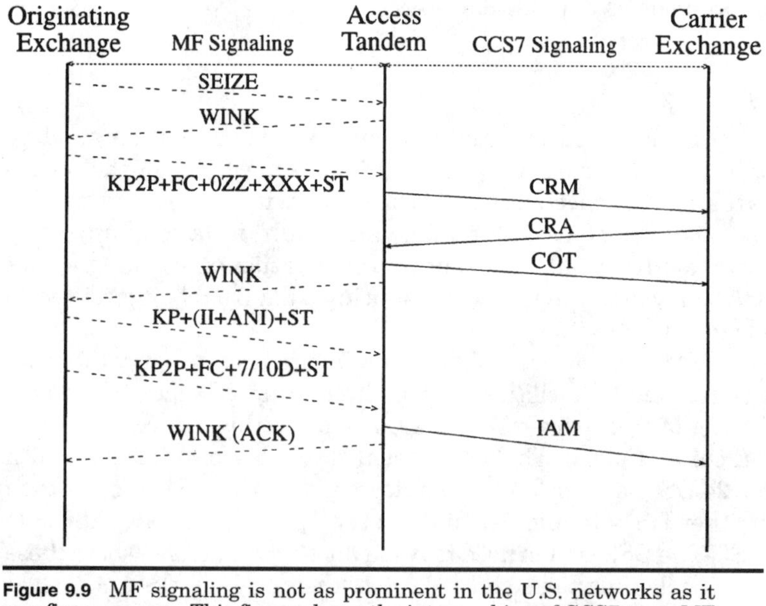

Figure 9.9 MF signaling is not as prominent in the U.S. networks as it was five years ago. This figure shows the interworking of CCS7 to an MF signaling network. Prior to the initial address message (IAM) is the continuity test (COT) procedure.

When it is determined that a continuity test is needed before a call can be set up, and the call is to another exchange using EAS or some other method of signaling other than SS7, a reservation message is sent to the adjacent exchange (exchange B). The Circuit Reservation Message (CRM) allows the voice circuit to be reserved without actually connecting the circuit. A request for a continuity test is sent in the CRM. The acknowledgment expected is the Circuit Reservation Acknowledgment (CRA).

The CRM also contains information regarding the nature of the connection to be established. This information includes satellite requirements (if any) as well as information regarding the usage of echo cancelers and end-to-end ISUP. The natures of connection parameters are also used in the IAM message, which follows the continuity test.

After the circuit has been reserved, the Continuity Test (COT) message is sent via SS7. The COT message indicates to the adjacent switch that the voice circuit should be tested for continuity using conventional testing methods (usually a loop-back test). The results of the COT test (if successful) are carried in the IAM.

In the event the circuit should fail the continuity test, the circuit is released, and a COT message with "test failed" is returned to the originating exchange (originating being the exchange which requested the continuity test to be performed). Another circuit is selected and the procedure begins again on the new circuit.

As can be seen in Fig. 9.9, the COT is sent in the same direction as the IAM. When the voice circuit is tested, the requester of the message will determine if the test was successful or not. This is done by sending a signal or current on the indicated circuit, and when the adjacent switch performs a "loop back," receiving the same signal back at the originating switch. If the signal is the same as what was sent, the continuity test is said to be successful.

The continuity test is not required on every call. Its usage is network dependent. Some networks will perform a continuity test on every circuit selected before every call, even when SS7 is used throughout the network. Other networks will perform a continuity test on any one circuit every 100 calls. The network operator must configure the voice switch to perform the continuity test at whatever intervals used in their network. An STP has no knowledge of these tests, as its function is to pass the information along to another SSP. This is an SSP function only.

Circuit validation test

The circuit validation test is typically used when a new facility or a new translation is added to the network. The purpose of this test is to validate the translation data and ensure that circuits between two exchanges can be selected by the routing function properly.

The translation is tested locally first, to validate the local routing entry. Once maintenance personnel have verified the circuit can be accessed locally by CLLI code, the technician generates a Circuit Validation Test Message (CVM) to verify whether the distant end can route to the new entry using the new translations.

The distant end will perform various tests upon receipt of this test message to verify that the physical port of a signaling link can be accessed for the new translation, and a CLLI code can be derived from the port. A response is sent to the originating exchange providing the results of the test by the Circuit Validation Response (CVR) message.

This test should be part of the routine maintenance check anytime translations are added to a signaling point. It can also be helpful in troubleshooting routing problems between two exchanges.

Functionality of the ISUP Protocol

The ISDN User Part (ISUP) protocol is a circuit-related protocol, used primarily for the establishment of connections between exchanges for the transmission of bearer traffic. The bearer traffic, which is usually generated by subscribers, can consist of voice, data, video, multimedia, or audio.

In today's network, only voice and data are achievable. Through the development of technologies such as ATM and broadband ISDN, video and multimedia will also be available through the Public Switched Telephone Network (PSTN).

Regardless of the technology, ISUP protocol provides the mechanism for establishing the connections from the originating exchange to the destination exchange, without using the bearer circuit itself.

In addition to connection establishment, ISUP also provides a means for passing information between exchanges associated with a call that is already in progress. However, the connection must already be established and the information must be related to that call's circuit or services.

Any information about the subscriber or network features, or anything that is not directly related to the circuit itself uses the Transaction Capabilities Application Part (TCAP). This protocol was established (as described in the previous chapter) for non-circuit-related messages.

The type of information provided by the ISUP protocol includes resource requirements for completion of the connection (resources such as ISUP all the way through the network and echo cancelers on the voice circuit). Bandwidth information and service information (call waiting and call forwarding, for example) are service related, and require that information be sent between both end-to-end exchanges.

Intermediate exchanges do not need to see this information, as it does not involve their interaction. The objective of the intermediate exchange is to provide the connection through their facilities to the next exchange, until an end-to-end path is established for the bearer traffic.

ISUP Services

The ISUP protocol provides two methods for reaching the end destination. As mentioned in the previous discussion about ISUP functions and call setups, there are two types of services provided by ISUP, Basic Service, and Supplementary Service.

In addition to these two types of services, there are two ways to reach the end destination. End-to-end signaling can use either the SCCP Method or the Pass Along Method. Even though these have been defined already, we will repeat them here as a reference.

Basic service

Basic service is defined as the setup and teardown of circuits in the telephone network. These circuits are used for voice, data, and video transmission. Currently, most networks are using some form of digital transmission for all transmission, regardless of the source.

This digital transmission is now being further enhanced by the addition of fiber optics into the network. Basic services will still be used for setting up and tearing down these connections as well.

As broadband technologies such as ATM and broadband ISDN are deployed, it will be basic services that will be used to control these circuits. The protocol does not care what is being transmitted, although it will carry some indication of the source and the type of transmission being carried.

Supplementary service

Supplementary service is defined, as all other services needed to support these circuits. Other services may include the sending of caller information from one end point to another while a call is in progress. This information may be feature related or caller related.

The Intelligent Network will rely on supplementary service to send information about established calls. This is different than the usage of TCAP, as TCAP is not circuit related.

End-to-end signaling

End-to-end signaling is defined as signaling information that must be sent from the originating exchange to the final destination exchange.

This information may be part of basic services or supplementary services.

End-to-end signaling will almost always involve intermediate signaling points, even though they are not concerned with the call itself. There are two ways to reach the final destination exchange for end-to-end signaling. These are explained below.

SCCP method

SCCP can be used to provide network (Layer 3) routing for messages, but it is not used in today's U.S. networks. The usage of SCCP for end-to-end signaling would be an enhancement over the current method, however. The SCCP protocol allows ISUP messages to be routed to the distant exchange using any route, but only for messages related to a circuit already connected end to end.

The method currently used requires the ISUP message to use intermediate switches, using the same path as the messages used to set-up the circuit associated with the information being sent.

Pass along method

The pass along method is widely used in today's network. This method uses the same path as the setup messages. This method works, but requires messages to make unnecessary stops at intermediate switches.

This is not necessary, since the information does not concern the intermediate switch. It would be much more efficient if the message could be sent directly to the exchange through STPs, using any available route.

Message Format

The ISUP protocol uses message types to indicate the type of message being carried as well as the format of the message (see Fig. 9.10). Each message type has a distinct format, with mandatory and optional parameters. The parameters depend on the message type.

As found in the TCAP protocol, the ISUP also uses mandatory fixed parts and mandatory variables. These are parameters that must always exist, depending on the type of ISUP message. Again, the parameters used will depend on the message type.

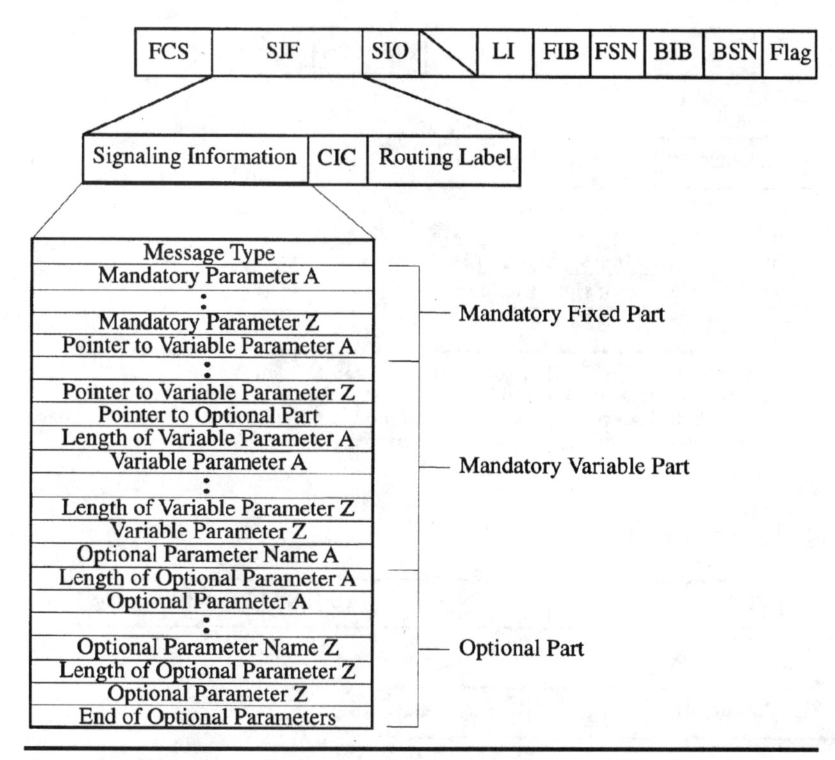

Figure 9.10 This figure shows the components used in an ISDN User Part (ISUP) message.

Circuit identification code (CIC)

The Circuit Identification Code (CIC) identifies the circuit being setup or released (see Fig. 9.11). The CIC may be a voice trunk or any other transmission medium in the Public Telephone Switched Network (PSTN).

Currently, there are no defined standards for allocating circuit identifiers. These are determined by agreement between the telephone companies. The CIC is provided to the originator of the ISUP message (SSP) by the end switch.

The end switch may be incorporated into the SSP, as many of these systems are fully integrated. This means that a voice subsystem provides the switching functionality for the voice circuits while the SS7 subsystem provides all the circuit control.

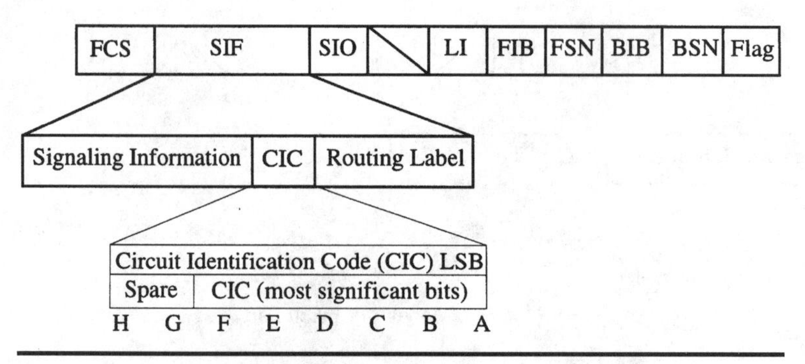

Figure 9.11 The circuit identification code (CIC) is used to identify the trunk circuit to be connected and associated with this message. The CIC is not used in broadband services. Instead, virtual paths and virtual connections are identified in the B-ISUP message content.

Message type codes for normal ISUP

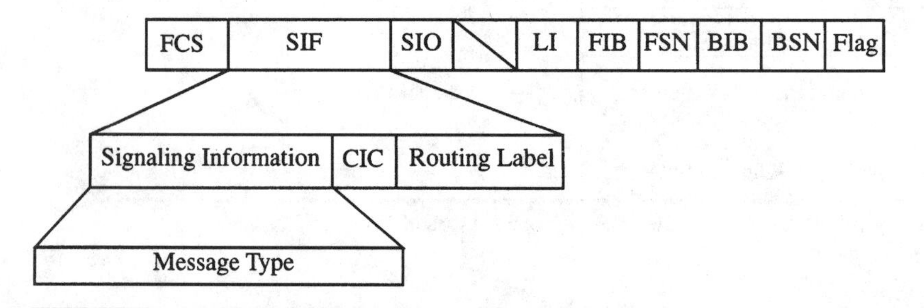

Figure 9.12 As found in TCAP and SCCP, the message type field identifies the nature of the message. Each message type consists of specific parameters (both mandatory and optional).

This is a one-octet field that is used to define the action to be taken by the exchange. In addition, the message type also implicitly defines the message structure. The parameters used will depend on the message type.

The message type is found in the mandatory fixed part of the message. The coding of the message type follows the same general rule used in other SS7 protocols. Any message type intended for national usage (internetworking) is to be coded using the upper range of the code. The upper range begins at 1111 1111, and works backward.

Following are the message types currently used in U.S. networks and defined in ANSI T1.113-1992 as well as BELLCORE GR-246-CORE, T1.113.3.

Message Types	H	G	F	E	D	C	B	A
Initial Address (IAM)	0	0	0	0	0	0	0	1
*Subsequent Address (SAM)	0	0	0	0	0	0	1	0
Information Request (INR)	0	0	0	0	0	0	1	1
Information (INF)	0	0	0	0	0	1	0	0
Continuity (COT)	0	0	0	0	0	1	0	1
Address Complete (ACM)	0	0	0	0	0	1	1	0
*Connect (CON)	0	0	0	0	0	1	1	1
Forward Transfer (FOT)	0	0	0	0	1	0	0	0
Answer (ANM)	0	0	0	0	1	0	0	1
Release (REL)	0	0	0	0	1	1	0	0
Suspend (SUS)	0	0	0	0	1	1	0	1
Resume (RES)	0	0	0	0	1	1	1	0
Release Complete (RLC)	0	0	0	1	0	0	0	0
Continuity Check Request (CCR)	0	0	0	1	0	0	0	1
Reset Circuit (RSC)	0	0	0	1	0	0	1	0
Blocking (BLO)	0	0	0	1	0	0	1	1
Unblocking (UBL)	0	0	0	1	0	1	0	0
Blocking Acknowledgment (BLA)	0	0	0	1	0	1	0	1
Unblocking Acknowledgment (UBA)	0	0	0	1	0	1	1	0
Circuit Group Reset (GRS)	0	0	0	1	0	1	1	1
Circuit Group Blocking (CGB)	0	0	0	1	1	0	0	0
Circuit Group Unblocking (CGU)	0	0	0	1	1	0	0	1
Circuit Group Blocking Acknowledgment (CGBA)	0	0	0	1	1	0	1	0
Circuit Group Unblocking Acknowledgment (CGUA)	0	0	0	1	1	0	1	1
*Call Modification Request (CMR)	0	0	0	1	1	1	0	0
*Call Modification Completed (CMC)	0	0	0	1	1	1	0	1
*Call Modification Reject (CMRJ)	0	0	0	1	1	1	1	0
*Facility Request (FAR)	0	0	0	1	1	1	1	1
*Facility Accepted (FAA)	0	0	1	0	0	0	0	0
*Facility Reject (FRJ)	0	0	1	0	0	0	0	1
Facility Deactivated (FAD)	0	0	1	0	0	0	1	0
Facility Information (FAI)	0	0	1	0	0	0	1	1
Loop-back Acknowledgment (LPA)	0	0	1	0	0	1	0	0
CUG Selection & Validation Request (CSVR)	0	0	1	0	0	1	0	1
CUG Selection & Validation Response (CSVS)	0	0	1	0	0	1	1	0
*Delayed Release (DRS)	0	0	1	0	0	1	1	1
Pass Along (PAM)	0	0	1	0	1	0	0	0
Circuit Group Reset Acknowledgment (GRA)	0	0	1	0	1	0	0	1
Circuit Query (CQM)	0	0	1	0	1	0	1	0
Circuit Query Response (CQR)	0	0	1	0	1	0	1	1
Call Progress (CPG)	0	0	1	0	1	1	0	0
*User-to-User Information (USR)	0	0	1	0	1	1	0	1

Unequipped Circuit Identification Code (UCIC)	0 0 1 0	1 1 1 0
Confusion (CFN)	0 0 1 0	1 1 1 1
*Overload (OLM)	0 0 1 1	0 0 0 0
*Charge Information (CRG)	0 0 1 1	0 0 0 1
Facility (FAC)	0 0 1 1	0 0 1 1
Circuit Reservation Acknowledgment (CRA)	1 1 1 0	1 0 0 1
**Circuit Reservation (CRM)	1 1 1 0	1 0 1 0
**Circuit Validation Response (CVR)	1 1 1 0	1 0 1 1
**Circuit Validation Test (CVT)	1 1 1 0	1 1 0 0
**Exit (EXM)	1 1 1 0	1 1 0 1

*ITU-TS only, not currently specified in ANSI or Bellcore networks.
**ANSI only, not currently specified in ITU-TS networks.

The above message types have been defined in the ANSI and Bellcore standards as well as ITU-TS. Each of these message types has a distinct structure, with parameters. Following are descriptions of each of the message types and the message structure that accompanies it.

The section that follows provides the one octet value which indicates the message type (in bold), followed by the parameter names and the one octet values for the parameter names. Each parameter may have several additional bits defining the actual parameter. This section only provides the values for the parameter names.

The parameters shown as optical are supported within each message type, but their usage is dependent on the network, and may or may not be used. Many of these parameters may be used in multiple message types, which is why they are discussed in the last section of this chapter.

The section below illustrates the message format and provides the basic structure for each message type. The next section following this will provide the detailed structure for each parameter type along with a description of the parameter. Refer to the last section for detailed information on parameters. All parameters with an asterisk are used in ITU ISUP only.

Message type structure

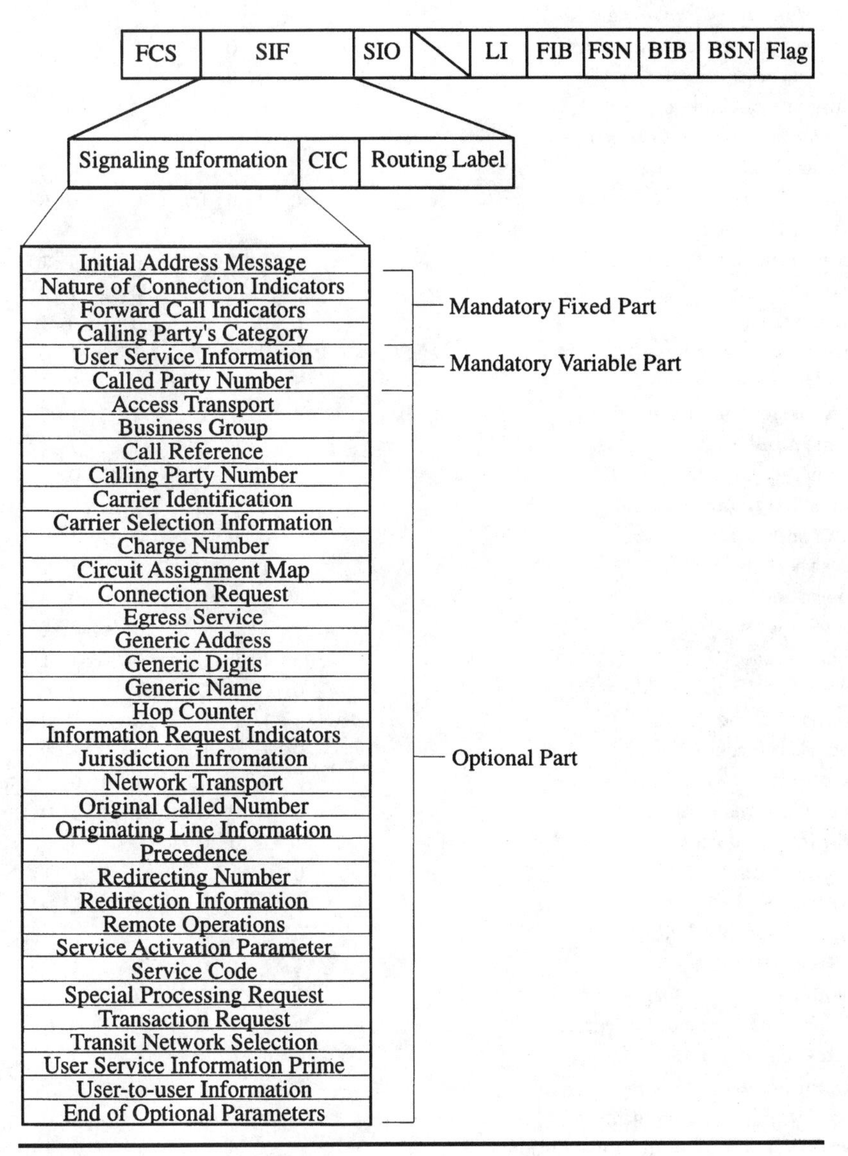

Figure 9.13

Initial Address Message	**0 0 0 0 0 0 0 1**
Fixed Mandatory Parameters	
Nature of connection indicators	0 0 0 0 0 1 1 0
Forward call indicators	0 0 0 0 0 1 1 1
Calling party's category	0 0 0 0 1 0 0 1
Variable Mandatory Parameters	
User service information	0 0 0 1 1 1 0 1
Called party number	0 0 0 0 0 1 0 0
Optional Parameters	
Access transport	0 0 0 0 0 0 1 1
Business group	1 1 0 0 0 1 1 0
Call reference	0 0 0 0 0 0 0 1
Calling party number	0 0 0 0 1 0 1 0
Carrier identification	1 1 0 0 0 1 0 1
Carrier selection information	1 1 1 0 1 1 1 0
Charge number	1 1 1 0 1 0 1 1
Circuit assignment map	0 0 1 0 0 1 0 1
Connection request	0 0 0 0 1 1 0 1
*CUG interlock code	0 0 0 1 1 0 1 0
Egress service	1 1 0 0 0 0 1 1
Generic address	1 1 0 0 0 0 0 0
Generic digits	1 1 0 0 0 0 0 1
Generic name	1 1 0 0 0 1 1 1
*Generic notification indicator	0 0 1 0 1 1 0 0
*Generic number	1 1 0 0 0 0 0 0
*Generic reference	0 1 0 0 0 0 1 0
Hop counter	0 0 1 1 1 1 0 1
Information request indicators	0 0 0 0 1 1 1 0
Jurisdiction information	1 1 0 0 0 1 0 0
*Location number	0 0 1 1 1 1 1 1
*MLPP precedence	0 0 1 1 1 0 1 0
*Network specific facility	0 0 1 0 1 1 1 1
Network transport	1 1 1 0 1 1 1 1
Operator services information	1 1 0 0 0 0 1 0
*Optional forward call indicators	0 0 0 0 1 0 0 0
Original called number	0 0 1 0 1 0 0 0
*Originating ISC point code	0 0 1 0 1 0 1 1
Originating line information	1 1 1 0 1 0 1 0
*Parameter compatibility information	0 0 1 1 1 0 0 1
Precedence	0 0 1 1 1 0 1 0
*Propagation delay counter	0 0 1 1 0 0 0 1
Redirecting number	0 0 0 0 1 0 1 1

Redirection information	0	0	0	1	0	0	1	1
Remote operations	0	0	1	1	0	0	1	0
Service activation parameter	1	1	1	0	0	0	1	0
Service code indicator	1	1	1	0	1	1	0	0
Special processing request	1	1	1	0	1	1	0	1
Transaction request	1	1	1	0	0	0	1	1
Transit network selection	0	0	1	0	0	0	1	1
*Transmission medium requirement	0	0	0	0	0	0	1	0
*Transmission medium requirement prime	0	0	1	1	1	1	1	0
User service information	0	0	0	1	1	1	0	1
User service information prime	0	0	1	1	0	0	0	0
User-to-user indicators	0	0	1	0	1	0	1	0
User-to-user information	0	0	1	0	0	0	0	0

This is the message used to establish a connection on a specified circuit. The IAM provides the circuit information which includes the carrier identification (long-distance carrier to be used for this call) and any special requirements to be considered in the handling of this call. The IAM message is by far the most comprehensive of the ISUP messages, with many parameters. Refer to the end of this chapter for the parameter values and definitions.

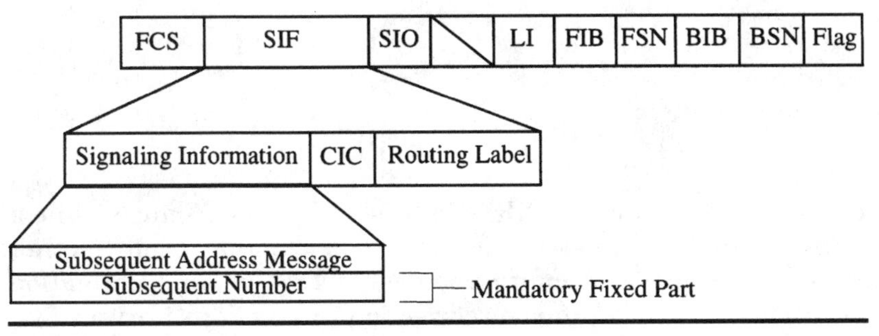

Figure 9.14

Subsequent Address (SAM) **0 0 0 0 0 0 1 0**

No procedure is presently defined for ANSI networks. This message is used in ITU-TS networks only. The structure of this message has been omitted from this section, as the scope of this book is ANSI and Bellcore networks. Refer to the ITU specifications for information regarding this and other messages.

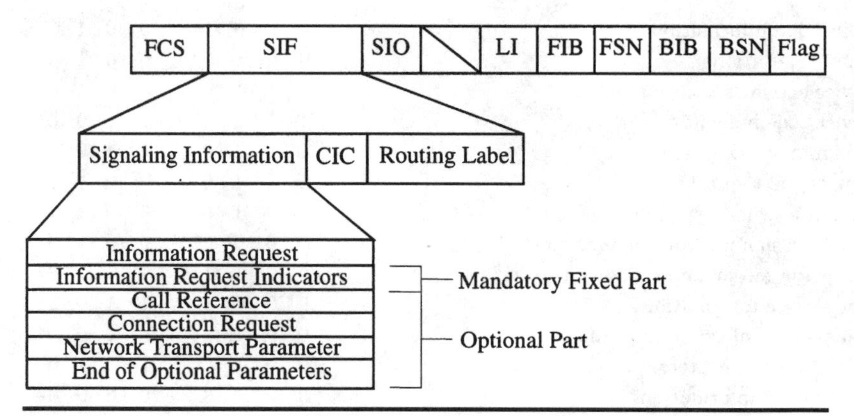

Figure 9.15

Information Request (INR)	**0 0 0 0**	**0 0 1 1**
Fixed Mandatory Parameters		
Information request indicators	0 0 0 0	1 1 1 0
Optional Parameters		
Call reference	0 0 0 0	0 0 0 1
Connection request	0 0 0 0	1 1 0 1
*Network specific facility	0 0 1 0	1 1 1 1
Network transport parameter	1 1 1 0	1 1 1 1
*Parameter compatibility information	0 0 1 1	1 0 0 1

The Information Request (INR) can be sent by an exchange while a call is in progress to request additional information from another exchange. The additional information is carried in an Information (INF) message, and may provide redirection instructions (forwarding) or other call-handling information.

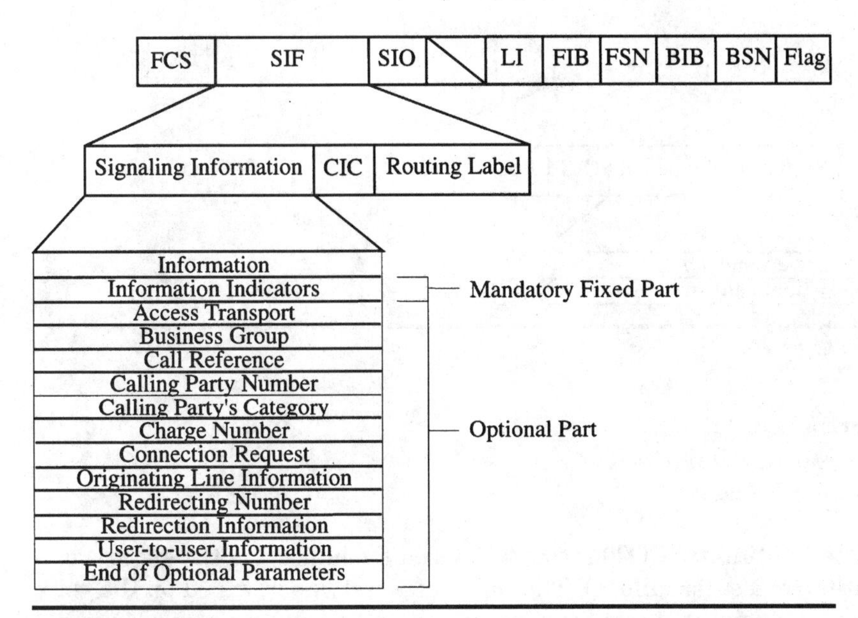

Figure 9.16

Information (INF)	**0 0 0 0**	**0 1 0 0**
Fixed Mandatory Parameters		
Information indicators	0 0 0 0	1 1 1 1
Optional Parameters		
Access transport	0 0 0 0	0 0 1 1
Business group	1 1 0 0	0 1 1 0
Call reference	0 0 0 0	0 0 0 1
Calling party number	0 0 0 0	1 0 1 0
Calling party's category	0 0 0 0	1 0 0 1
Charge number	1 1 1 0	1 0 1 1
Connection request	0 0 0 0	1 1 0 1
Originating line information	1 1 1 0	1 0 1 0
Redirection information	0 0 0 1	0 0 1 1
User-to-user information	0 0 1 0	0 0 0 0

The Information (INF) message is used to pass additional information about a call upon request from the distant exchange. The information is requested from an exchange using the Information Request (INR) message, and the reply is carried in this INF message. The type of information is usually call-handling information, such as the number to forward a call to or a billing number.

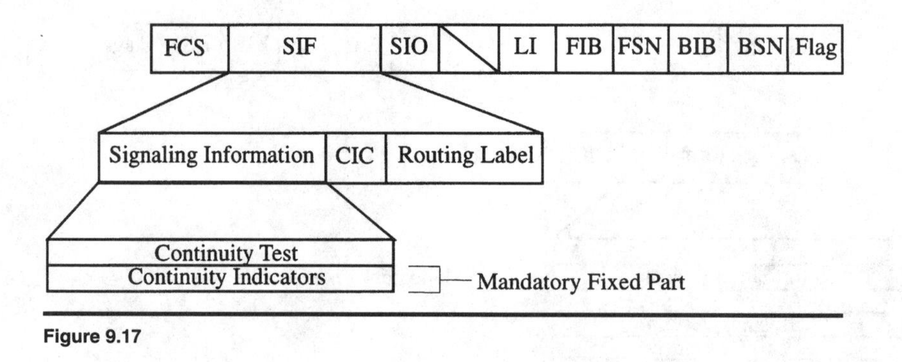

Figure 9.17

Continuity (COT) **0 0 0 0 0 1 0 1**

Fixed Mandatory Parameters

Continuity indicators 0 0 0 1 0 0 0 0

The Continuity (COT) message is used for indicating the success of a continuity test (or failure). The continuity test is performed on the voice circuit depending on criteria set by the network operator at time of deployment. The COT is used to indicate the status of the preceding circuit and the circuit selected in the forward direction to the next exchange.

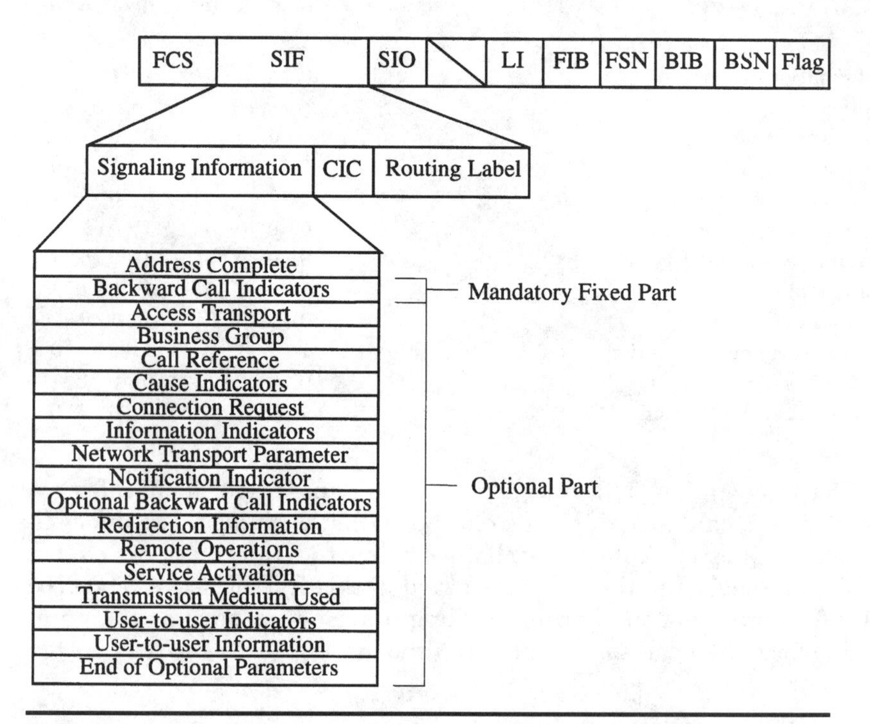

Figure 9.18

Address Complete (ACM)	**0 0 0 0**	**0 1 1 0**
Mandatory Fixed Parameters		
Backward call indicators	0 0 0 1	0 0 0 1
Optional Parameters		
*Access delivery information	0 0 1 0	1 1 1 0
Access transport	0 0 0 0	0 0 1 1
Business group	1 1 0 0	0 1 1 0
*Call diversion information	0 0 1 1	0 1 1 0
Call reference	0 0 0 0	0 0 0 1
Cause indicators	0 0 0 1	0 0 1 0
Connection request	0 0 0 0	1 1 0 1
*Echo control information	0 0 1 1	0 1 1 1
*Generic notification indicators	0 0 1 0	1 1 0 0
Information indicators	0 0 0 0	1 1 1 1
*Network specific facility	0 0 1 0	1 1 1 1
Network transport parameter	1 1 1 0	1 1 1 1
Notification indicator	1 1 1 0	0 0 0 1
Optional backward call indicators	0 0 1 0	1 0 0 1
*Parameter compatibility information	0 0 1 1	1 0 0 1
Redirection information	0 0 0 1	0 0 1 1
*Redirection number	0 0 0 0	1 1 0 0
*Redirection number restriction	0 1 0 0	0 0 0 0
Remote operations	0 0 1 1	0 0 1 0
Service activation	0 0 1 1	0 0 1 1
Transmission medium used	0 0 1 1	0 1 0 1
User-to-user indicators	0 0 1 0	1 0 1 0
User-to-user information	0 0 1 0	0 0 0 0

This message (ACM) is sent by a distant exchange upon receipt of all address signals (IAM and any subsequent information sent) needed to establish a connection on a circuit between the two exchanges. The ACM indicates that the call is being processed, and the distant exchange is checking the availability of the called party. This could mean the called party's telephone is being signaled (ringing if analog or "alerting message" if ISDN). In some networks, cut-through on the voice circuit can take place after receipt of the ACM.

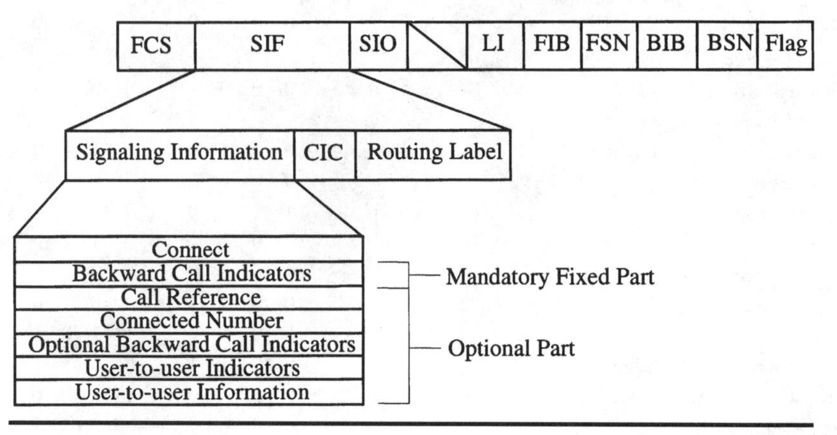

Figure 9.19

Connect (CON) **0 0 0 0 0 1 1 1**

The Connect (CON) message is defined for use in International networks but not ANSI networks.

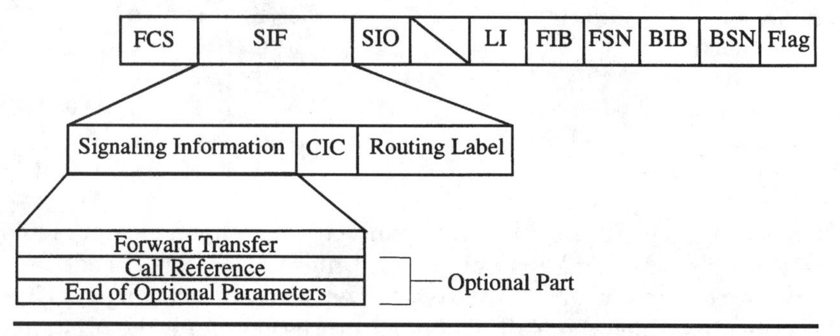

Figure 9.20

Forward Transfer (FOT) **0 0 0 0 1 0 0 0**

Optional Parameters

Call reference 0 0 0 0 0 0 0 1

The Forward Transfer (FOT) is used in conjunction with operator services. In exchanges where telephone calls are setup automatically (which is the case in all of North America today), an operator is only needed in certain circumstances. This message is sent in the forward direction to bring an operator into the circuit when operator assistance is required to complete the call. When the call has been completed, the operator can be recalled to terminate the call or initiate another call for the same calling party.

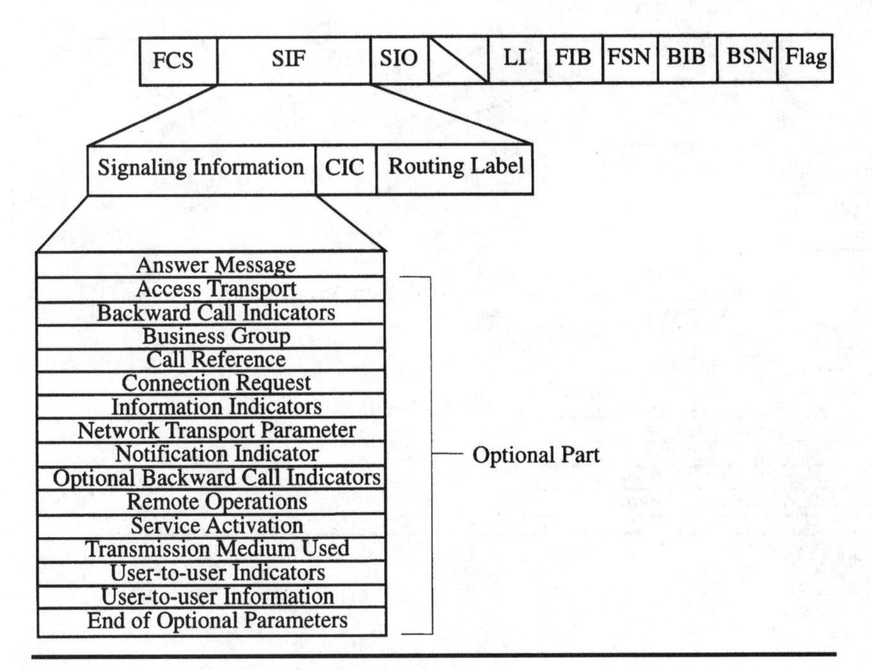

Figure 9.21

Answer (ANM)	**0 0 0 0**	**1 0 0 1**
Optional Parameters		
Access transport	0 0 0 0	0 0 1 1
Backward call indicators	0 0 0 1	0 0 0 1
Business group	1 1 0 0	0 1 1 0
Call reference	0 0 0 0	0 0 0 1
Connection request	0 0 0 0	1 1 0 1
Information indicators	0 0 0 0	1 1 1 1
Network transport parameter	1 1 1 0	1 1 1 1
Optional backward call indicators	0 0 1 0	1 0 0 1
Remote operations	0 0 1 1	0 0 1 0
Service activation	0 0 1 1	0 0 1 1
Transmission medium used	0 0 1 1	0 1 0 1
User-to-user indicators	0 0 1 0	1 0 1 0
User-to-user information	0 0 1 0	0 0 0 0

This message (ANM) is sent in the backward direction to indicate the called party has answered the call. The usage of this parameter is really two-fold. In semi-automatic networks, this parameter is used for call supervision. In automatic networks, the ANM message is used to begin metering the call for billing purposes. Metering of domestic calls and international calls can be activated using this parameter.

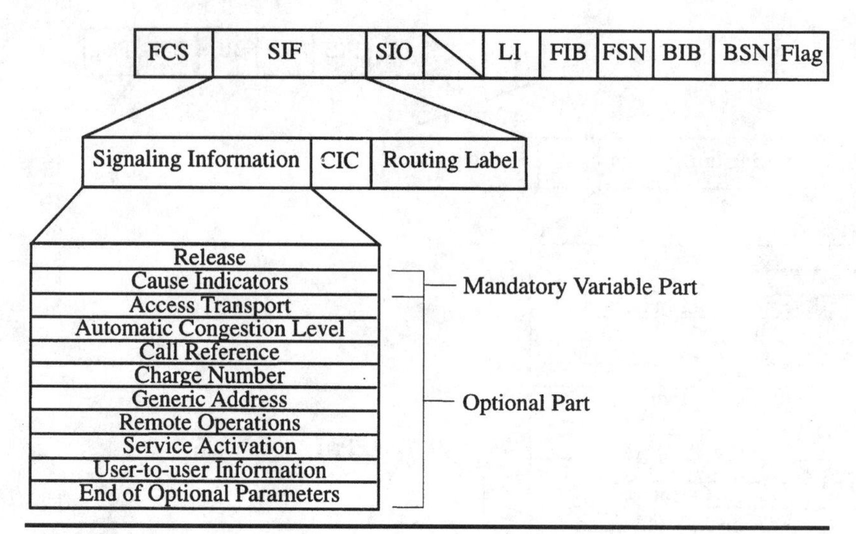

Figure 9.22

Release (REL)	**0 0 0 0**	**1 1 0 0**
Mandatory Variable Parameters		
Cause indicators	0 0 0 1	0 0 1 0
Optional Parameters		
Access transport	0 0 0 0	0 0 1 1
Automatic congestion level	0 0 1 0	0 1 1 1
Call reference	0 0 0 0	0 0 0 1
Charge number	1 1 1 0	1 0 1 1
Generic address	1 1 0 0	0 0 0 0
Service activation	0 0 1 1	0 0 1 1
User-to-user information	0 0 1 0	0 0 0 0

The Release (REL) message is sent in either direction indicating that either one of the parties (called or calling) has gone on-hook and the call is being terminated. The REL message does not return the circuit back to its idle state, however. A Release Complete (RLC) must be received before the circuit is returned to idle.

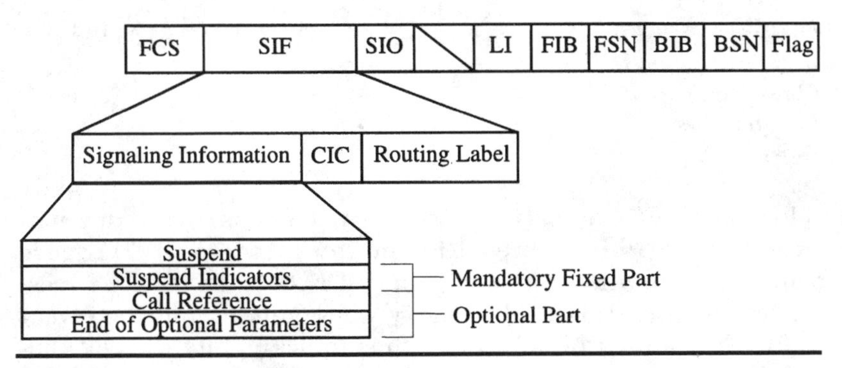

Figure 9.23

Suspend (SUS)	**0 0 0 0 1 1 0 1**
Mandatory Fixed Parameters	
Suspend/resume indicators	0 0 1 0 0 0 0 1
Optional Parameters	
Call reference	0 0 0 0 0 0 0 1

This message is used when a non-ISDN party returned to an on-hook state. When an ISDN party returned on-hook, only the Release (REL) message is used, but with non-ISDN the SUS is sent first, followed by the Release (REL) and the Release Complete (RLC). For complete call setup and teardown information, review the previous sections in this chapter on call setup and teardown.

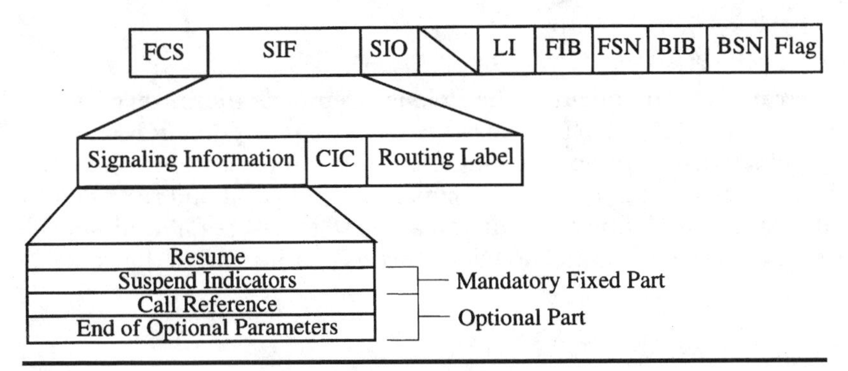

Figure 9.24

Resume (RES)	**0 0 0 0**	**1 1 1 0**
Mandatory Fixed Parameters		
Suspend/resume indicators	0 0 1 0	0 0 1 0
Optional Parameters		
Call reference	0 0 0 0	0 0 0 1

The Resume (RES) message is used in two circumstances. In a network where interworking is used, RES indicates the interworking node has reanswered. In a network with non-ISDN circuits, the RES message indicates a non-ISDN called party went on-hook, but then went back off-hook again within a certain time (quickly) and the call connection should remain established. Had the called party stayed on-hook past the specified time (network dependent) the SUS message would have been sent in the backward direction to begin releasing the circuit.

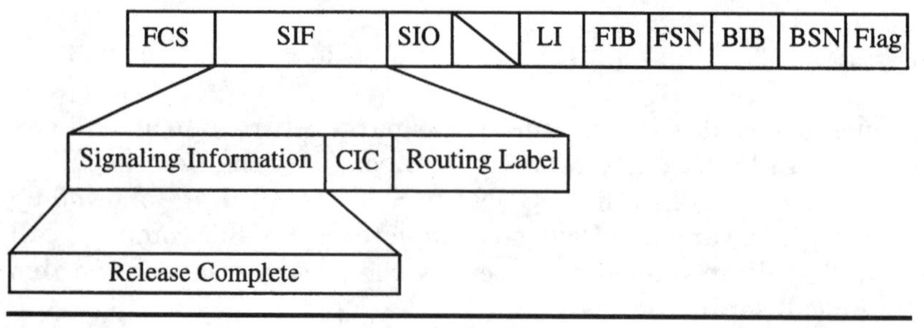

Figure 9.25

Release Complete (RLC) **0 0 0 1 0 0 0 0**

No parameters are given in the Release Complete message, only the message type field. The RLC is used to indicate receipt of an REL message, and serves as an acknowledgment of the release. Once the RLC has been received, the indicated circuit can be released and returned to its idle state. The CIC is sent with this message, but is not an integral part of the message itself. The CIC is presented just after the routing label.

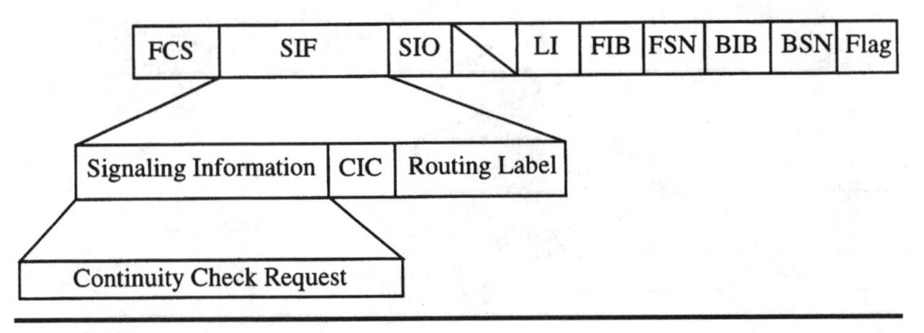

Figure 9.26

Continuity Check Request (CCR) **0 0 0 1 0 0 0 1**

No parameters are given in the Continuity Check Request (CCR). The CCR is used to request continuity check equipment to be attached to the circuit indicated in the CIC field of the message. The equipment is attached to the voice circuit for loopback testing. Once loop-back has been detected the status of "successful" is sent through the SS7 network using the IAM message or the Continuity (COT) message, depending on the network and the circumstances.

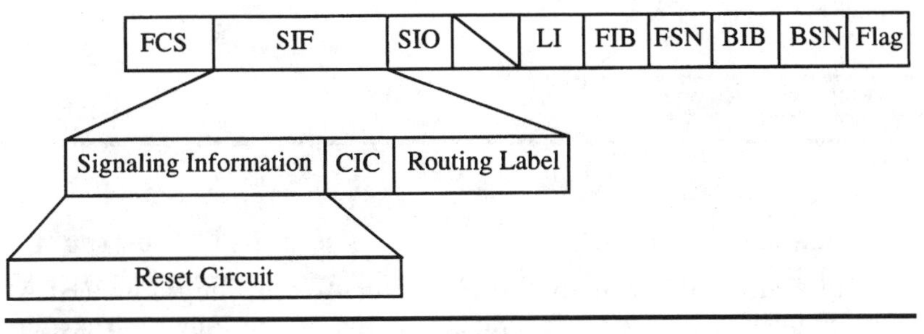

Figure 9.27

Reset Circuit (RSC) **0 0 0 1 0 0 1 0**

No parameters are given in the Reset Circuit (RSC) message. The purpose of this message is to allow an exchange to reset a circuit to the state that exchange thinks the circuit should be in. This occurs when a memory error occurs at an exchange, and it no longer knows the state of the circuit in question. To restart from scratch, the RSC message is sent. Any calls in progress or blocked conditions are released and the circuit is returned to an idle state after an alignment procedure (not to be confused with the alignment procedure used on SS7 links).

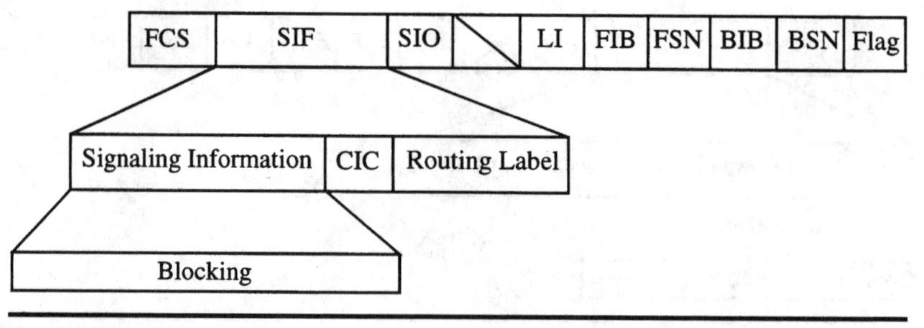

Figure 9.28

Blocking (BLO) **0 0 0 1 0 0 1 1**

No parameters are given in the Blocking (BLO) message. This message allows one exchange to block a voice circuit at a remote exchange, preventing voice calls from being reserved on the voice circuit from the remote end. The circuit is identified in the CIC field.

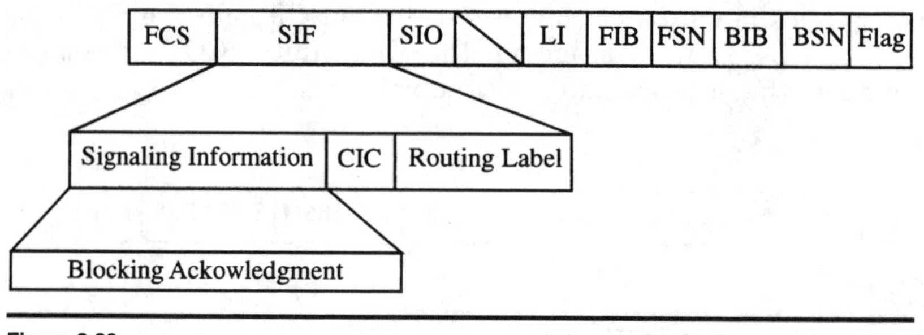

Figure 9.29

Blocking Acknowledgment (BLA) **0 0 0 1 0 1 0 1**

No parameters are given in the Blocking Acknowledgment (BLA) message. This message acknowledges receipt of the Blocking (BLO) message, and indicates that the circuit has been blocked. The voice circuit is identified in the CIC field.

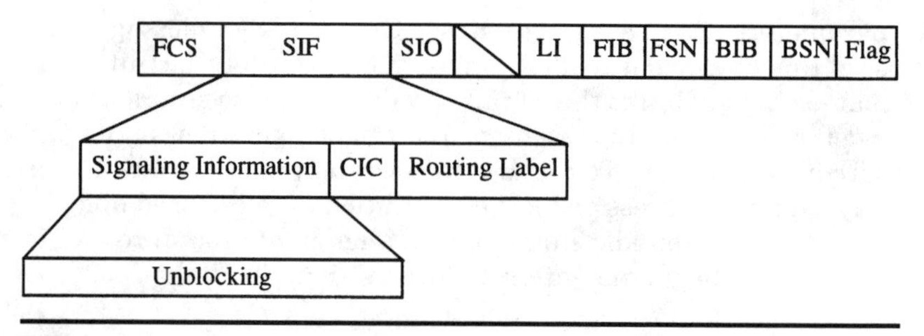

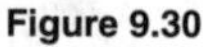
Figure 9.30

Unblocking (UBL) **0 0 0 1 0 1 0 0**

No parameters are given in the Unblocking (UBL) message. This message is sent by an exchange to remove a blocking condition at a remoter exchange. The circuit being unblocked is indicated in the CIC field.

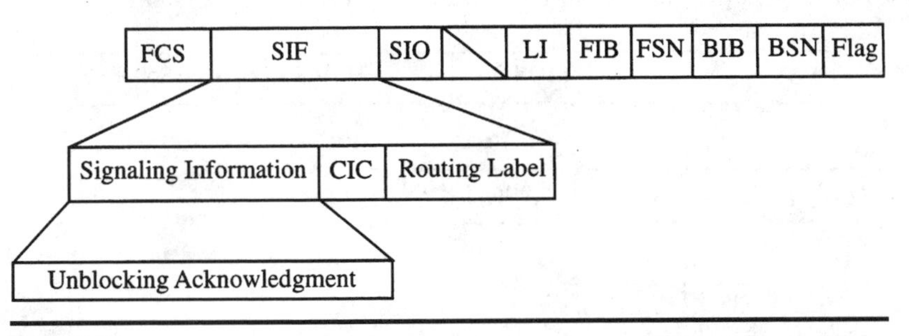

Figure 9.31

Unblocking Acknowledgment (UBA) **0 0 0 1 0 1 1 0**

No parameters are given in the Unblocking Acknowledgment (UBA) message. This message is sent to acknowledge receipt of the Unblocking (UBL) message. The acknowledgment also indicates that the circuit has been unblocked. The circuit is identified in the CIC field.

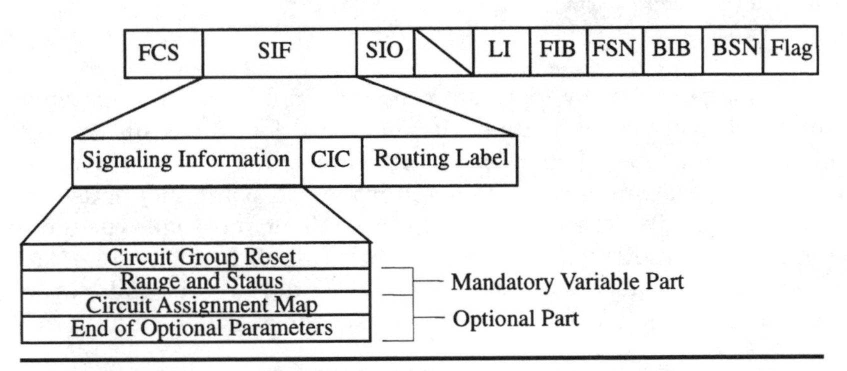

Figure 9.32

Circuit Group Reset (GRS) **0 0 0 1 0 1 1 1**

Mandatory Variable Parameters

Range and status 0 0 0 1 0 1 1 0

Optional Parameters

Circuit assignment map 0 0 1 0 0 1 0 1

This message is used to reset a group of voice circuits when the exchange no longer knows the status of the voice circuits. This could be the result of memory malfunction or some other error that caused it to lose track of the circuits status. The range parameter is used to identify the range of voice circuits to be reset. Any calls in progress or blocked conditions will be canceled and the voice circuits indicated released. However, they must go through diagnostics and alignment procedures (voice alignment, not SS7 alignment) before becoming available for calls again.

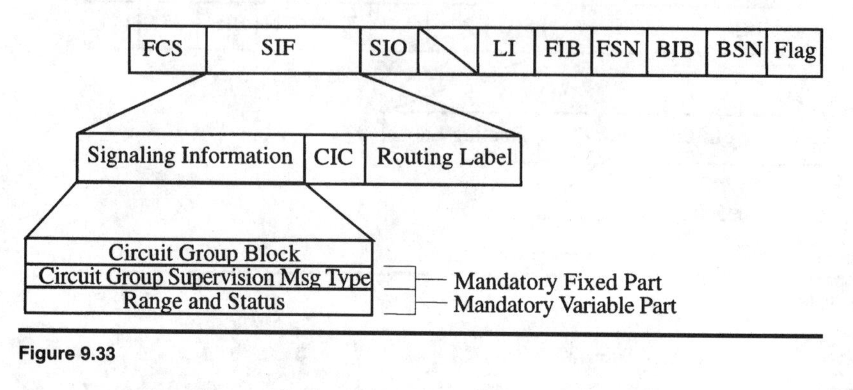

Figure 9.33

Circuit Group Blocking (CGB)	**0 0 0 1**	**1 0 0 0**
Fixed Mandatory Parameters		
Circuit group supervision message type indicator	0 0 0 1	0 1 0 1
Mandatory Variable Parameters		
Range and status	0 0 0 1	0 1 1 0

This message is sent by maintenance personnel from an operations terminal to block voice circuits from being used for voice calls during maintenance routines. The voice circuits are manually busied until maintenance procedures are completed; at which point they must be unblocked manually. The circuit group supervision message type indicator parameter indicates what type of blocking to invoke, while the range and status indicate what range of circuits to block, and status indicates the status (blocked or unblocked) of the specified circuits.

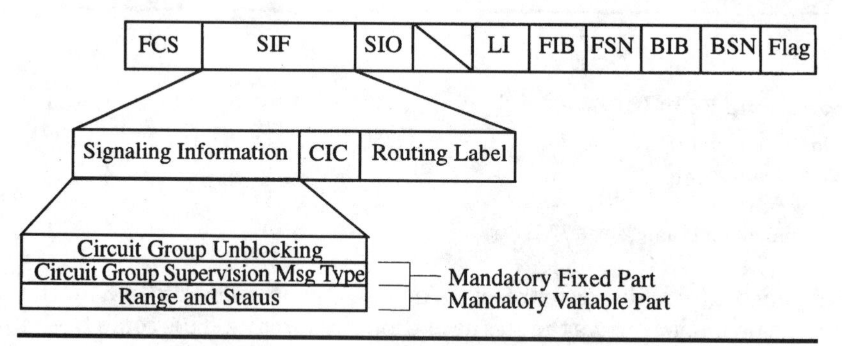

Figure 9.34

Circuit Group Unblocking (CGU)	**0 0 0 1**	**1 0 0 1**
Fixed Mandatory Parameters		
Circuit group supervision message type indicator	0 0 0 1	0 1 0 1
Mandatory Variable Parameters		
Range and status	0 0 0 1	0 1 1 0

This message is sent by maintenance personnel from an operations terminal to unblock voice circuits that were previously blocked for maintenance purposes. The circuit group supervision message type indicator parameter indicates what type of unblocking to invoke, while the range and status indicate what range of circuits to unblock, and status indicates the status (blocked or unblocked) of the specified circuits.

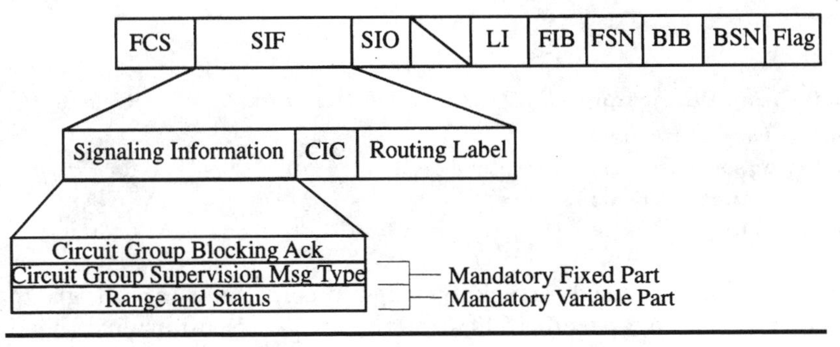

Figure 9.35

Circuit Group Blocking Acknowledgment (CGBA)	**0 0 0 1**	**1 0 1 0**
Fixed Mandatory Parameters		
Circuit group supervision message type indicator	0 0 0 1	0 1 0 1
Mandatory Variable Parameters		
Range and status	0 0 0 1	0 1 1 0

This message is used to acknowledge receipt of a circuit group blocking message, and indicates that the circuits have been blocked. The supervision message type indicator shows the type of blocking invoked while the range and status parameter shows the range of circuits that were blocked and their present status (blocked).

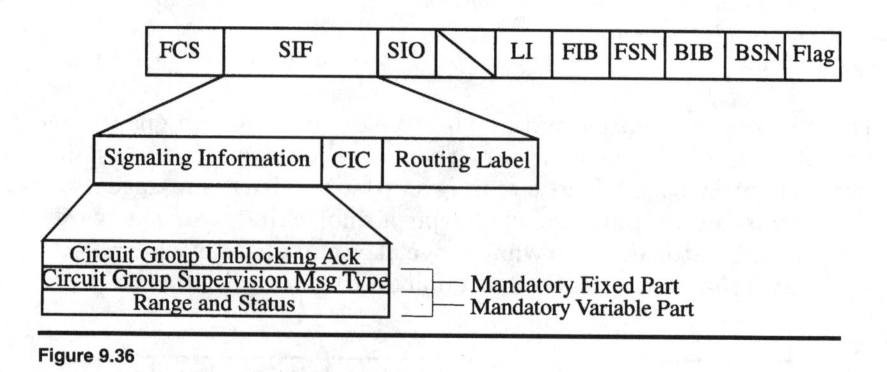

Figure 9.36

Circuit Group Unblocking Ack (CGUA)	**0 0 0 1**	**1 0 1 1**
Fixed Mandatory Parameters		
Circuit group supervision message type indicator	0 0 0 1	0 1 0 1
Mandatory Variable Parameters		
Range and status	0 0 0 1	0 1 1 0

This message is used to acknowledge receipt of a circuit group unblocking message, and indicates that the circuits have been unblocked. The supervision message type parameter indicates the type of unblocking used while the range and status indicates the range of circuits that have been unblocked and the status of those circuits (unblocked).

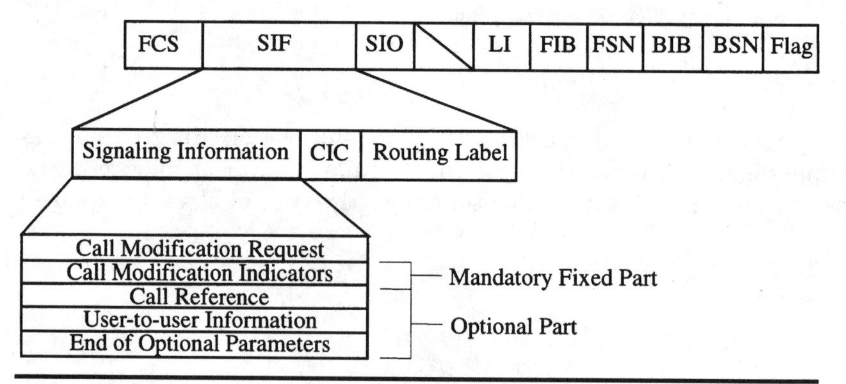

Figure 9.37

Call Modification Request (CMR) 0 0 0 1 1 1 0 0

There are currently no procedures written for this in ANSI networks. This message is only found in ITU-TS networks.

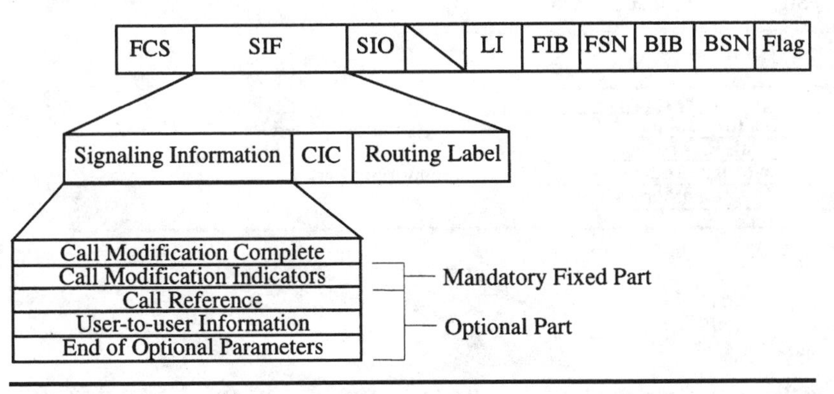

Figure 9.38

Call Modification Completed (CMC) 0 0 0 1 1 1 0 1

There are currently no procedures written for this in ANSI networks. This message is only found in ITU-TS networks.

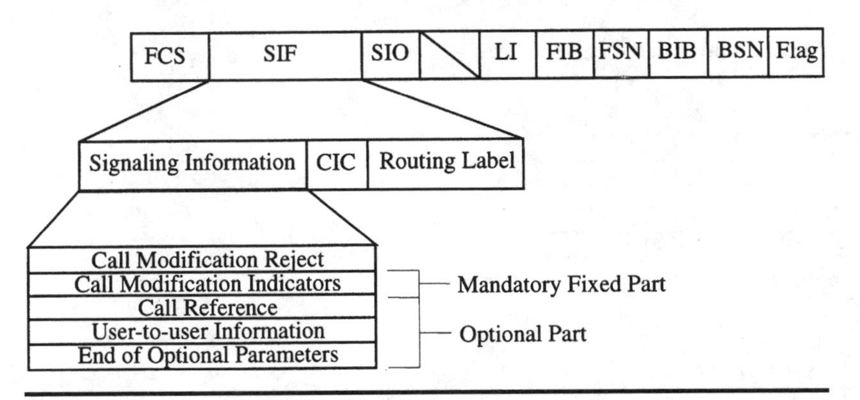

Figure 9.39

Call Modification Reject (CMRJ) 0 0 0 1 1 1 1 0

There are currently no procedures written for this in ANSI networks. This message is only found in ITU-TS networks.

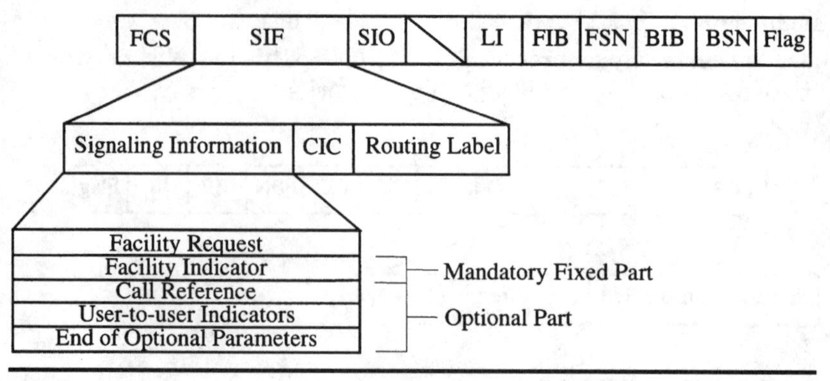

Figure 9.40

Facility Request (FAR) **0 0 0 1 1 1 1 1**

There are currently no procedures written for this in ANSI networks. This message is only found in ITU-TS networks.

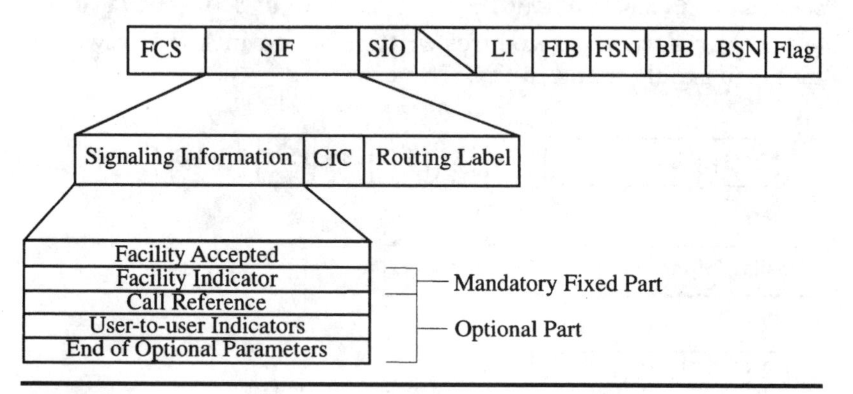

Figure 9.41

Facility Accepted (FAA) **0 0 1 0 0 0 0 0**

There are currently no procedures written for this in ANSI networks. This message is only found in ITU-TS networks.

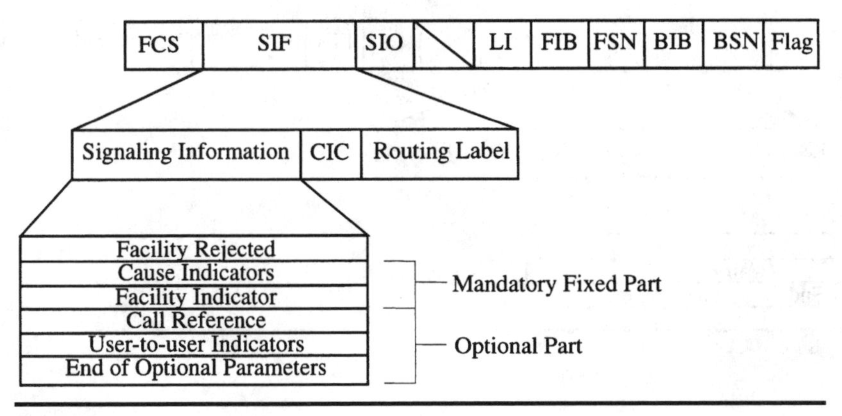

Figure 9.42

Facility Reject (FRJ) **0 0 1 0 0 0 0 1**

There are currently no procedures written for this in ANSI networks. This message is only found in ITU-TS networks.

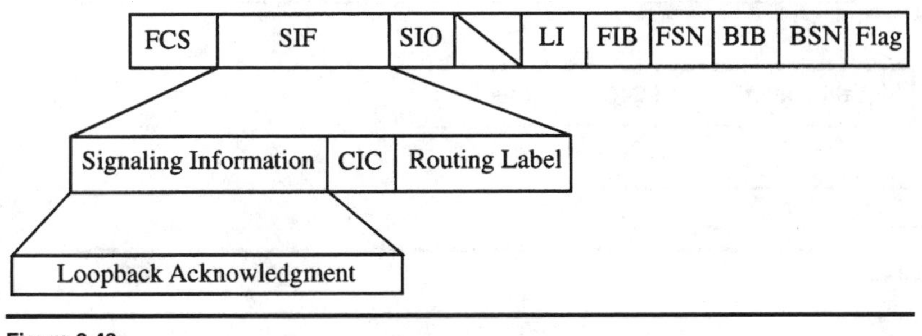

Figure 9.43

Loop Back Acknowledgment (LPA) **0 0 1 0 0 1 0 0**

No parameters are given in this message. This message is used to indicate that loop-back equipment has been connected in response to a Continuity Check Request message, and loop-back testing is being performed. The voice circuit identification is provided in the CIC field.

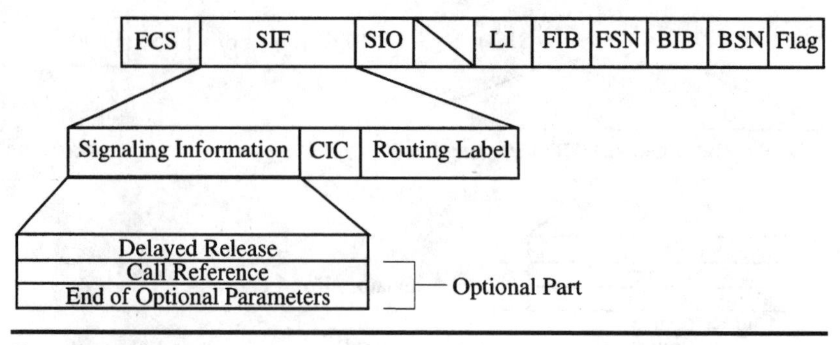

Figure 9.44

Delayed Release (DRS) 0 0 1 0 0 1 1 1

There are currently no procedures written for this in ANSI networks. This message is only found in ITU-TS networks.

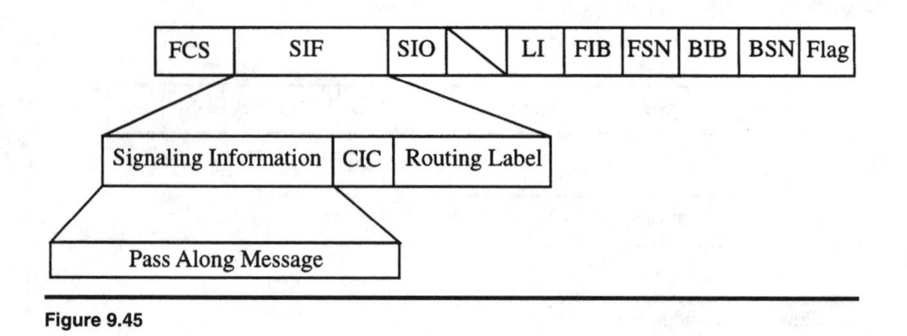

Figure 9.45

Pass Along (PAM) 0 0 1 0 1 0 0 0

There are no specific parameters associated with this command, however, when the pass along message type is given, another message type is normally contained within (as if parameters). This allows a message to be routed to the exchange associated with the specified voice circuit connection so that information may be passed along using the same path as that used for the call setup messages.

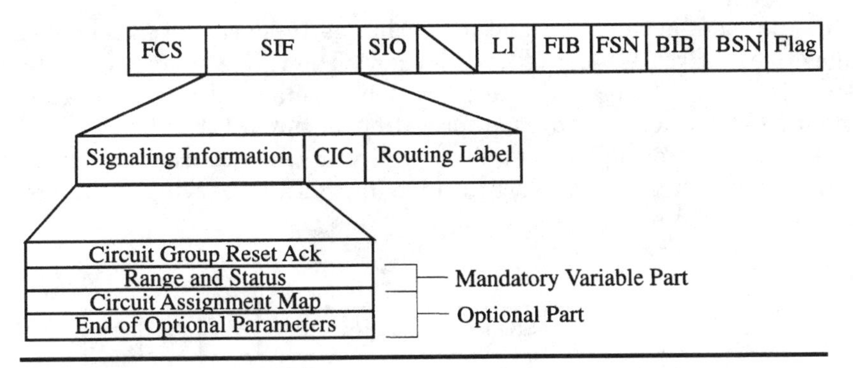

Figure 9.46

Circuit Group Reset Acknowledgment (GRA)	**0 0 1 0**	**1 0 0 1**
Mandatory Variable Parameters		
Range and status	0 0 0 1	0 1 1 0
Optional Parameters		
Circuit assignment map	0 0 1 0	0 1 0 1

This message is used to indicate receipt of a circuit group reset message. This message also indicates that the reset has been performed on the circuits identified in the range parameter. The status parameter indicates the current status of those circuits.

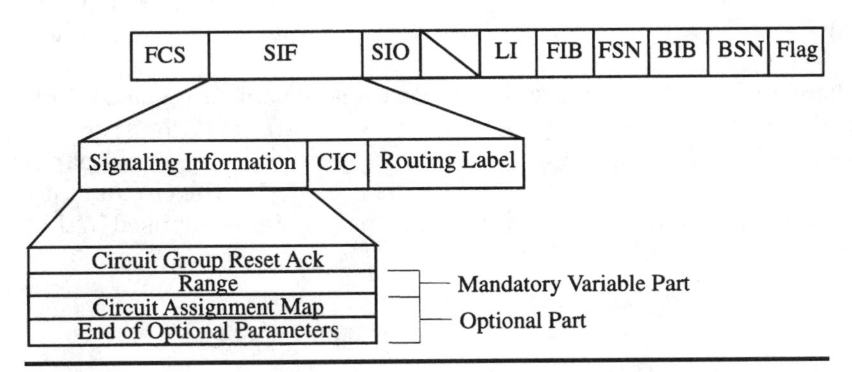

Figure 9.47

Circuit Query (CQM)	**0 0 1 0**	**1 0 1 0**
Mandatory Variable Parameters		
Range and status	0 0 0 1	0 1 1 0
Optional Parameters		
Circuit assignment map	0 0 1 0	0 1 0 1

This message is sent to a distant exchange to learn the status of a range of voice circuits (blocked, unblocked). The range of voice circuits is specified in the range parameter, which normally also has a status subfield. However, the status information is not returned with this message, therefore the status field is not used (set to zeros). A circuit query response message (CQR) is used to inform the querying exchange the status information.

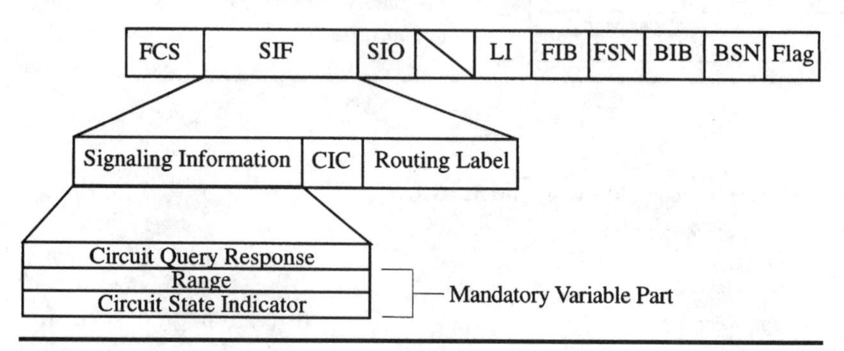

Figure 9.48

Circuit Query Response (CQR)	**0 0 1 0**	**1 0 1 1**
Mandatory Variable Parameters		
Range and status	0 0 0 1	0 1 1 0
Circuit state indicator	0 0 1 0	0 1 1 0

The circuit query response (CQR) message is sent in response to a circuit query (CQM) message, and provides the status of the specified voice circuits. The range of voice circuits is specified in the range parameter, while the status of those circuits is provided in the circuit state indicator. The status subfield of the range parameter is not used in this message (set to zeros).

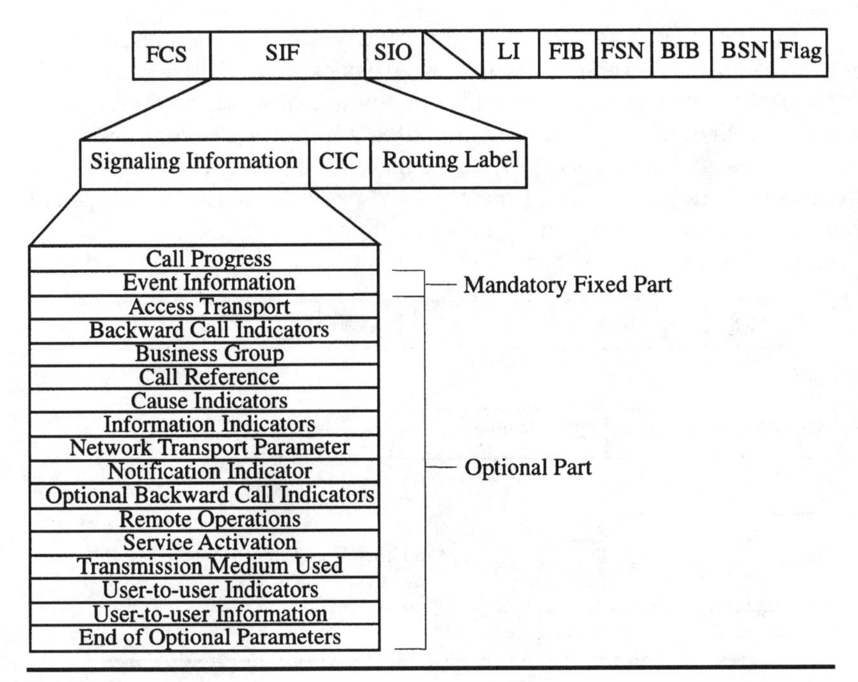

Figure 9.49

Call Progress (CPG)	**0 0 1 0 1 1 0 0**
Mandatory Fixed Parameters	
Event information	0 0 1 0 0 1 0 0
Optional Parameters	
Access transport	0 0 0 0 0 0 1 1
Backward call indicators	0 0 0 1 0 0 0 1
Business group	1 1 0 0 0 1 1 0
Call reference	0 0 0 0 0 0 0 1
Cause indicators	0 0 0 1 0 0 1 0
Information indicators	0 0 0 0 1 1 1 1
Network transport parameter	1 1 1 0 1 1 1 1
Notification indicator	1 1 1 0 0 0 0 1
Optional backward call indicators	0 0 1 0 1 0 0 1
Redirecting number	0 0 0 0 1 0 1 1
Remote operations	0 0 1 1 0 0 1 0
Service activation	0 0 1 1 0 0 1 1
Transmission medium used	0 0 1 1 0 1 0 1
User-to-user indicators	0 0 1 0 1 0 1 0
User-to-user information	0 0 1 0 0 0 0 0

The call progress message is used to notify a distant exchange that some event has occurred during the progress of a call. The event is not a catastrophic event, or an error-related event, but a call-related event. The event information parameter indicates what type of event occurred (alerting message was received, the call was forwarded because of a busy, etc.) while the optional parameters provide additional support information required depending on the event.

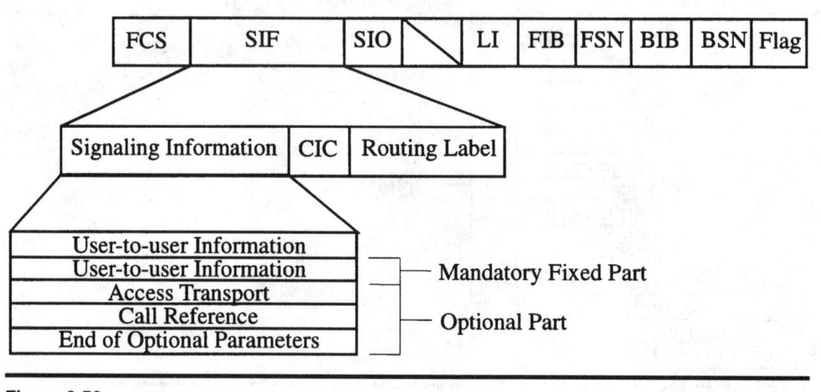

Figure 9.50

User-to-User Information (USR) **0 0 1 0 1 1 0 1**

This message does not have any definition in ANSI networks, and is used only in international ITU-TS networks.

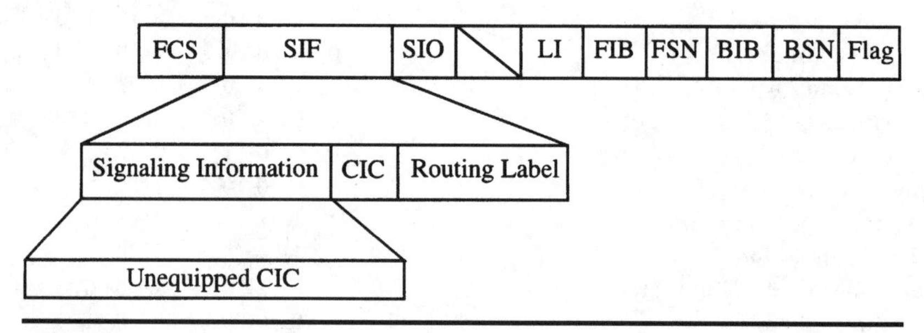

Figure 9.51

Unequipped Circuit Identification Code (UCIC) **0 0 1 0 1 1 1 0**

There are no parameters in this message. This message is used to notify a distant exchange that is the originator of an ISUP Initial Address Message (IAM) that the Circuit Identification Code (CIC) it has requested to be connected is not equipped. Upon receipt of this message, the exchange that originated the IAM must select a different CIC, while marking the first CIC as unavailable.

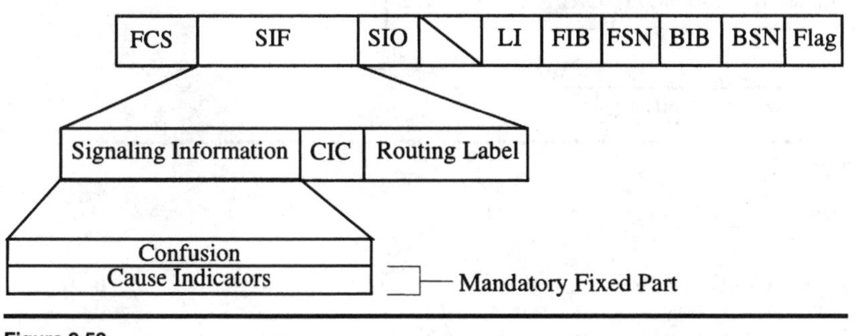

Figure 9.52

Confusion (CFN)	**0 0 1 0**	**1 1 1 1**
Mandatory Variable Parameters		
Cause indicators	0 0 0 1	0 0 1 0

The confusion message indicates that the exchange has received a message it does not recognize, and it does not know how to handle the message. The confusion message is sent to the originator of the ISUP message. This only applies to ISUP messages, and does not apply to TCAP, SCCP, or any other protocol message other than ISUP. The cause indicators parameter indicates where the confusion message was originated, as well as why the message is being sent.

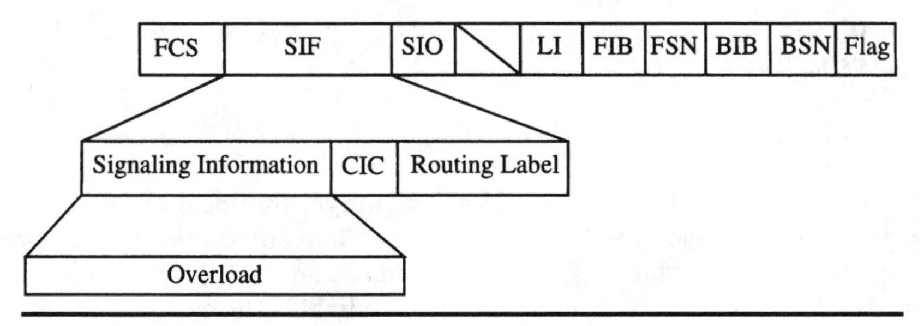

Figure 9.53

Overload (OLM)	**0 0 1 1**	**0 0 0 0**

This message is not supported in ANSI networks, and is used only in international ITU-TS networks.

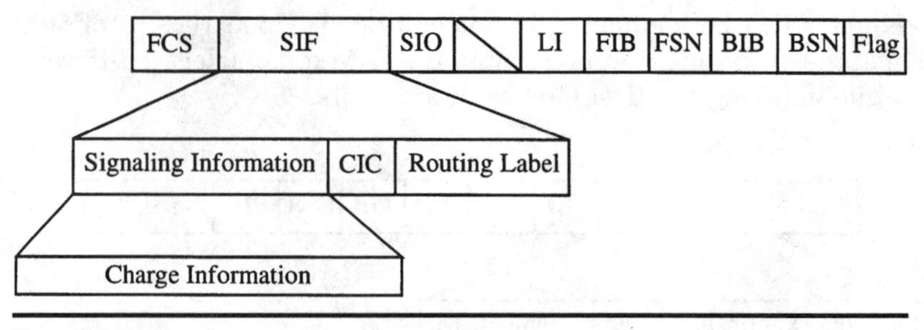

Figure 9.54

Charge Information (CRG) **0 0 1 1 0 0 0 1**

This message is not supported in ANSI networks, and is used only in international ITU-TS networks.

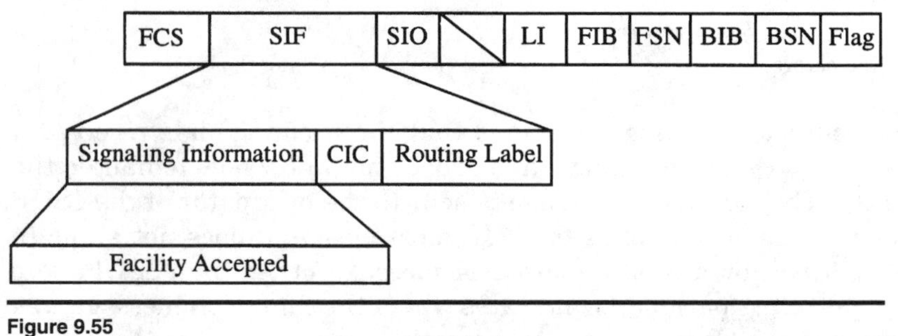

Figure 9.55

Facility (FAC)	**0 0 1 1**	**0 0 1 1**
Optional Parameters		
Remote operations	0 0 1 1	0 0 1 0
Service activation	0 0 1 1	0 0 0 0

This message may be sent by either exchange, the local or the distant, to request an action at that exchange. The same message may also be used as an acknowledgment that the action was performed successfully. The service activation parameter indicates the type of service that is being requested (or has been invoked in the case of an acknowledgment). Call waiting is defined in the Bellcore standard, but all other codes are considered as network specific and are undefined in the standards.

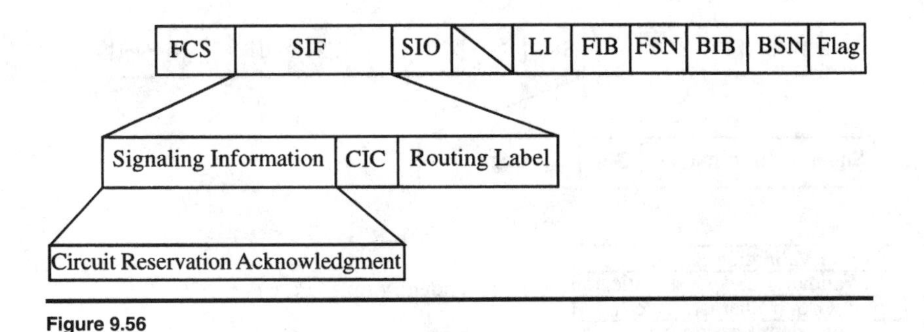

Figure 9.56

Circuit Reservation Acknowledgment (CRA) **1 1 1 0 1 0 0 1**

There are no parameters in this message. This is sent to an exchange after receipt of a Circuit Reservation Message (CRM) as an acknowledgment that the circuit has been reserved for a call. This message only applies when the circuit reservation procedure is incorporated.

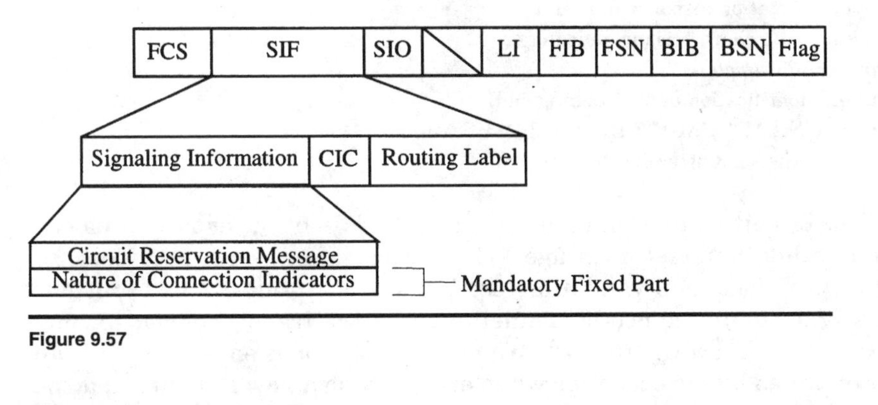

Figure 9.57

Circuit Reservation (CRM) **1 1 1 0 1 0 1 0**

Mandatory Fixed Parameters

Nature of connection indicators 0 0 0 0 0 1 1 0

This is used only interworking with a non-SS7 network, such as an analog network using MF signaling. This is typically the case in rural areas here in the U.S., although rural areas are quickly being converted to SS7 networks. Where SS7 is not available, the exchanges rely on conventional signaling methods (such as MF), which require special handling by the SS7 network. Circuit reservation allows the voice circuit to be reserved for a call. See the section on interworking earlier in this chapter to understand how this works. This procedure is only used in ANSI networks.

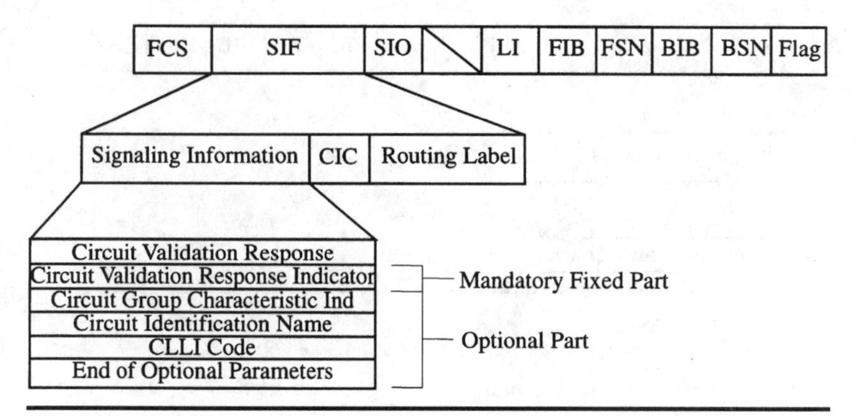

Figure 9.58

Circuit Validation Response (CVR)	**1 1 1 0 1 0 1 1**
Mandatory Fixed Parameters	
Circuit validation response indicator	1 1 1 0 0 1 1 0
Circuit group characteristic indicators	1 1 1 0 0 1 0 1
Optional Parameters	
Circuit identification name (sending end)	1 1 1 0 1 0 0 0
COMMON LANGUAGE™ Location Identification Code (for the sending end)	1 1 1 0 1 0 0 1

The circuit validation response message is sent in response to a circuit validation test message. The response message provides the results of the circuit validation test. The circuit validation test provides a way for maintenance personnel to check the translations at far-end exchanges and verify that when a new translation is entered, a CIC can be obtained by the exchange when establishing a new call, and that the physical port associated with that CIC can be seized. The response message will indicate whether this test passed or failed. This procedure is only used in ANSI networks.

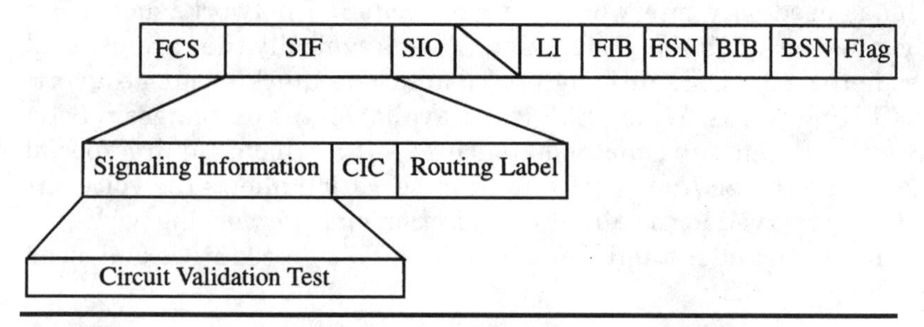

Figure 9.59

Circuit Validation Test (CVT) **1 1 1 0 1 1 0 0**

There are no parameters for this message. This message is used to initiate a translations test at a distant exchange. The circuit validation response message provides the results of the circuit validation test. The primary purpose of this test is to verify that new translations were entered properly, and a physical port can be connected between two exchanges.

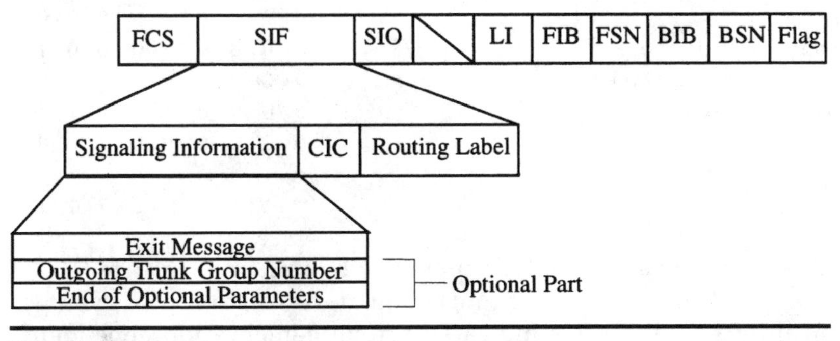

Figure 9.60

Exit (EXM) **1 1 1 0 1 1 0 1**

Optional Parameters

Outgoing trunk group number 1 1 1 0 0 1 1 1

The Exit Message (EXM) is used only when interworking with another network. When an IAM is sent to a gateway to establish a connection in another network, the EXM is sent in the backward direction to indicate that the IAM has passed the gateway and is being forwarded to the other network.

ISUP Parameters

The previous section defined all the message types and their structures. Each of the parameters possible with any one message type was outlined, and the parameter name value provided. However, parameters contain additional information besides the parameter name.

Because parameters can be used in multiple message types, it is easier to list them all here in alphabetical order. In this section, all parameters for both ANSI and ITU-TS are listed, along with illustrations showing the message structure of each parameter. To find this information in any of the standards publications you would have to refer to several sections, because of the way the information is segregated. All of the information regarding parameters has been grouped here in one section for easy reference.

Access transport

The access transport parameter provides the parameters per the ANSI T1.607 standard. The method of transport parameter is found in the IAM message as well as the ACM.

Automatic Congestion Level	**H G F E**	**D C B A**
Spare	0 0 0 0	0 0 0 0
Automatic Congestion Level 1	0 0 0 0	0 0 0 1
Automatic Congestion Level 2	0 0 0 0	0 0 1 0
Automatic Congestion Level 3	0 0 0 0	0 0 1 1
Spare	0 0 0 0	0 1 0 0
	to	
	1 1 1 1	1 1 1 1

This is an optional parameter that may be sent in a release (REL) message to indicate the level of congestion at the originating exchange. The meanings of the levels indicated are implementation dependent. The action taken by the receiving exchange is also implementation dependent. This parameter simply provides an alerting mechanism to show some level of congestion exists in another exchange.

Backward Call Indicators	**H G**	**F E**	**D C**	**B A**
First Octet				
Charge Indicators				
No indication				0 0
No charge				0 1
*Charge				1 0
Spare				1 1
Called party's status indicator				
No indication			0 0	
Subscriber free			0 1	
*Connect when free			1 0	
Excessive delay			1 1	
Called party's category indicator				
No indication		0 0		
Ordinary subscriber		0 1		
*Payphone		1 0		
Spare		1 1		
End-to-end method indicator				
No end-to-end method available	0 0			
Pass Along method available	0 1			
SCCP method available	1 0			
Pass Along and SCCP methods available	1 1			

Second Octet	H	G	F	E	D	C	B	A
Interworking indicator								
No interworking encountered								0
Interworking encountered								1
IAM Segmentation Indicator								
No indication							0	
Additional info has been received and added to IAM							1	
ISDN User Part Indicator								
ISUP not used all the way (end to end)						0		
ISUP used all the way (end to end)						1		
Holding Indicator								
Holding not required					0			
*Holding required					1			
ISDN Access Indicator								
Terminating access non-ISDN				0				
Terminating access ISDN				1				
Echo Control Device Indicator								
Incoming half echo control device not included			0					
Incoming half echo control device included			1					
SCCP Method Indicator								
No indication	0	0						
*Connectionless method available	0	1						
*Connection-oriented method available	1	0						
*Connectionless and connection-oriented available	1	1						

The backward call indicators parameter is sent in the backward direction (back to the originating exchange) to provide information regarding charging, status of the called party, and various other forms of information which may be needed to complete processing of a call.

The charge indicator indicates whether or not the call is a chargeable call. If the call is chargeable, then the number to which the call is to be charged is also provided. The status parameter indicates whether or not the called party is available. If not, then no indication is given in the charge field and the calling party is returned a busy tone.

The category indicator indicates which category (pay phone, etc.) the called party fits. This information is used during call processing and may imply special handling of the call (for example, a pay phone may require special handling for billing and operator assistance).

The end-to-end indicator is used by the protocol to designate the method of ISUP signaling available for this call. The protocol can then use this information to make a decision as to which method will be used. In ANSI networks the pass along method is most widely used. The SCCP method is not supported in ANSI networks.

Interworking encountered means that another signaling system other than SS7 will be encountered for this call. For example, an exchange may still be using MF signaling, and is needed to complete the path for this call. In this case, the SS7 interworking procedures discussed early in this chapter will have to be used.

The ISUP indicator will tell whether or not ISUP is used through all segments of the network involved with this call. If this field shows that ISUP is used all the way, then there will be no interworking encountered, and vice versa.

This parameter also indicates if ISDN is the interface to the subscriber or not. If ISDN is the subscriber interface, then special handling is required at the interface to signal the subscriber being called. There is direct mapping for SS7 signaling to ISDN, as the ISDN protocol was developed as an extension of SS7.

Echo cancelers are indicated in the echo control device indicator. This information may be used by the originating switch when encoding voice signals to prevent original signals from being mistaken for noise, and may trigger special digitization of the voice.

The last field is the SCCP method indicator, which indicates the type of SCCP method available for this call. Since SCCP services are not supported with ISUP in ANSI networks, this should only be found in private and international networks.

Business Group

Octet 1	H	G	F	E	D	C	B	A
Party Selector								
No indication					0	0	0	0
Calling party number					0	0	0	1
Called party number					0	0	1	0
Connected party number					0	0	1	1
Redirecting number					0	1	0	0
Original called number					0	1	0	1
Spare					0	1	1	0
						to		
					1	1	1	1
Line privilege information indicator								
Fixed line privileges				0				
Customer defined line privileges				1				
Business group identifier type								
Multilocation business group identifier			0					
Interworking with private networks identifier			1					
Attendant status								
No indication		0						
Attendant line		1						
Spare	0							

Octets 2, 3, and 4	H	G	F	E	D	C	B	A
Business group identifier								

	HGFE	DCBA
No indication	0 0 0 0	0 0 0 0
Public network	0 0 0 0	0 0 0 1
Network dependent	0 0 0 0	0 0 1 0
	to	
	1 1 1 1	1 1 1 1

Binary representation of the actual number assigned may be three octets long.

Octets 5 and 6	**H G F E**	**D C B A**
Subgroup identifier		
No subgroups	0 0 0 0	0 0 0 0

All other values represent the number assigned to any subgroups and may be two octets long.

Octet 7	**H G F E**	**D C B A**
If line privileges information indicator = 0		
Terminating line privileges		
Unrestricted		0 0 0 0
Semirestricted		0 0 0 1
Fully restricted		0 0 1 0
Fully restricted, intra-switch		0 0 1 1
Denied		0 1 0 0
Spare		0 1 0 1
		to
		1 1 1 1
Originating restrictions		
Unrestricted	0 0 0 0	
Semirestricted	0 0 0 1	
Fully restricted	0 0 1 0	
Fully restricted, intra switch	0 0 1 1	
Denied	0 1 0 0	
Spare	0 1 0 1	
	to	
	1 1 1 1	
If line privileges information indicator = 1		
Customer defined line privilege codes	0 0 0 0	0 0 0 0
	to	
	1 1 1 1	1 1 1 1

The business group parameter identifies the properties of a group of subscriber lines that belong to a common subscriber (such as Centrex services). The business group is assigned specific features and restrictions, which must be identified through the protocol during call setup and information sharing between exchanges.

The party indicator identifies the type of number, called or calling, original called (in the case of forwarded numbers), and connected numbers. The fixed line privileges indicates the type of restrictions to be applied, those already defined (fixed) by the protocol, or those defined by the customer.

The business group identifier indicates whether the business group is one of multiple locations or if the group requires interworking with a private network. The attendant status is also indicated when the business group attendant places calls through the network.

The business group identifier and subgroup identifiers are the actual numbers assigned to these groups by the network provider. The service provider allocates the business group numbers.

Call reference

The call reference is a number assigned to a call for tracking of messages and for use as a reference for additional information exchanged between two offices during the duration of the call. The first octet is the call identity number, which uniquely identifies each call. The number is of local significance only. The second octet bears the point code that assigned the call identity number. The point code indicated would be the only point code for which the identity number is of any significance.

Called Party Number

Octet 1	H G F E	D C B A
nature of address indicator		
Spare	0 0 0	0 0 0 0
Subscriber number	0 0 0	0 0 0 1
Spare, reserved for national use	0 0 0	0 0 1 0
National significant number	0 0 0	0 0 1 1
International number	0 0 0	0 1 0 0
Spare	0 0 0	0 1 0 1
		to
	1 1 1	0 0 0 0
Subscriber number, operator requested	1 1 1	0 0 0 1
National number, operator requested	1 1 1	0 0 1 0
International number, operator requested	1 1 1	0 0 1 1
No number present, operator requested	1 1 1	0 1 0 0
No number present, cut-through call to carrier	1 1 1	0 1 0 1
950 + call from local exchange carrier public station, hotel/motel, or nonexchange access end-office	1 1 1	0 1 1 0
Test line code	1 1 1	0 1 1 1
Reserved for network specific use	1 1 1	1 0 0 0
		to
	1 1 1	1 1 1 0

	H	G	F	E	D	C	B	A
Spare		1	1	1	1	1	1	1
Odd/even bits								
Even number of address signals	0							
Odd number of address signals	1							
Octet 2	**H**	**G**	**F**	**E**	**D**	**C**	**B**	**A**
Reserved					0	0	0	0
Numbering Plan								
Unknown numbering plan		0	0	0				
ISDN numbering plan (Rec. E.164, E.163)		0	0	1				
Spare		0	1	0				
Reserved ITU-TS Data Numbering Plan		0	1	1				
Reserved ITU-TS Telex Numbering Plan		1	0	0				
Private numbering plan		1	0	1				
Spare		1	1	0				
Spare		1	1	1				
Spare	0							
Octets 3–n	**H**	**G**	**F**	**E**	**D**	**C**	**B**	**A**
Address Signal—1st address								
Digit 0					0	0	0	0
Digit 1					0	0	0	1
Digit 2					0	0	1	0
Digit 3					0	0	1	1
Digit 4					0	1	0	0
Digit 5					0	1	0	1
Digit 6					0	1	1	0
Digit 7					0	1	1	1
Digit 8					1	0	0	0
Digit 9					1	0	0	1
Spare					1	0	1	0
Code 11					1	0	1	1
Code 12					1	1	0	0
Spare					1	1	0	1
Spare					1	1	1	0
End of pulse signal					1	1	1	1
Address Signal—2nd address								
Digit 0	0	0	0	0				
Digit 1	0	0	0	1				
Digit 2	0	0	1	0				
Digit 3	0	0	1	1				
Digit 4	0	1	0	0				
Digit 5	0	1	0	1				
Digit 6	0	1	1	0				
Digit 7	0	1	1	1				
Digit 8	1	0	0	0				

Digit 9	1 0 0 1
Spare	1 0 1 0
Code 11	1 0 1 1
Code 12	1 1 0 0
Spare	1 1 0 1
Spare	1 1 1 0
End of pulse signal	1 1 1 1

The called party address can be as large as needed to accommodate the dialed digits. The above only defines the first three octets, although the actual parameter will be larger. Each of the dialed digits uses 4 bits, as shown above. The purpose of this parameter is to provide the distant exchange the number of the called party (dialed digits), or the location routing number (LRN) associated with the called party. This parameter can be used in a number of message types, and is usually used when establishing a connection from one exchange to another. If the dialed digits are to a ported number, the nature of address indicator will indicate a ported number. If this parameter contains the LRN to be used for routing, the dialed digits are placed in the GAP parameter.

Calling Party Number

Octet 1	**H**	**G F E**	**D C B A**
Nature of address indicator			
Spare		0 0 0	0 0 0 0
Unique subscriber number		0 0 0	0 0 0 1
Spare, reserved for national use		0 0 0	0 0 1 0
Unique national significant number		0 0 0	0 0 1 1
Unique international number		0 0 0	0 1 0 0
Spare		0 0 0	0 1 0 1
		to	
		1 1 1	0 0 0 0
Non-unique subscriber number		1 1 1	0 0 0 1
Spare, reserved for national use		1 1 1	0 0 1 0
Non-unique national number		1 1 1	0 0 1 1
Non-unique international number		1 1 1	0 1 0 0
Spare		1 1 1	0 1 0 1
Spare		1 1 1	0 1 1 0
Test line code		1 1 1	0 1 1 1
Reserved for network-specific use		1 1 1	1 0 0 0
		to	
		1 1 1	1 1 1 0
Spare		1 1 1	1 1 1 1
Odd/even indicator			
Even number of address signals	0		
Odd number of address signals	1		

Octet 2	H	G	F	E	D	C	B	A
Screening								
Reserved (for user provided, not screened)							0	1
User provided, screening passed							1	0
Network provided							1	1
Address Presentation								
Presentation allowed					0	0		
Presentation restricted					0	1		
Spare					1	0		
Spare					1	1		
Numbering Plan								
Unknown numbering plan		0	0	0				
ISDN numbering plan (Rec. E.164, E.163)		0	0	1				
Spare		0	1	0				
Reserved ITU-TS Data Numbering Plan		0	1	1				
Reserved ITU-TS Telex Numbering Plan		1	0	0				
Private numbering plan		1	0	1				
Spare		1	1	0				
Spare		1	1	1				
Spare	0							

Octets 3–n	H	G	F	E	D	C	B	A
Address Signal—1st address								
Digit 0					0	0	0	0
Digit 1					0	0	0	1
Digit 2					0	0	1	0
Digit 3					0	0	1	1
Digit 4					0	1	0	0
Digit 5					0	1	0	1
Digit 6					0	1	1	0
Digit 7					0	1	1	1
Digit 8					1	0	0	0
Digit 9					1	0	0	1
Spare					1	0	1	0
Code 11					1	0	1	1
Code 12					1	1	0	0
Spare					1	1	0	1
Spare					1	1	1	0
End of pulse signal					1	1	1	1
Address Signal—2nd address								
Digit 0	0	0	0	0				
Digit 1	0	0	0	1				
Digit 2	0	0	1	0				
Digit 3	0	0	1	1				
Digit 4	0	1	0	0				
Digit 5	0	1	0	1				

Digit 6	0 1 1 0
Digit 7	0 1 1 1
Digit 8	1 0 0 0
Digit 9	1 0 0 1
Spare	1 0 1 0
Code 11	1 0 1 1
Code 12	1 1 0 0
Spare	1 1 0 1
Spare	1 1 1 0
End of pulse signal	1 1 1 1

The calling party's address is much the same as the called party's address, for the exception of the second octet. Some differences also exist in the nature of address indicator.

The screening indicator is used to indicate who provided the dialed digits in the address. The digits could have been provided by the subscriber (originating party) or by the network (global title translation). The purpose of this parameter is to provide a means for the network to determine the origin of the digits and whether or not this number is viewed by the network as the true called party number or as an alias. In the case of ported numbers, the parameter will not contain the LRN of the calling party. The LRN of the calling party is found in the JIP parameter.

The address presentation parameter serves the same purpose as in the calling party address. The default for this parameter is to be "presentation restricted." The presentation of the dialed digits allows called parties to identify who is calling them (caller party ID). When presentation is restricted, the number can still be presented to another network, but it is restricted from presentation to an end user.

Calling Party's Category	**H G F E**	**D C B A**
Calling party's category unknown (default)	0 0 0 0	0 0 0 0
*French language operator	0 0 0 0	0 0 0 1
*English language operator	0 0 0 0	0 0 1 0
*German language operator	0 0 0 0	0 0 1 1
*Russian language operator	0 0 0 0	0 1 0 0
*Spanish language operator	0 0 0 0	0 1 0 1
Reserved for network-dependent language selection	0 0 0 0	0 1 1 0
	0 0 0 0	0 1 1 1
	0 0 0 0	1 0 0 0
National networks—operator service	0 0 0 0	1 0 0 1
Ordinary calling subscriber	0 0 0 0	1 0 1 0
*Calling subscriber with priority	0 0 0 0	1 0 1 1
*Data call (voiceband data)	0 0 0 0	1 1 0 0
Test call	0 0 0 0	1 1 0 1

Spare	0	0	0	0	1	1	1	0
*Pay phone	0	0	0	0	1	1	1	1
Spare (ITU-TS)	0	0	0	1	0	0	0	0
				to				
	1	1	0	1	1	1	1	1
Emergency service call in progress	1	1	1	0	0	0	0	0
High-priority call indication	1	1	1	0	0	0	0	1
National Security & Emergency Preparedness Call	1	1	1	0	0	0	1	0
Spare (ANSI)	1	1	1	0	0	0	1	1
				to				
	1	1	1	0	1	1	1	1
Network-specific use	1	1	1	1	0	0	0	0
				to				
	1	1	1	1	1	1	1	0
Reserved for expansion	1	1	1	1	1	1	1	1

The calling party's category indicates the type of subscriber originating the call. In the case of a special-language operator, the originator of the call (the operator services) will require special handling. The same is true of the pay phone, which may require special operator assistance.

The test call indicator in this parameter is used for remote testing of translations (as discussed earlier in this chapter). A technician may initiate a test call from a remote maintenance center terminal to test and verify newly added translations or routing information. The IAM would then include the calling party's category parameter with the test call value.

Carrier Identification

Octet 1	**H**	**G**	**F**	**E**	**D**	**C**	**B**	**A**
Network identification plan								
Unknown					0	0	0	0
3-digit carrier identification code					0	0	0	1
4-digit carrier identification code					0	0	1	0
Spare					0	0	1	1
							to	
					1	1	1	1
Type of network identification		0	0	0				
Spare		0	0	0				
Spare		0	0	1				
National network identification		0	1	0				
Spare		0	1	1				
			to					
		1	1	1				
Spare	0							

Octet 2	H	G	F	E	D	C	B	A
Digit One								
Digit 0					0	0	0	0
Digit 1					0	0	0	1
Digit 2					0	0	1	0
Digit 3					0	0	1	1
Digit 4					0	1	0	0
Digit 5					0	1	0	1
Digit 6					0	1	1	0
Digit 7					0	1	1	1
Digit 8					1	0	0	0
Digit 9					1	0	0	1
Spare					1	0	1	0
Code 11					1	0	1	1
Code 12					1	1	0	0
Spare					1	1	0	1
Spare					1	1	1	0
End of pulse signal					1	1	1	1
Digit Two								
Digit 0	0	0	0	0				
Digit 1	0	0	0	1				
Digit 2	0	0	1	0				
Digit 3	0	0	1	1				
Digit 4	0	1	0	0				
Digit 5	0	1	0	1				
Digit 6	0	1	1	0				
Digit 7	0	1	1	1				
Digit 8	1	0	0	0				
Digit 9	1	0	0	1				
Spare	1	0	1	0				
Code 11	1	0	1	1				
Code 12	1	1	0	0				
Spare	1	1	0	1				
Spare	1	1	1	0				
End of pulse signal	1	1	1	1				

Octet 3	H	G	F	E	D	C	B	A
Digit Three								
Digit 0					0	0	0	0
Digit 1					0	0	0	1
Digit 2					0	0	1	0
Digit 3					0	0	1	1
Digit 4					0	1	0	0
Digit 5					0	1	0	1
Digit 6					0	1	1	0

Digit 7		0 1 1 1
Digit 8		1 0 0 0
Digit 9		1 0 0 1
Spare		1 0 1 0
Code 11		1 0 1 1
Code 12		1 1 0 0
Spare		1 1 0 1
Spare		1 1 1 0
End of pulse signal		1 1 1 1
Digit Four (if four-digit code is used)		
Digit 0	0 0 0 0	
Digit 1	0 0 0 1	
Digit 2	0 0 1 0	
Digit 3	0 0 1 1	
Digit 4	0 1 0 0	
Digit 5	0 1 0 1	
Digit 6	0 1 1 0	
Digit 7	0 1 1 1	
Digit 8	1 0 0 0	
Digit 9	1 0 0 1	
Spare	1 0 1 0	
Code 11	1 0 1 1	
Code 12	1 1 0 0	
Spare	1 1 0 1	
Spare	1 1 1 0	
End of pulse signal	1 1 1 1	

Carrier Selection Information	**H G F E**	**D C B A**
Carrier selection information		
No indication (default)	0 0 0 0	0 0 0 0
Subscribers designated (preselected) carrier	0 0 0 0	0 0 0 1
Subscribers designated carrier as input by caller	0 0 0 0	0 0 1 0
Subscribers designated carrier (undetermined)	0 0 0 0	0 0 1 1
Carrier designated by caller at time of call	0 0 0 0	0 1 0 0
Spare	0 0 0 0	0 1 0 1
	to	
	1 1 1 1	1 1 1 0
Reserved	1 1 1 1	1 1 1 1

The carrier identification parameter is used to identify the carrier selected by the caller. The selection can be accomplished in one of several ways. All subscribers have preselected carriers listed in the Line Information Database (LIDB) as a part of their customer record. This is the carrier to be used for all long-distance calls, except when another carrier is selected manually.

Any carrier can be selected by dialing the carrier access code. The carrier access code is the 10xxx number assigned to all long-distance carriers. This number allows a caller to use any long distance provider for that call. After the call is finished the next call defaults back to the preselected carrier.

A subscriber may also be calling from a pay phone or some other phone where a long-distance carrier has not been preselected, but the caller has entered in a 10xxx code for carrier selection. This is a rarity in U.S. networks, as equal access has required that all subscribers have a designated carrier preselected for every line.

Cause Indicators

Octet 1	H	G	F	E	D	C	B	A
Location								
User					0	0	0	0
Local private network					0	0	0	1
Local local network					0	0	1	0
Transit network					0	0	1	1
Remote local network					0	1	0	0
Remote private network					0	1	0	1
Local interface controlled by this signaling link					0	1	1	0
International network					0	1	1	1
Network beyond interworking point					1	0	1	0
Spare				0				
Coding Standard								
ITU-TS Standard (default)		0	0					
Reserved for other international standards		0	1					
ANSI standard		1	0					
Reserved		1	1					
Extension Bit								
Parameter continues to next octet	0							
Last octet	1							

Octet 2

Cause Value—The cause value provides the reason for the message failure. These cause codes are grouped as ITU-TS and ANSI codes. All cause codes are divided into two parts; bits ABCD represent the cause while bits EFG represent the class.

ITU-TS cause codes (coding standard 0 0)

Class 0 0 0 and 0 0 1, normal event	H	G	F	E	D	C	B	A
Unallocated (unassigned) number		0	0	0	0	0	0	1
No route to specified transit network		0	0	0	0	0	1	0
No route to destination		0	0	0	0	0	1	1

Send special information tone	0 0 0	0 1 0 0
*Misdialed trunk prefix	0 0 0	0 1 0 1
Preemption	0 0 0	1 0 0 0
Preemption—circuit reserved for reuse	0 0 0	1 0 0 1
Normal clearing	0 0 1	0 0 0 0
User busy	0 0 1	0 0 0 1
No user responding	0 0 1	0 0 1 0
No answer from user (user alerted)	0 0 1	0 0 1 1
Subscriber absent	0 0 1	0 1 0 0
Call rejected	0 0 1	0 1 0 1
Number changed	0 0 1	0 1 1 0
Redirect to new destination	0 0 1	0 1 1 1
Destination out of order	0 0 1	1 0 1 1
Address incomplete	0 0 1	1 1 0 0
Facility rejected	0 0 1	1 1 0 1
Normal—unspecified (default)	0 0 1	1 1 1 1

Class 0 1 0, resource unavailable	**H G F E**	**D C B A**
No circuit/channel available	0 1 0	0 0 1 0
Network out of order	0 1 0	0 1 1 0
Temporary failure	0 1 0	1 0 0 1
Switching equipment congestion	0 1 0	1 0 1 0
Access information discarded	0 1 0	1 0 1 1
Requested circuit/channel not available	0 1 0	1 1 0 0
Precedence call blocked	0 1 0	1 1 1 0
Resource unavailable—unspecified (default)	0 1 0	1 1 1 1

Class 0 1 1, service or option not available	**H G F E**	**D C B A**
Requested facility not subscribed	0 1 1	0 0 1 0
*Outgoing calls barred within Closed User Group	0 1 1	0 1 0 1
*Incoming calls barred within Closed User Group	0 1 1	0 1 1 1
Bearer capability not authorized	0 1 1	1 0 0 1
Bearer capability not presently available	0 1 1	1 0 1 0
Inconsistency in designated outgoing access and subscriber class	0 1 1	1 1 1 0
Service option not available—unspecified (default)	0 1 1	1 1 1 1

Class 1 0 0, service or option not implemented	**G G F E**	**D C B A**
Bearer capability not implemented	1 0 0	0 0 0 1
Requested facility not implemented	1 0 0	0 1 0 1
Only restricted digital info bearer capability available	1 0 0	0 1 1 0
Service not implemented—unspecified (default)	1 0 0	1 1 1 1

Class 1 0 1, invalid message	**H G F E**	**D C B A**
*User not member of Closed User Group	1 0 1	0 1 1 1
Incompatible destination	1 0 1	1 0 0 0

	HGFE	DCBA
*Nonexistent Closed User Group	1 0 1	1 0 1 0
Invalid transit network selection	1 0 1	1 0 1 1
Invalid message, unspecified (default)	1 0 1	1 1 1 1
Class 1 1 0, protocol error (i.e., unknown message)	**H G F E**	**D C B A**
Message type nonexistent or not implemented	1 1 0	0 0 0 1
Information element parameter nonexistent/not implemented	1 1 0	0 0 1 1
Recovery on timer expiration	1 1 0	0 1 1 0
Parameter nonexistent/not implemented—passed on	1 1 0	0 1 1 1
Message with unrecognized parameter discarded	1 1 0	1 1 1 0
Protocol error, unspecified (default)	1 1 0	1 1 1 1
Class 1 1 1, interworking class	**H G F E**	**D C B A**
Interworking, unspecified (default)	1 1 1	1 1 1 1

ANSI cause codes (coding standard 1 0)

Class 0 0 0 and 0 0 1, normal event	**H G F E**	**D C B A**
Unallocated destination number	0 0 1	0 1 1 1
Unknown business group	0 0 1	1 0 0 0
Exchange routing error	0 0 1	1 0 0 1
Misrouted call to a ported number	0 0 1	1 0 1 0
Number portability Query on Release (QoR) number not found	0 0 1	1 0 1 1
Class 0 1 0, resource unavailable		
Preemption	0 1 0	1 1 0 1
Precedence call blocked	0 1 0	1 1 1 0
Class 0 1 1, service or option not available	**H G F E**	**D C B A**
Call type incompatibility with service request	0 1 1	0 0 1 1
Call blocked due to group restrictions	0 1 1	0 1 1 0
Extension Bit	0	
Octet 3	**H G F E**	**D C B A**
Diagnostics (if applicable)	0 0 0 0	0 0 0 0

The diagnostics field is dependent on the cause value. Not all cause codes will require a diagnostics field afterward. The diagnostic field uses the same format as the specified parameters (i.e., called party number). The following lists the cause codes that generate a diagnostics field, and the parameter structure used for the diagnostic value.

Cause Code	Diagnostic	Structure
0 0 1 0 1 1 0	Called party number (new)	See called party parameter
0 1 0 0 1 1 0	Transit network identity	Transit network selection
0 1 0 1 0 1 0	Transit network identity	Transit network selection
0 1 1 1 0 0 1	Attribute identity	See below
0 1 1 1 0 1 0	Attribute identity	See below
1 0 0 0 0 0 1	Attribute identity	See below

Attribute identity	H G F E	D C B A
Information transfer capability	0 1 1	0 0 0 1
Information transfer mode	0 1 1	0 0 1 0
Information transfer rate	0 1 1	0 0 1 1
Structure	0 1 1	0 1 0 0
Configuration	0 1 1	0 1 0 1
Establishment	0 1 1	0 1 1 0
Symmetry	0 1 1	0 1 1 1
Information transfer rate (destination to origination)	0 1 1	1 0 0 0
Layer identification and corresponding user info	0 1 1	1 0 0 1

This rather lengthy parameter is full of variables, and is dependent on the cause as to the full contents of the parameter. The parameter can be found in Release (REL) messages, Address Complete Messages (ACM), or Confusion (CON) messages. The purpose is to identify the cause for the failure or disconnect or message rejection. Appendix C provides full explanations for all cause codes and diagnostics, to save space here.

Charge Number

Octet 1	H G F E	D C B A
Nature of address indicator		
Spare	0 0 0	0 0 0 0
ANI of the calling party; subscriber number	0 0 0	0 0 0 1
ANI not available or not provided	0 0 0	0 0 1 0
ANI of the calling party; national number	0 0 0	0 0 1 1

	H G F E	D C B A
Spare	0 0 0	0 1 0 0
ANI of the called party; subscriber number	0 0 0	0 1 0 1
ANI of the called party; no number present	0 0 0	0 1 1 0
ANI of the called party; national number	0 0 0	0 1 1 1
Spare	0 0 0	1 0 0 0
	to	
	1 1 1	0 1 1 1
Reserved for network-specific use	1 1 1	1 0 0 0
	to	
	1 1 1	1 1 1 0
Spare	1 1 1	1 1 1 1
Odd/Even bit		
Even number of address signals	0	
Odd number of address signals	1	
Octet 2	**H G F E**	**D C B A**
Reserved		0 0 0 0
Numbering plan		
Unknown	0 0 0	
ISDN numbering plan (Rec E.164, E.163)	0 0 1	
Spare	0 1 0	
Reserved (ITU data numbering plan)	0 1 1	
Reserved (ITU Telex numbering plan)	1 0 0	
Private numbering plan	1 0 1	
Spare	1 1 0	
Spare	1 1 1	
Spare	0	
Octet 3	**H G F E**	**D C B A**
1st address signal		
Digit 0		0 0 0 0
Digit 1		0 0 0 1
Digit 2		0 0 1 0
Digit 3		0 0 1 1
Digit 4		0 1 0 0
Digit 5		0 1 0 1
Digit 6		0 1 1 0
Digit 7		0 1 1 1
Digit 8		1 0 0 0
Digit 9		1 0 0 1
Spare		1 0 1 0
Code 11		1 0 1 1
Code 12		1 1 0 0
2nd address signal		
Digit 0	0 0 0 0	

Digit 1	0 0 0 1
Digit 2	0 0 1 0
Digit 3	0 0 1 1
Digit 4	0 1 0 0
Digit 5	0 1 0 1
Digit 6	0 1 1 0
Digit 7	0 1 1 1
Digit 8	1 0 0 0
Digit 9	1 0 0 1
Spare	1 0 1 0
Code 11	1 0 1 1
Code 12	1 1 0 0
Spare	1 1 0 1
Spare	1 1 1 0
End of pulse signal	1 1 1 1

Circuit Assignment Map

Octet 1	**H**	**G**	**F**	**E**	**D**	**C**	**B**	**A**
Map Type								
Spare			0	0	0	0	0	0
DS1 map format			0	0	0	0	0	1
Spare			0	0	0	0	1	0
					to			
			1	1	1	1	1	1
Spare	0	0						

Octets 2–4	**H**	**G**	**F**	**E**	**D**	**C**	**B**	**A**
Map	x	x	x	x	x	x	x	x
0 = 64-kbps circuit is not used								
1 = 64-kbps circuit is used								

The map portion of this parameter provides a 1-bit representation for each circuit. If the value of the circuit bit is a 1, then the 64-kbps circuit is used. If the value is a 0, it is not used. Up to 24 circuits may be represented in the map fields. All 24 circuits are represented, but set to 0 if the circuits are not used. This means that this parameter is always a fixed length and not variable.

This parameter is only used when DS0 circuits that are not contiguous are used. Noncontiguous means that a full DS1 is not being utilized for these circuits. There may be a partial DS1 or the various DS0s may be split between multiple DS1s. Either way, the map indicates which circuits are or are not used within a DS1.

Circuit Group Supervision Msg Type Indicator

	H	G	F	E	D	C	B	A
Circuit group blocking type indicator								
Block without release							0	0
Block with immediate release							0	1
Reserved for national use							1	0
Spare							1	1
Spare	0	0	0	0	0	0		

This parameter provides instructions to another exchange on the method of circuit blocking to be implemented on the designated circuit. Blocking allows craft personnel to perform tests on a circuit without the change of the circuit being seized for another call.

There are two methods of blocking, blocking with a release (which disconnects any call in progress) and blocking without release (which will not send a release message through the network).

Circuit Identification Name

Octet 1	H	G	F	E	D	C	B	A
Trunk number (first digit)		x	x	x	x	x	x	x
Spare	0							

Octets 2–4	H	G	F	E	D	C	B	A
Trunk number (digits two through four)		x	x	x	x	x	x	x
Spare	0							

Octet 5	H	G	F	E	D	C	B	A
CLLI code—office A		x	x	x	x	x	x	x
Spare	0							

Octets 6–26	H	G	F	E	D	C	B	A
CLLI code—office Z		x	x	x	x	x	x	x
Spare	0							

The circuit identification name parameter is used to identify to a distant exchange the CLLI of a specific trunk. This parameter is coded using IA5 characters, with each character using one octet.

Office A is designated as follows:

- If the trunk is a one-way trunk group, then the office which originates the calls for this trunk is office A.
- If the trunk is a two-way trunk, the office with the lower alphanumeric CLLI code is office A.

The same rules apply to subgroups that can be one-way or two-way trunks. This parameter allows two exchanges to exchange information regarding their identity for usage in routing tables.

Circuit State Indicator	**H G F E**	**D C B A**
Transient	0 0 0 0	0 0 0 0
Spare	0 0 0 0	0 0 0 1
Spare	0 0 0 0	0 0 1 0
Unequipped	0 0 0 0	0 0 1 1
Incoming circuit busy, active	0 0 0 0	0 1 0 0
Incoming circuit busy, locally blocked	0 0 0 0	0 1 0 1
Incoming circuit busy, remotely blocked	0 0 0 0	0 1 1 0
Incoming circuit busy, locally and remotely blocked	0 0 0 0	0 1 1 1
Outgoing circuit busy, active	0 0 0 0	1 0 0 0
Outgoing circuit busy, locally blocked	0 0 0 0	1 0 0 1
Outgoing circuit busy, remotely blocked	0 0 0 0	1 0 1 0
Outgoing circuit busy, locally and remotely blocked	0 0 0 0	1 0 1 1
Idle	0 0 0 0	1 1 0 0
Idle, locally blocked	0 0 0 0	1 1 0 1
Idle, remotely blocked	0 0 0 0	1 1 1 0
Idle, locally and remotely blocked	0 0 0 0	1 1 1 1
Spare	0 0 0 1	0 0 0 0
	to	
	1 1 1 1	1 1 1 1

This parameter allows exchanges to send status information regarding specific trunk circuits, providing the status of the circuit in the distant exchange's perspective. The circuit identification code is carried in the field after the routing label.

This parameter may be from one octet up to *n* octets in length. There may be situations when a two-way trunk will have more than one status indicator. For example, a two-way trunk can be incoming circuit busy, active, while also being outgoing circuit busy, locally blocked.

Circuit Validation Response Indicator	**H G F E**	**D C B A**
Successful	0 0 0 0	0 0 0 0
Failure (default)	0 0 0 0	0 0 0 1
Spare	0 0 0 0	0 0 1 0
	to	
	1 1 1 1	1 1 1 1

The circuit validation response indicator provides the results of a circuit validation in response to a distant exchange request.

COMMON LANGUAGE™ Location Identifier (CLLI)

	H	G	F	E	D	C	B	A
Octet 1	**H**	**G**	**F**	**E**	**D**	**C**	**B**	**A**
Town (1st character)		0	0	0	0	0	0	0
Spare	0							
Octets 2–4	**H**	**G**	**F**	**E**	**D**	**C**	**B**	**A**
Town (characters two thru four)		x	x	x	x	x	x	x
Spare	0							
Octet 5	**H**	**G**	**F**	**E**	**D**	**C**	**B**	**A**
State (1st character)		x	x	x	x	x	x	x
Spare	0							
Octet 6	**H**	**G**	**F**	**E**	**D**	**C**	**B**	**A**
State (2nd character)		x	x	x	x	x	x	x
Spare	0							
Octet 7	**H**	**G**	**F**	**E**	**D**	**C**	**B**	**A**
Building (1st character)		x	x	x	x	x	x	x
Spare	0							
Octet 8	**H**	**G**	**F**	**E**	**D**	**C**	**B**	**A**
Building (2nd character)		x	x	x	x	x	x	x
Spare	0							
Octet 9	**H**	**G**	**F**	**E**	**D**	**C**	**B**	**A**
Building subdivision (1st character)		x	x	x	x	x	x	x
Spare	0							
Octet 10	**H**	**G**	**F**	**E**	**D**	**C**	**B**	**A**
Building subdivision (2nd character)		x	x	x	x	x	x	x
Spare	0							
Octet 11	**H**	**G**	**F**	**E**	**D**	**C**	**B**	**A**
Building subdivision (3rd character)		x	x	x	x	x	x	x
Spare	0							

The COMMON LANGUAGE™ Location Identifier (CLLI) parameter is used during circuit validation to identify an exchange. All signaling points in the ANSI SS7 network must have a CLLI code. This provides the means for identifying by location where a particular signaling point is located. A typical CLLI may look like "RLGHNCXA03W." All characters are IA5 characters.

Connection request

The connection request parameter is sent in the forward direction for the SCCP function. This allows the ISUP protocol to establish an end-to-end connection on which the SCCP may send TCAP or ISUP messages (if ISUP is using the service of SCCP).

The local reference number is the number assigned for the specific call, and is used as a reference in the originating exchange. The reference number allows the exchange to monitor all messages and associate them to their proper calls.

The protocol class field identifies the protocol class to be used on this end-to-end connection. The protocol class is directly related to the SCCP protocol (refer to the chapter on SCCP for more details on protocol class). The protocol class specifies whether or not the services on this connection are to be connection-oriented or connectionless.

The credit field is used for changing the window size of the exchange during the connection. This is only valid if class 3 or 4 is specified (connection-oriented services).

Continuity Indicators	**H**	**G**	**F**	**E**	**D**	**C**	**B**	**A**
Continuity indicator								
Continuity check failed								0
Continuity check successful								1
Spare		0	0	0	0	0	0	

The continuity indicator parameter indicates whether or not a continuity test was successful or not. The continuity check may be requested by an originating exchange according to predetermined criteria. The criteria for conducting a continuity check are found in the continuity check requirements indicator field of the circuit group characteristic indicator parameter.

Egress service

This parameter is used to send network-specific information regarding a terminating exchange such as the interexchange carrier, the type of terminating access service, and the point of interconnection. This information is sent in the forward direction by the first incoming exchange to the terminating exchange.

End of Optional Parameter Fields Indicator	**H**	**G**	**F**	**E**	**D**	**C**	**B**	**A**
End of optional parameters	0	0	0	0	0	0	0	0

The end of optional parameters parameter is the last octet in a message containing any optional parameters.

Event information	**H G F E**	**D C B A**
Event indicator		
Spare	0 0 0	0 0 0 0
ALERTing	0 0 0	0 0 0 1
PROGress	0 0 0	0 0 1 0
In-band info or appropriate pattern now available	0 0 0	0 0 1 1
*Call forwarded on busy	0 0 0	0 1 0 0
*Call forwarded on no reply	0 0 0	0 1 0 1
*Call forwarded unconditional	0 0 0	0 1 1 0
Call deflected	0 0 0	0 1 1 1
Notification for supplementary service	0 0 0	1 0 0 0
Spare	0 0 0	1 0 0 1
	to	
	1 1 0	1 1 1 0
Service information included	1 1 0	1 1 1 1
Spare	1 1 1	0 0 0 0
	to	
	1 1 1	1 1 1 0
Reserved	1 1 1	1 1 1 1

	H G F E	**D C B A**
Event presentation restricted indicator (restrict)		
No indication	0	
Presentation restricted	1	

Forward Call Indicators

Octet 1

	H G F E	**D C B A**
Incoming international call indicator		
Not an incoming international call		0
Incoming international call		1

	H G F E	**D C B A**
End-to-end method indicator		
No end-to-end method available		0 0
Pass along method available		0 1
SCCP method available		1 0
Pass along and SCCP methods available		1 1

	H G F E	**D C B A**
Interworking indicator		
No interworking encountered (SS7 all the way)		0
Interworking encountered		1

	H	G	F	E	D	C	B	A
IAM segmentation indicator								
No indication				0				
Additional info being sent by unsolicited info msg				1				

	H	G	F	E	D	C	B	A
ISDN User Part indicator								
ISUP not used all the way			0					
ISUP used all the way			1					

	H	G	F	E	D	C	B	A
ISDN User Part preference indicator								
ISUP preferred all the way (default)	0	0						
ISUP not required all the way	0	1						
ISUP required all the way	1	0						
Spare	1	1						

Octet 2

	H	G	F	E	D	C	B	A
ISDN access indicator								
Originating access non-ISDN								0
Originating access ISDN								1

	H	G	F	E	D	C	B	A
SCCP method indicator								
No indication						0	0	
*Connectionless method available						0	1	
*Connection-oriented method available						1	0	
*Connectionless and connection-oriented available						1	1	
Spare					0			

	H	G	F	E
Ported number translation indicator				
Number not translated				0
Number translated				1
No QoR routing attempt in progress			0	
QoR routing attempt in progress			1	
Reserved for national use			0	0

The forward call indicators are sent with an IAM to alert the distant exchange of the services required for the call. The international call indicator identifies international calls that have entered through a gateway STP. Without this indicator, it would be difficult for the distant exchange to know if the call was international (mapping of the dialed digits to a conversion table would be necessary).

In addition to identifying international calls, the type of end-to-end signaling method available is also indicated through the end-to-end method indicator. This field identifies the method of signaling available for use; pass along method, SCCP method, or both are available. In U.S. networks, only the pass along method is currently supported.

The interworking indicator identifies any networks encountered along the way which are not SS7 networks. The location of this network is not provided (as that is of no importance). The distant exchange only needs to be aware of its existence.

The IAM segmentation indicator shows when an IAM has been divided into separate messages, because of length or any other reason. IAM information can be sent in an additional signal unit after the initial IAM.

The ISUP indicators are used to indicate whether or not ISUP is used end to end, whether it is required end to end, and whether or not the subscriber interface at the originating exchange is ISDN.

An SCCP indicator is also provided for those networks using the services of SCCP for supporting the ISUP. SCCP method of end-to-end signaling is not supported in U.S. networks.

The ported number translation indicator is used with the LNP application. It is used to indicate when a specific number has been looked up in the LNP database, to prevent unnecessary queries.

Generic Address Parameter

Octet 1	**H G F E**	**D C B A**
Type of address	x x x x	x x x x
Dialed number	0 0 0 0	0 0 0 0
Destination number	0 0 0 0	0 0 0 1
Supplemental user provided calling address—failed network screening	0 0 0 0	0 0 1 0
Supplemental user provided calling address—not screened	0 0 0 0	0 0 1 1
Completion number	0 0 0 0	0 1 0 0
ITU spare	0 0 0 0	0 1 0 1
	to	
	0 1 1 1	1 1 1 1
Network specific use	1 0 0 0	0 0 0 0
	to	
	1 0 1 1	1 1 1 1
Ported number	1 1 0 0	0 0 0 0
ANSI spare	1 1 0 0	0 0 0 1
	to	
	1 1 1 1	0 1 1 1
Transfer number 6	1 1 1 1	1 0 0 0
Transfer number 5	1 1 1 1	1 0 0 1

Transfer number 4	1	1	1	1	1	0	1	0
Transfer number 3	1	1	1	1	1	0	1	1
Transfer number 2	1	1	1	1	1	1	0	0
Transfer number 1	1	1	1	1	1	1	0	1
Callers Emergency Service Identification (CESID)	1	1	1	1	1	1	1	0
Reserved for expansion	1	1	1	1	1	1	1	1

Octet 2

(For type dialed digits and destination number type of address)

	H	G	F	E	D	C	B	A
Nature of address indicator								
Spare		0	0	0	0	0	0	0
Subscriber number		0	0	0	0	0	0	1
Spare reserved, for national use		0	0	0	0	0	1	0
National (significant number)		0	0	0	0	0	1	1
International number		0	0	0	0	1	0	0
Spare		0	0	0	0	1	0	1
Abbreviated number		0	0	0	0	1	1	0
Spare		0	0	0	0	1	1	1
				to				
		1	1	1	1	1	1	1

(For type supplemental user provided calling address)

	H	G	F	E	D	C	B	A
Nature of address indicator								
Spare		0	0	0	0	0	0	0
Unique subscriber number		0	0	0	0	0	0	1
Spare, reserved for national use		0	0	0	0	0	1	0
Unique national significant number		0	0	0	0	0	1	1
Unique international number		0	0	0	0	1	0	0
Spare		0	0	0	0	1	0	1
				to				
		1	1	1	0	0	0	0
Non-unique subscriber number		1	1	1	0	0	0	1
Spare, reserved for national use		1	1	1	0	0	1	0
Non-unique national number		1	1	1	0	0	1	1
Non-unique international number		1	1	1	0	1	0	0
Spare		1	1	1	0	1	0	1
Spare		1	1	1	0	1	1	0
Test line code		1	1	1	0	1	1	1
Reserved for network-specific use		1	1	1	1	0	0	0
				to				
		1	1	1	1	1	1	0
Spare		1	1	1	1	1	1	1

(For completion number type of address)

	H	G	F	E	D	C	B	A
Nature of address indicator								
Spare		0	0	0	0	0	0	0
Subscriber number		0	0	0	0	0	0	1
Spare, reserved for national use		0	0	0	0	0	1	0
National significant number		0	0	0	0	0	1	1
International number		0	0	0	0	1	0	0
Spare		0	0	0	0	1	0	1
					to			
		1	1	1	0	0	0	0
Subscriber number, operator requested		1	1	1	0	0	0	1
National number, operator requested		1	1	1	0	0	1	0
International number, operator requested		1	1	1	0	0	1	1
No number present, operator requested		1	1	1	0	1	0	0
No number present, cut-through call to carrier		1	1	1	0	1	0	1
950 + call from local exchange carrier public station, hotel/motel, or nonexchange access end office		1	1	1	0	1	1	0
Test line code		1	1	1	0	1	1	1
Reserved for network-specific use		1	1	1	1	0	0	0
					to			
		1	1	1	1	1	1	0
Spare		1	1	1	1	1	1	1
Odd/even indicator								
Even number of address signals	0							
Odd number of address signals	1							

Octet 3	H	G	F	E	D	C	B	A
Reserved							0	0
Presentation								
Presentation allowed					0	0		
Presentation restricted					0	1		
Spare					1	0		
Spare					1	1		
Numbering Plan								
Unknown numbering plan		0	0	0				
ISDN numbering plan (Rec. E.164, E.163)		0	0	1				
Spare		0	1	0				
Reserved ITU-TS Data Numbering Plan		0	1	1				
Reserved ITU-TS Telex Numbering Plan		1	0	0				
Private numbering plan		1	0	1				
Spare		1	1	0				
Spare		1	1	1				
Spare	0							

Octet 4	H	G	F	E	D	C	B	A
Address Signal—1st address								
Digit 0					0	0	0	0

Digit 1		0 0 0 1
Digit 2		0 0 1 0
Digit 3		0 0 1 1
Digit 4		0 1 0 0
Digit 5		0 1 0 1
Digit 6		0 1 1 0
Digit 7		0 1 1 1
Digit 8		1 0 0 0
Digit 9		1 0 0 1
Spare		1 0 1 0
Code 11		1 0 1 1
Code 12		1 1 0 0
Spare		1 1 0 1
Spare		1 1 1 0
End of pulse signal		1 1 1 1
Address Signal—2nd address		
Digit 0	0 0 0 0	
Digit 1	0 0 0 1	
Digit 2	0 0 1 0	
Digit 3	0 0 1 1	
Digit 4	0 1 0 0	
Digit 5	0 1 0 1	
Digit 6	0 1 1 0	
Digit 7	0 1 1 1	
Digit 8	1 0 0 0	
Digit 9	1 0 0 1	
Spare	1 0 1 0	
Code 11	1 0 1 1	
Code 12	1 1 0 0	
Spare	1 1 0 1	
Spare	1 1 1 0	
End of pulse signal	1 1 1 1	

Octets 5–n	**H G F E**	**D C B A**
Address Signal—1st address		
Digit 0		0 0 0 0
Digit 1		0 0 0 1
Digit 2		0 0 1 0
Digit 3		0 0 1 1
Digit 4		0 1 0 0
Digit 5		0 1 0 1
Digit 6		0 1 1 0
Digit 7		0 1 1 1
Digit 8		1 0 0 0
Digit 9		1 0 0 1
Spare		1 0 1 0
Code 11		1 0 1 1

Code 12		1 1 0 0
Spare		1 1 0 1
Spare		1 1 1 0
End of pulse signal		1 1 1 1
Address Signal—2nd address		
Digit 0	0 0 0 0	
Digit 1	0 0 0 1	
Digit 2	0 0 1 0	
Digit 3	0 0 1 1	
Digit 4	0 1 0 0	
Digit 5	0 1 0 1	
Digit 6	0 1 1 0	
Digit 7	0 1 1 1	
Digit 8	1 0 0 0	
Digit 9	1 0 0 1	
Spare	1 0 1 0	
Code 11	1 0 1 1	
Code 12	1 1 0 0	
Spare	1 1 0 1	
Spare	1 1 1 0	
End of pulse signal	1 1 1 1	

The generic address parameter identifies the type of address (dialed digits, etc.) being presented in a call setup. It also indicates the numbering plan used in the address and the actual address. When LNP is provided, the GAP provides the actual dialed digits for a ported number. The called party address is then used for the location routing number (LRN).

The GAP parameter is also used in LNP applications. When a number has been ported, the dialed digits are placed in the GAP parameter. The called party address contains the LN used to route the call to the proper exchange. See Chapter 10 for more details on LNP.

The nature of address indicator is dependent on the type of address provided. All of the options are shown above for clarity.

Generic Digits Parameter

Octet 1	**H G F E**	**D C B A**
Type of digits		
Account code	0	0 0 0 0
Authorization code	0	0 0 0 1
Private network traveling class mark	0	0 0 1 0
ANSI Spare	0	0 0 1 1
	to	
	0	1 1 0 1
Originating party service provider	0	1 1 0 1

Bill to number				0	1	1 1 1	
Reserved for network-specific use				1	0	0 0 0	
					to		
				1	1	1 1 0	
Reserved for extension				1	1	1 1 1	
Encoding scheme							
BCD even	0	0	0				
BCD odd	0	0	1				
IA5	0	1	0				
Binary	0	1	1				
Spare	1	0	0				
		to					
	1	1	1				

Octets 2–n

Digits	(Encoded in the format specified above)

The generic digits parameter provides additional numeric data pertaining to supplementary services such as authorization code, PIN number, or account code.

The type of digits is found in the first octet. These types are related to PBX features and/or business group features. The transfer of this data is possible from an ISDN-compatible PBX to another ISDN-compatible PBX through the Public Switched Telephone Network (PSTN) using the SS7 ISUP protocol and parameters such as this one.

Generic Name Parameter	**H**	**G**	**F**	**E**	**D**	**C**	**B**	**A**
Presentation								
Presentation allowed							0	0
Presentation restricted							0	1
Blocking toggle							1	0
No indication							1	1
Spare					0	0		
Availability								
Name available/unknown				0				
Name not available				1				
Type of name								
Spare	0	0	0					
Calling name	0	0	1					
Original called name	0	1	0					
Redirecting name	0	1	1					
Connected name	1	0	0					
Spare	1	0	1					
		to						
	1	1	1					

The Generic Name parameter contains information to be used for name display features. In ANSI networks, Calling Name (CNAM) display requires the terminating exchange to access a database and search for the name information. In some networks (outside the United States) the name information is actually carried in the IAM from the originating exchange. The name information is found in this parameter in the IA5 format. Up to 15 characters may be sent.

Hop Count Parameter	**H**	**G**	**F**	**E**	**D**	**C**	**B**	**A**
Hop Counter				x	x	x	x	x
Spare	0	0	0					

The hop counter is used to ensure that ISUP looping does not occur. The initial message is sent in the forward direction with the maximum value allowed (network dependent). As the message is passed through each circuit, the counter is decremented by one. When the counter reaches zero, the message is discarded. The counter value is the number of contiguous SS7 circuits that this message must pass to reach its destination, and is provided in binary form.

Information Indicators

Octet 1	**H**	**G**	**F**	**E**	**D**	**C**	**B**	**A**
Calling party address response indicator								
Calling party address not included							0	0
Calling party address not available							0	1
Spare							1	0
Calling party address included, hold not provided							1	1
Hold provided indicator								
Hold not provided (default)						0		
*Hold provided						1		
Spare				0	0			

	H	**G**	**F**	**E**	**D**	**C**	**B**	**A**
Calling party's category response indicator								
Calling party's category not included			0					
*Calling party's category included			1					
Charge information response indicator								
Charge information not included		0						
*Charge information included		1						
Solicited information indicator								
Solicited	0							
Unsolicited	1							

Octet 2	**H**	**G**	**F**	**E**	**D**	**C**	**B**	**A**
Spare		0	0	0	0	0	0	0

	H	G	F	E	D	C	B	A
Multilocation business group info response indicator								
Multilocation business group info not included	0							
Multilocation business group info included	1							

This parameter provides additional information related to a call in progress. It can be sent in either direction, and can be a solicited message or unsolicited. Solicited indicates that the distant exchange requested information (billing information, for example) and the receiving exchange is replying to that request.

This parameter does not provide the requested information, it simply indicates that the information is in this message.

Information Request Indicators

Octet 1	H	G	F	E	D	C	B	A
Calling party address request indicator								
Calling party address not requested								0
Calling party address requested								1

	H	G	F	E	D	C	B	A
Holding indicator								
Holding not requested							0	
*Holding requested							1	
Spare						0		

	H	G	F	E	D	C	B	A
Calling party's category request indicator								
Calling party's category not requested					0			
*Calling party's category requested					1			

	H	G	F	E	D	C	B	A
Charge information request indicator								
Charge information not requested				0				
Charge information requested				1				
Spare		0	0					

	H	G	F	E	D	C	B	A
Malicious call identification request indicator								
Malicious call identification not requested	0							
*Malicious call identification requested	1							

Octet 2	H	G	F	E	D	C	B	A
Spare		0	0	0	0	0	0	0
Multilocation business group info indicator								
Multilocation business group info not requested	0							
Multilocation business group info requested	1							

The information request parameter is used to request specific information regarding a call already in progress. The response to the request parameter is the information indicators, with the appropriate parameters providing the actual data.

This information is used in call processing and billing for the call. Not all of these procedures are currently in use in U.S. networks, yet the functionality is defined in both ITU and ANSI standards.

Jurisdiction Information Parameter (JIP)	**H**	**G**	**F**	**E**	**D**	**C**	**B**	**A**
Address Signal—1st address								
Digit 0					0	0	0	0
Digit 1					0	0	0	1
Digit 2					0	0	1	0
Digit 3					0	0	1	1
Digit 4					0	1	0	0
Digit 5					0	1	0	1
Digit 6					0	1	1	0
Digit 7					0	1	1	1
Digit 8					1	0	0	0
Digit 9					1	0	0	1
Spare					1	0	1	0
Code 11					1	0	1	1
Code 12					1	1	0	0
Spare					1	1	0	1
Spare					1	1	1	0
End of pulse signal					1	1	1	1
Address Signal—2nd address								
Digit 0	0	0	0	0				
Digit 1	0	0	0	1				
Digit 2	0	0	1	0				
Digit 3	0	0	1	1				
Digit 4	0	1	0	0				
Digit 5	0	1	0	1				
Digit 6	0	1	1	0				
Digit 7	0	1	1	1				
Digit 8	1	0	0	0				
Digit 9	1	0	0	1				
Spare	1	0	1	0				
Code 11	1	0	1	1				
Code 12	1	1	0	0				
Spare	1	1	0	1				
Spare	1	1	1	0				
End of pulse signal	1	1	1	1				

The JIP is used in LNP applications to support billing systems. When an originating number has been ported, billing systems may not be able to determine the correct billing for the call (since one billion systems are still based on the North American Numbering Plan). The JIP provides the LRN assigned to the originating number, which is then used for determining proper billing for the call. The calling party number is no longer applicable for billing of a call in the case of ported numbers.

This parameter is a variable length parameter consisting of address signals only. The parameter provides numerical data indicating the geographic origination of the call.

Nature of Connection Indicators	**H**	**G**	**F**	**E**	**D**	**C**	**B**	**A**
Satellite indicator								
No satellite circuit in the connection							0	0
One satellite circuit in the connection							0	1
Two satellite circuits in the connection							1	0
Three or more satellite circuits in the connection							1	1
Continuity check indicator								
Continuity check not required					0	0		
Continuity check required on this circuit					0	1		
Continuity check performed on a previous circuit					1	0		
Spare					1	1		

	H	**G**	**F**	**E**	**D**	**C**	**B**	**A**
Echo control device indicator								
Outgoing half echo control device not included				0				
Outgoing half echo control device included				1				
Spare	0	0	0					

The nature of connection indicators is sent in the forward direction to provide information regarding the circuit connection specified in the CIC parameter of the message. The values within this parameter allow intermediate exchanges to determine how to handle the processing of this message.

Network transport

As seen in the figure above, this parameter consists of other ISUP parameters. It is used to send ISUP parameters through the network transparently, without involving a call setup or other mechanism. The objective is to send parameters end-to-end through the network, without the intermediate exchanges having to process the message.

Notification Indicator

	H G F E	D C B A
Notification indicator		
Spare	0 0 0	0 0 0 0
	to	
	0 0 0	0 0 1 1
Call completion delay	0 0 0	0 1 0 0
Spare	0 0 0	0 1 0 1
	to	
	1 0 0	0 0 0 1
Conference established (multiparty call)	1 0 0	0 0 1 0
Conference disconnected	1 0 0	0 0 1 1
Other party added to conference	1 0 0	0 1 0 0
Isolated	1 0 0	0 1 0 1
Reattached	1 0 0	0 1 1 0
Other part isolated	1 0 0	0 1 1 1
Other party reattached	1 0 0	1 0 0 0
Other part split	1 0 0	1 0 0 1
Other part disconnected	1 0 0	1 0 1 0
Conference floating	1 0 0	1 0 1 1
Spare	1 0 0	1 1 0 0
Spare	1 0 0	1 1 0 1
Spare	1 0 0	1 1 1 0
Conference floating, served user preempted	1 0 0	1 1 1 1
Spare	1 0 1	0 0 0 0
	to	
	1 0 1	1 1 1 1
Call is a waiting call	1 1 0	0 0 0 0
Reserved for transfer in progress	1 1 0	0 0 0 1
Reserved for call isolated from conference call	1 1 0	0 0 1 0
Reserved for call split from conference call	1 1 0	0 0 1 1
Reserved for call reattached to conference call	1 1 0	0 1 0 0
Reserved for call added to conference call	1 1 0	0 1 0 1
Spare	1 1 0	0 1 1 0
	to	
	1 1 0	1 0 0 0
Call transfer, alerting	1 1 0	1 0 0 1
Call transfer, active	1 1 0	1 0 1 0
Spare	1 1 0	1 0 1 1
	to	
	1 1 1	1 0 0 0
Remote hold	1 1 1	1 0 0 1
Remote hold released	1 1 1	1 0 1 0
Call is forwarded/deflected	1 1 1	1 0 1 1
Spare	1 1 1	1 1 0 0
	to	

	H	G	F	E	D	C	B	A
		1	1	1	1	1	1	0
Reserved		1	1	1	1	1	1	1
Extension indicator								
Octet continues through the next octet	0							
Last octet	1							

This parameter provides information regarding supplementary services. These services are related to business group services (such as Centrex) which provide PBX-like features to a group of business lines within a group. This parameter provides information regarding the nature of the calling party and its origin (forwarded call, call from hold, etc.).

Optional Backward Call Indicators	**H**	**G**	**F**	**E**	**D**	**C**	**B**	**A**
In-band information indicator								
No indication								0
In-band info or an appropriate pattern is now avail								1
Call forwarding may occur indicator								
No indication							0	
*Call forwarding may occur							1	
Spare					0	0		
Reserved for national use					0	0		
Network excessive delay indicator								
No indication		0						
Network excessive delay encountered		1						
User-network interaction indicator								
No indication	0							
User-network interaction occurs, cut-through in both directions	1							

This parameter is sent in the backward direction providing information to another exchange regarding a call in progress. The in-band information indicator is used to alert the distant exchange that in-band information such as a recording or a service tone is present on the voice circuit.

The call forwarding may occur parameter is used to indicate that the call may be forwarded if the called party does not answer. There presently are no national ANSI procedures for this field.

The user-network interaction indicator is sent from the originating exchange to indicate that additional information is being gathered from the caller (such as PIN number, or maybe special code) before routing the call.

Operator Services Information	**H**	**G**	**F**	**E**	**D**	**C**	**B**	**A**
Original access prefix	0	0	0	1				
Unknown					0	0	0	0
1 + or 011+					0	0	0	1

0 + or 01+		0 0 1 0
0–		0 0 1 1
Spare		0 1 0 0 to 1 0 0 1
Reserved for national use		1 0 1 0 to 1 1 1 1
Bill to information entry type and handling type	0 0 1 0	
Info entry unknown, unknown handling		0 0 0 0
Info entry manual by operator, station handling		0 0 0 1
Info entry manual by operator, person handling		0 0 1 0
Info entry automated by tone input, station handling		0 0 1 1
Info entry unknown, station handling		0 1 0 0
Info entry unknown, person handling		0 1 0 1
Info entry manual by operator, unknown handling		0 1 1 0
Info entry automated by tone input, unknown handling		0 1 1 1
Info entry automated by tone input, person handling		1 0 0 0
Info entry automated by spoken input, unknown handling		1 0 0 1
Info entry automated by spoken input, station handling		1 0 1 0
Info entry automated by spoken input, person handling		1 0 1 1
Spare		1 1 0 0 to 1 1 0 1
Reserved for network-specific use		1 1 1 0 to 1 1 1 1
Bill-to type	0 0 1 1	
Unknown		0 0 0 0
Calling card—14-digit format		0 0 0 1
Calling card—89C format		0 0 1 0
Calling card—other format		0 0 1 1
Collect		0 1 0 0
Third part number billing		0 1 0 1
Sent paid (prepaid calling card)		0 1 1 0
Spare		0 1 1 1 to 1 0 1 0
Reserved for network-specific use		1 1 1 1
Bill-to specific information	0 1 0 0	
Spare		0 0 0 0
NIDB authorizes		0 0 0 1

NIDB reports, verify by automated means		0 0 1 0
NIDB reports, verify by operator		0 0 1 1
No NIDB query		0 1 0 0
No NIDB response		0 1 0 1
NIDB reports unavailable		0 1 1 0
No NIDB response—timeout		0 1 1 1
No NIDB response—reject component		1 0 0 0
No NIDB response—ACG in effect		1 0 0 1
No NIDB response—SCCP failure		1 0 1 0
Spare		1 0 1 1
Reserved for network-specific use		1 1 0 0 to 1 1 1 1
Special handling	0 1 0 1	
Unknown		0 0 0 0
Call completion		0 0 0 1
Rate information		0 0 1 0
Trouble reporting		0 0 1 1
Time and charges		0 1 0 0
Credit reporting		0 1 0 1
General assistance		0 1 1 0
Spare		0 1 1 1 to 1 0 1 0
Reserved for network-specific use		1 0 1 1 to 1 1 1 1
Spare	0 1 1 0	
Accessing signaling	0 1 1 1	
Unknown		0 0 0 0
Dial pulse		0 0 0 1
Dualtone multifrequency (DTMF)		0 0 1 0
Spare		0 0 1 1 to 1 0 0 1
Reserved for network-specific use		1 0 1 0 to 1 1 1 1

The operator services parameter is used to identify how a caller accessed an operator for assistance, and how the call should be billed. For example, if a caller dialed 01 + the number, special handling may be required. Billing may also be affected based on the caller's interaction with the operator (the caller may request the call to be billed to a calling card for example). The values in this parameter are used to identify how the call should be handled in the billing system.

Original Called Number	H	G	F	E	D	C	B	A
Octet 1								
Nature of address indicator								
Spare		0	0	0	0	0	0	0
Unique subscriber number		0	0	0	0	0	0	1
Spare, reserved for national use		0	0	0	0	0	1	0
Unique national significant number		0	0	0	0	0	1	1
Unique international number		0	0	0	0	1	0	0
Spare		0	0	0	0	1	0	1
				to				
		1	1	1	0	0	0	0
Non-unique subscriber number		1	1	1	0	0	0	1
Spare, reserved for national use		1	1	1	0	0	1	0
Non-unique national significant number		1	1	1	0	0	1	1
Non-unique international number		1	1	1	0	1	0	0
Spare		1	1	1	0	1	0	1
Spare		1	1	1	0	1	1	0
Test line test code		1	1	1	0	1	1	1
Reserved for network specific use		1	1	1	1	0	0	0
				to				
		1	1	1	1	1	1	0
Spare		1	1	1	1	1	1	1
Odd/even indicator								
Even number of address signals	0							
Odd number of address signals	1							

Octet 2	H	G	F	E	D	C	B	A
Reserved							0	0
Address presentation								
Presentation					0	0		
Presentation restricted (default)					0	1		
Spare					1	0		
Spare					1	1		
Numbering plan								
Unknown		0	0	0				
ISDN numbering plan (E.164, E.163)		0	0	1				
Spare		0	1	0				
Reserved (ITU: Data numbering plan)		0	1	1				
Reserved (ITU: Telex numbering plan)		1	0	0				
Private numbering plan		1	0	1				
Spare		1	1	0				
Spare		1	1	1				
Spare	0							

Octet 3	H	G	F	E	D	C	B	A
1st address								
Digit 0					0	0	0	0
Digit 1					0	0	0	1
Digit 2					0	0	1	0
Digit 3					0	0	1	1
Digit 4					0	1	0	0
Digit 5					0	1	0	1
Digit 6					0	1	1	0
Digit 7					0	1	1	1
Digit 8					1	0	0	0
Digit 9					1	0	0	1
Spare					1	0	1	0
Code 11					1	0	1	1
Code 12					1	1	0	0
Spare					1	1	0	1
Spare					1	1	1	0
End of pulse signal					1	1	1	1
2nd address								
Digit 0	0	0	0	0				
Digit 1	0	0	0	1				
Digit 2	0	0	1	0				
Digit 3	0	0	1	1				
Digit 4	0	1	0	0				
Digit 5	0	1	0	1				
Digit 6	0	1	1	0				
Digit 7	0	1	1	1				
Digit 8	1	0	0	0				
Digit 9	1	0	0	1				
Spare	1	0	1	0				
Code 11	1	0	1	1				
Code 12	1	1	0	0				
Spare	1	1	0	1				
Spare	1	1	1	0				
End of pulse signal	1	1	1	1				

This parameter is used when call redirecting (forwarding) occurs. The parameter identifies the address of the party that initiated the redirection. This parameter also provides information regarding the presentation of the calling party number, used by the end exchange when connecting the call to its destination.

Only the first and second address signals are shown here (for brevity), but there can be additional address signals. The address signals are typically the telephone number assigned to the subscriber who initiated the redirecting.

Originating Line Information	**H G F E**	**D C B A**
Binary equivalent of the II digits	0 0 0 0	0 0 0 0
	to	
	0 1 1 0	0 0 1 1
Reserved for future expansion	0 1 1 0	0 1 0 0
	to	
	1 1 1 1	1 1 1 1

This information is sent in the forward direction representing a toll class of service for the call.

Outgoing trunk group number

This parameter provides the trunk number used for an interworking call. Interworking means that the call was sent to another exchange in another network. The trunk number represents the circuit used at the gateway switch into the other network. This parameter is only found in instances where internetworking occurs.

Precedence

Octet 1	**H**	**G**	**F**	**E**	**D**	**C**	**B**	**A**
Precedence level								
Flash override (0)					0	0	0	0
Flash (1)					0	0	0	1
Immediate (2)					0	0	1	0
Priority (3)					0	0	1	1
Routine (4)					0	1	0	0
Spare					0	1	0	1
						to		
					1	1	1	1
Spare				0				
Look ahead for busy (LFB)								
Look ahead for busy allowed		0	0					
Look ahead for busy not allowed		1	0					
Path reserved		0	1					
Spare		1	1					
Extension bit								
Octet continues through the next octet	0							
Last octet	1							

Network identity (coded in BCD), the telephone country code is placed in the second to fourth digits of this parameter. The first digit is set to zero.

Octet 2	H	G	F	E	D	C	B	A
Multilevel Precedence and Preemption (MLPP)								
Defense switched network		0	0	0	0	0	0	0
Spare		0	0	0	0	0	0	1
					to			
		1	1	1	1	1	1	1
Extension								
Octet continues through the next octet	0							
Last octet	1							

Precedence is a feature provided in defense networks, where an individual of higher rank is given priority for outgoing trunks over someone of lower rank. This feature is typically found in AUTOVON systems, but is now being offered through the central office services.

This parameter identifies the level of precedence allowed and whether or not the "look ahead for busy" feature is allowed.

Range and Status

Octet 1

Range

This field of the parameter is a binary representation of the range of circuits that are affected by the status field that follows it. Because this is a zero-based number, the actual circuit number is the binary representation plus one.

In national circuits (ANSI only), the range is from 0 to 23. International circuits range from 0 to 31.

Octet 2

Status

The status field provides the status of the circuits indicated in the range parameter. The status bits are numbered from 0 to 23, or 0 to 31 in international circuits. There is a direct correlation between the range and status fields. The number of status bits is equal to the value of the range field plus 1.

Status bit 0 is located in the first bit position of the status octet. Other status bits follow in numerical order. More than one octet can exist. In the range field, if the range is coded as zero, the status bit is not provided.

The status bits also depend on the message type for their value. For instance, in a Circuit Group Blocked message, the status bits are different than in the Circuit Group Unblocking message. The status bit values are as follows:

In Circuit Group Blocking messages:	
No blocking	0
Blocking	1
In Circuit Group Blocking Acknowledgment messages:	
No blocking acknowledgment	0
Blocking acknowledgment	1
In Circuit Group Unblocking messages:	
No unblocking	0
Unblocking	1
In Circuit Group Unblocking Acknowledgment messages:	
No unblocking acknowledgment	0
Unblocking acknowledgment	1
In Circuit Group Reset Acknowledgment messages:	
No blocking	0
Blocked	1

The range and status parameter is found in circuit group supervision messages to indicate the range of circuits that are affected by the status indicator, and the status of those circuits. The status field correlates to the type of circuit group message. The status bits are provided in numerical order, with each status bit corresponding with the CIC of the affected circuit.

As seen in the range field, the range field identifies the actual circuit number (or range of circuits), which is used to also identify which status bits are required for each circuit. Status bit 0 is the first status bit, found in the first bit location of the first octet in the status field.

Redirecting Number

Octet 1	H G F E	D C B A
Nature of address indicator		
Spare	0 0 0	0 0 0 0
Subscriber number	0 0 0	0 0 0 1
Spare, reserved for national use	0 0 0	0 0 1 0
National significant number	0 0 0	0 0 1 1
International number	0 0 0	0 1 0 0
Spare	0 0 0	0 1 0 1
	to	
	1 1 1	0 0 0 0
Subscriber number, operator requested	1 1 1	0 0 0 1
National number, operator requested	1 1 1	0 0 1 0
International number, operator requested	1 1 1	0 0 1 1
No number present, operator requested	1 1 1	0 1 0 0

	H	G	F	E	D	C	B	A
No number present, cut-through call to carrier		1	1	1	0	1	0	1
950 + call from local exchange carrier public station, hotel/motel, or nonexchange access end-office		1	1	1	0	1	1	0
Test line code		1	1	1	0	1	1	1
Reserved for network specific use		1	1	1	1	0	0	0
				to				
		1	1	1	1	1	1	0
Spare		1	1	1	1	1	1	1
Odd/even bits								
Even number of address signals	0							
Odd number of address signals	1							
Octet 2	**H**	**G**	**F**	**E**	**D**	**C**	**B**	**A**
Reserved							0	0
Address presentation								
Presentation allowed					0	0		
Presentation restricted					0	1		
Spare					1	0		
Spare					1	1		
Numbering Plan								
Unknown numbering plan		0	0	0				
ISDN numbering plan (Rec. E.164, E.163)		0	0	1				
Spare		0	1	0				
Reserved ITU-TS Data Numbering Plan		0	1	1				
Reserved ITU-TS Telex Numbering Plan		1	0	0				
Private numbering plan		1	0	1				
Spare		1	1	0				
Spare		1	1	1				
Spare	0							
Octet 3	**H**	**G**	**F**	**E**	**D**	**C**	**B**	**A**
Address Signal—1st address								
Digit 0					0	0	0	0
Digit 1					0	0	0	1
Digit 2					0	0	1	0
Digit 3					0	0	1	1
Digit 4					0	1	0	0
Digit 5					0	1	0	1
Digit 6					0	1	1	0
Digit 7					0	1	1	1
Digit 8					1	0	0	0
Digit 9					1	0	0	1
Spare					1	0	1	0
Code 11					1	0	1	1
Code 12					1	1	0	0
Spare					1	1	0	1

	H	G	F	E	D	C	B	A
Spare					1	1	1	0
End of pulse signal					1	1	1	1
Address Signal—2nd address								
Digit 0	0	0	0	0				
Digit 1	0	0	0	1				
Digit 2	0	0	1	0				
Digit 3	0	0	1	1				
Digit 4	0	1	0	0				
Digit 5	0	1	0	1				
Digit 6	0	1	1	0				
Digit 7	0	1	1	1				
Digit 8	1	0	0	0				
Digit 9	1	0	0	1				
Spare	1	0	1	0				
Code 11	1	0	1	1				
Code 12	1	1	0	0				
Spare	1	1	0	1				
Spare	1	1	1	0				
End of pulse signal	1	1	1	1				

When call forwarding is applied to a call, this parameter is used to indicate the telephone number from which the called number was last forwarded. Call forwarding can be invoked from any telephone, hence the need to identify from where the telephone number was last forwarded.

The address signals, as is the case in all parameters with this field, can consist of several digits, even though only one octet is shown here (for brevity). At least four octets are needed to represent a telephone number. Any odd number of address signals requires a filler for the last half of the octet. The odd/even address indicator identifies those addresses that do not require a full octet, and contain a filler at the end.

Redirect Capability	**H**	**G**	**F**	**E**	**D**	**C**	**B**	**A**
Redirection possibility indicator								
Not used						0	0	0
Redirection possible before ACM						0	0	1
Redirection possible before ANM						0	1	0
Redirection possible at any time during the call						0	1	1
Spare						1	0	0
							to	
						1	1	1
Spare		0	0	0	0			
Extension indicator								
Next octet	0							
Last octet	1							

This information is sent in the forward direction. It indicates that the succeeding exchange is allowed to initiate redirection of a call, and indicates when the redirection may take place (in relation to the call processing procedures).

Redirect Counter	**H**	**G**	**F**	**E**	**D**	**C**	**B**	**A**
Counter				x	x	x	x	x

This parameter indicates the number of times a call has been redirected within the network. This may be used to control the maximum number of times a number can be redirected as part of a service offering. The number is represented in binary.

Redirection Information	**H**	**G**	**F**	**E**	**D**	**C**	**B**	**A**
Octet 1								
Spare						0	0	0
Reserved					0			
Original redirecting reason								
Unknown/not available (default)	0	0	0	0				
User busy	0	0	0	1				
No reply	0	0	1	0				
Unconditional	0	0	1	1				
Deflection	0	1	0	0				
Spare	0	1	0	1				
		to						
	1	1	1	0				
Reserved	1	1	1	1				

Octet 2	**H**	**G**	**F**	**E**	**D**	**C**	**B**	**A**
Redirection counter								
No redirection has occurred					0	0	0	0
Redirected 1 time					0	0	0	1
Redirected 2 times					0	0	1	0
Redirected 3 times					0	0	1	1
Redirected 4 times					0	1	0	0
Redirected 5 times					0	1	0	1
Redirected 6 times					0	1	1	0
Redirected 7 times					0	1	1	1
Redirected 8 times					1	0	0	0
Redirected 9 times					1	0	0	1
Redirected 10 times					1	0	1	0
Redirected 11 times					1	0	1	1
Redirected 12 times					1	1	0	0
Redirected 13 times					1	1	0	1
Redirected 14 times					1	1	1	0

Redirected 15 times					1	1	1	1
Redirecting reason								
Unknown/not available (default)	0	0	0	0				
User busy	0	0	0	1				
No reply	0	0	1	0				
Unconditional	0	0	1	1				
Spare	0	1	0	0				
			to					
	1	1	1	1				

This parameter is used with calls where call forwarding has been invoked. The information provided indicates the original reason for the forwarding, and in the case where the call has undergone more than one forward, the reason for the subsequent forwardings.

It is quite possible for a call to be forwarded a number of times, without the knowledge of the caller. For example, a telephone number may be forwarded to another telephone number. When a caller tries to reach that number, they are forwarded to the second number. The second number may also be forwarded, resulting in the call being redirected again. This parameter indicates the number of times the call was redirected and the reason for the redirecting.

Redirection number

This parameter identifies the number to which the called number is to be redirected. The values contained in this parameter include the nature of address indicator, numbering plan, and the telephone number the call is to be redirected to. The actual values have not been shown here, because they are repeated in several other places in this book.

Remote Operations	**H**	**G**	**F**	**E**	**D**	**C**	**B**	**A**
Protocol profile								
Spare				0	0	0	0	0
				to				
				1	0	0	0	0
Remote operations protocol				1	0	0	0	1
Spare				1	0	0	1	0
				to				
				1	1	1	1	1
Spare		0	0					
Extension bit								
Next octet	0							
Last octet	1							
Component (follows same format as TCAP component)								

Service Activation	**H G F E**	**D C B A**
Feature code indicators		
Reserved for international use	0 0 0 0	0 0 0 0
	to	
	0 1 1 1	1 0 1 1
Call waiting originating invoked	0 1 1 1	1 1 0 0
Dial call waiting invoked	0 1 1 1	1 1 0 1
Complete call req, ISUP used all the way	0 1 1 1	1 1 1 0
Complete call req, ISUP not used all the way	0 1 1 1	1 1 1 1
Network service attached	1 0 0 0	0 0 0 0
Network service released	1 0 0 0	0 0 0 1
Coin collect	1 0 0 0	0 0 1 0
Coin return	1 0 0 0	0 0 1 1
Network service recall	1 0 0 0	0 1 0 0
Billing verification	1 0 0 0	0 1 0 1
Hold available	1 0 0 0	0 1 1 0
Hold not available	1 0 0 0	0 1 1 1
Hold request	1 0 0 0	1 0 0 0
Hold acknowledge	1 0 0 0	1 0 0 1
Hold release request	1 0 0 0	1 0 1 0
Hold release acknowledge	1 0 0 0	1 0 1 1
Hold continuation request	1 0 0 0	1 1 0 0
Disconnect request	1 0 0 0	1 1 0 1
Reconnect request	1 0 0 0	1 1 1 0
Spare	1 0 0 0	1 1 1 1
	to	
	1 0 0 1	0 0 1 0
Resume operator services	1 0 0 1	0 0 1 1
Spare	1 0 0 1	0 1 0 0
	to	
	1 0 1 1	1 1 1 1
Network specific use	1 1 0 0	0 0 0 0
	to	
	1 1 1 1	1 1 1 0
Reserved for network specific use	1 1 1 1	1 1 1 1

This parameter is used to invoke supplementary services from another exchange. Presently, there are not a lot of features called for in this parameter, but there is room for expansion.

Service code

The service code field is a one-octet field representing the service code as assigned by the North American Numbering Plan Administration at Bellcore. Presently, this parameter is under further study, but can be used to identify a specific type of service, which a subscriber can invoke either in real time or otherwise. The number in this parameter is a binary number representing the decimal equivalent of the service code.

Special Processing Request

	H G F E	D C B A
Special processing request		
Spare	0 0 0 0	0 0 0 0
Reserved for international use	0 0 0 0	0 0 0 1
	to	
	0 0 0 0	1 1 1 1
Reserved for national use	0 0 0 1	0 0 0 0
	to	
	0 1 1 1	1 1 1 0
Service processing requested	0 1 1 1	1 1 1 1
Reserved for network-specific use	1 0 0 0	0 0 0 0
	to	
	1 1 1 1	1 1 1 0
Spare	1 1 1 1	1 1 1 1

In the event that a call originates in a private network, there may be a need for special number translation or authorization code verification. This parameter indicates the special processing requirements of just such a call. The receiver of this message is a service node in the Public Switched Telephone Network (PSTN) from a service node in the private network.

Suspend/Resume Indicators

	H G F E	D C B A
Suspend/Resume indicator		
ISDN subscriber initiated[		0
Network initiated (default)		1
Spare	0 0 0 0	0 0 0

The suspend/resume indicator is sent in the forward direction to indicate the originator of a suspend or a resume. There are only two options, either an ISDN subscriber or the network initiated the message.

Transaction request

This parameter follows the same message structure as seen in the Transaction Capabilities Application Part (TCAP), providing the transaction ID and the SCCP address for those messages used to carry information regarding a call in progress. This parameter can only be used for calls that are already in progress, and allows the ISUP protocol to use the services of the TCAP protocol to deliver service information relating to a call.

This parameter is carried in the Initial Address Message (IAM) during the circuit connection establishment. The receiving exchange then uses this parameter for all subsequent messages related to the call on that circuit, such as feature invocation or call hand-off procedures.

Transit Network Selection

Octet 1	H G F E	D C B A
Network identification plan (National ANSI networks)		
Unknown		0 0 0 0
3-digit carrier identification with circuit code		0 0 0 1
4-digit carrier identification with circuit code		0 0 1 0
Reserved		0 0 1 1
		to
		0 1 1 1
Reserved for network-specific use		1 0 0 0
		to
		1 1 1 1
Network identification plan (International networks)		
Unknown		0 0 0 0
Public data network identification code		0 0 1 1
Public land mobile network ID code		0 1 1 0
Type of network identification		
ITU standardized identification	0 0 0	
National network identification	0 1 0	
Spare	0	

Octet 2	H G F E	D C B A
Digit One		
Digit 0		0 0 0 0
Digit 1		0 0 0 1
Digit 2		0 0 1 0
Digit 3		0 0 1 1
Digit 4		0 1 0 0
Digit 5		0 1 0 1
Digit 6		0 1 1 0
Digit 7		0 1 1 1
Digit 8		1 0 0 0
Digit 9		1 0 0 1
Spare		1 0 1 0
Code 11		1 0 1 1
Code 12		1 1 0 0

	HGFE	DCBA
Spare		1 1 0 1
Spare		1 1 1 0
End of pulse signal		1 1 1 1
Digit Two		
Digit 0	0 0 0 0	
Digit 1	0 0 0 1	
Digit 2	0 0 1 0	
Digit 3	0 0 1 1	
Digit 4	0 1 0 0	
Digit 5	0 1 0 1	
Digit 6	0 1 1 0	
Digit 7	0 1 1 1	
Digit 8	1 0 0 0	
Digit 9	1 0 0 1	
Spare	1 0 1 0	
Code 11	1 0 1 1	
Code 12	1 1 0 0	
Spare	1 1 0 1	
Spare	1 1 1 0	
End of pulse signal	1 1 1 1	

Octet 3	**H G F E**	**D C B A**
Digit Three		
Digit 0		0 0 0 0
Digit 1		0 0 0 1
Digit 2		0 0 1 0
Digit 3		0 0 1 1
Digit 4		0 1 0 0
Digit 5		0 1 0 1
Digit 6		0 1 1 0
Digit 7		0 1 1 1
Digit 8		1 0 0 0
Digit 9		1 0 0 1
Spare		1 0 1 0
Code 11		1 0 1 1
Code 12		1 1 0 0
Spare		1 1 0 1
Spare		1 1 1 0
End of pulse signal		1 1 1 1
Circuit Code		
Unspecified	0 0 0 0	
International call, no operator requested	0 0 0 1	
International call, operator requested	0 0 1 0	
Spare	0 0 1 1 to 0 1 1 1	

Reserved for network-specific use	1 0 0 0
	to
	1 1 1 1

This parameter is sent in the forward direction to indicate the long-distance carrier or transit network to be used to carry this call. This is used whenever the call is an inter-LATA call or international call.

The carrier ID is a three- or four-digit code administered by Bellcore that uniquely identifies each of the long-distance carriers. This is the same code dialed when using calling cards (10xxx, where xxx equals the carrier code).

Presently, there are no implications allowing for the use of ISUP to international networks such as the public data network or the public land mobile network. These are for further study.

Transmission Medium Used

Octet 1	H G F E	D C B A
Transmission medium used		
Speech	0 0 0 0	0 0 0 0
Spare	0 0 0 0	0 0 0 1
Reserved for 64-kbps unrestricted	0 0 0 0	0 0 1 0
3.1-kHz audio	0 0 0 0	0 0 1 1
Reserved	0 0 0 0	0 1 0 0
Reserved	0 0 0 0	0 1 0 1
Reserved for 64-kbps preferred	0 0 0 0	0 1 1 0
Reserved	0 0 0 0	0 1 1 1
	to	
	0 0 0 0	1 0 1 0
Spare	0 0 0 0	1 0 1 1
	to	
	1 1 1 1	1 1 1 1

The transmission used in a call setup is sent in the backward direction in the event the original circuit requested could not be used. This parameter then identifies the circuit type and is carried in the ANM, ACM, and CPG messages.

User Service Information

Octet 1	H G F E	D C B A
Information transfer capability		
Speech	0	0 0 0 0
Unrestricted digital information	0	1 0 0 0
Restricted digital information	0	1 0 0 1
3.1-kHz audio	1	0 0 0 0

7-kHz audio				1	0	0	0	1
Coding standard								
ITU standardized coding		0	0					
National standard		1	0					
Extension bit								
Octet continues through the next octet	0							
Last octet	1							

*Note: Only permitted when 64-kbps information transfer rate is used.

Octet 2	**H**	**G**	**F**	**E**	**D**	**C**	**B**	**A**
Information transfer rate								
Code for packet mode calls				0	0	0	0	0
64 kbps				1	0	0	0	0
384 kbps				1	0	0	1	1
1472 kbps (national ANSI only)				1	0	1	0	0
1536 kbps				1	0	1	0	1
1920 kbps				1	0	1	1	1
Multirate (64 kbps based)				1	1	0	0	0
Transfer mode								
Circuit mode		0	0					
Packet mode		1	0					
Extension bit								
Octet continues through the next octet	0							
Last octet	1							

Octet 2a	**H**	**G**	**F**	**E**	**D**	**C**	**B**	**A**
Establishment								
Demand (default)							0	0
Configuration								
Point-to-point (default)					0	0		
Structure								
Default (see description)		0	0	0				
8-kHz integrity		0	0	1				
Service data unit integrity		1	0	0				
Unstructured		1	1	1				
Extension bit								
Octet continues through the next octet	0							
Last octet	1							

Octet 2b	**H**	**G**	**F**	**E**	**D**	**C**	**B**	**A**
Information transfer rate (destination to origination)								
Code for packet mode calls				0	0	0	0	0
64 kbps				1	0	0	0	0
384 kbps				1	0	0	1	1
1472 kbps				1	0	1	0	0

	H	G	F	E	D	C	B	A
1536 kbps				1	0	1	0	1
1920 kbps				1	0	1	1	1
Multirate (64-kbps based)				1	1	0	0	0
Symmetry								
Bidirectional symmetric (default)		0	0					
Extension bit								
Octet continues through the next octet	0							
Last octet	1							

Octet 2.1	**H**	**G**	**F**	**E**	**D**	**C**	**B**	**A**
(***present if octet 2 indicates multirate 64-kbps base rate***)								
Rate multiplier								
Reserved (0)		0	0	0	0	0	0	0
One (1) to thirty (30)		0	0	0	0	0	0	1
				to				
		0	0	1	1	1	1	0
Thirty-one (31) to one hundred twenty seven (127)		0	0	1	1	1	1	1
				to				
		1	1	1	1	1	1	1
Extension bit								
Octet continues through the next octet	0							
Last octet	1							

Octet 3	**H**	**G**	**F**	**E**	**D**	**C**	**B**	**A**
User information layer 1 protocol								
ITU standardized rate adaption V.110/X.30				0	0	0	0	1
Recommendation G.711 u-law speech				0	0	0	1	0
Recommendation G.722 and G.725 7 kHz audio				0	0	1	0	1
Non-ITU standardized rate adaption				0	0	1	1	1
ITU standardized rate adaption V.120				0	1	0	0	0
ITU standardized rate adaption X.31 HDLC flag stuffing				0	1	0	0	1
Layer 1 identification		0	1					
Extension bit								
Octet continues through the next octet	0							
Last octet	1							

Octet 3a	**H**	**G**	**F**	**E**	**D**	**C**	**B**	**A**
(***present if ITU standardized rate adaption V110/V120***)								
User rate				x	x	x	x	x
Rate is indicated by E-bits specified in Rec. I.460				0	0	0	0	0
0.6 kbps Recommendations V.6 & X.1				0	0	0	0	1
1.2 kbps Recommendations V.6				0	0	0	1	0
2.4 kbps Recommendations V.6 & X.1				0	0	0	1	1
3.6 kbps Recommendations V.6				0	0	1	0	0
4.8 kbps Recommendations V.6 & X.1				0	0	1	0	1

7.2 kbps Recommendations V.6				0	0	1	1	0
8.0 kbps Recommendations I.460				0	0	1	1	1
9.6 kbps Recommendations V.6 & X.1				0	1	0	0	0
14.4 kbps Recommendations V.6				0	1	0	0	1
16.0 kbps Recommendations I.460				0	1	0	1	0
19.2 kbps Recommendations V.6				0	1	0	1	1
32.0 kbps Recommendations I.460				0	1	1	0	0
48.0 kbps Recommendations V.6 & X.1				0	1	1	1	0
56.0 kbps Recommendations V.6				0	1	1	1	1
64.0 kbps Recommendations X.1				1	0	0	0	0
0.1345 kbps Recommendations X.1				1	0	1	0	1
0.100 kbps Recommendations X.1				1	0	1	1	0
0.075/1.2 kbps Recommendations V.6 & X.1				1	0	1	1	1
1.2/0.075 kbps Recommendations V.6 & X.1				1	1	0	0	0
0.050 kbps Recommendations V.6 & X.1				1	1	0	0	1
0.075 kbps Recommendations V.6 & X.1				1	1	0	1	0
0.110 kbps Recommendations V.6 & X.1				1	1	0	1	1
0.150 kbps Recommendations V.6 & X.1				1	1	1	0	0
0.200 kbps Recommendations V.6 & X.1				1	1	1	0	1
0.300 kbps Recommendations V.6 & X.1				1	1	1	1	0
12 kbps Recommendations V.6				1	1	1	1	1
Negotiation								
In-band negotiation not possible			0					
In-band negotiation possible			1					
Synchronous/asynchronous								
Synchronous at the "R" interface		0						
Asynchronous at the "R" interface		1						
Extension bit								
Octet continues through the next octet	0							
Last octet	1							

*Note: The first value is the transmit rate in the forward direction of the call, while the second value is the transmit rate in the backward direction of the call.

Octet 3b	**H**	**G**	**F**	**E**	**D**	**C**	**B**	**A**
(***present if ITU standardized rate adaption V110***)								
Spare								0
Flow control on receive								
Cannot accept data with flow control mechanism							0	
Can accept data with flow control mechanism							1	
Flow control on transmit								
Not required to send data with flow control mechanism						0		
Required to send data with flow control mechanism						1		
Network independent clock on receive								
Cannot accept data with independent clock					0			
Can accept data with independent clock					1			

Network-independent clock on transmit								
Not required to send data with networkindependent clock				0				
Required to send data with network-independent clock				1				
Intermediate rate								
Not used		0	0					
8 kbps		0	1					
16 kbps		1	0					
32 kbps		1	1					
Extension bit								
Octet continues through the next octet	0							
Last octet	1							

Octet 3c	**H**	**G**	**F**	**E**	**D**	**C**	**B**	**A**
(***present if ITU standardized rate adaption V.120***)								
Spare								0
In-band/out-band								
Not applicable to this standard							0	
Negotiation is done in-band using logical link zero							1	
Assignor/assignee								
Message originator is “Default Assignee”						0		
Message originator is “Assignor Only”						1		
Logical link identifier (LLI) negotiation								
Default LLI = 256					0			
LLI negotiation					1			
Mode of operation								
Bit transparent mode of operation				0				
Protocol sensitive mode of operation				1				
Multiframe								
Multiframe establishment not supported			0					
Multiframe establishment supported			1					
Rate adaption Header/no header								
Rate adaption header not included		0						
Rate adaption header included		1						
Extension bit								
Octet continues through the next octet	0							
Last octet	1							

Octet 3d	**H**	**G**	**F**	**E**	**D**	**C**	**B**	**A**
(***present if ITU standardized rate adaption V110/V120***)								
Parity						x	x	x
Odd						0	0	0
Even						0	1	0
None						0	1	1
Forced to 0						1	0	0
Forced to 1						1	0	1
Number of data bits excluding parity bit								

Not used				0	0			
5 data bits				0	1			
7 data bits				1	0			
8 data bits				1	1			
Number of stop bits								
Not used		0	0					
1 stop bit		0	1					
1.5 stop bits		1	0					
2 stop bits		1	1					
Extension bit								
Octet continues through the next octet	0							
Last octet	1							

Octet 3e	**H**	**G**	**F**	**E**	**D**	**C**	**B**	**A**
(***present if ITU standardized rate adaption V110/V120***)								
Modem type			x	x	x	x	x	x
Coded according to network-specific rules								
Duplex mode								
Half duplex		0						
Full duplex		1						
Extension bit								
Octet continues through the next octet	0							
Last octet	1							

Octet 4	**H**	**G**	**F**	**E**	**D**	**C**	**B**	**A**
User information (layer 2 protocol)								
ANSI T1.602				0	0	0	1	0
Recommendation X.25 link level				0	0	1	1	0
Layer 2 identifier		1	0					
Extension bit								
Octet continues through the next octet	0							
Last octet	1							

Octet 5	**H**	**G**	**F**	**E**	**D**	**C**	**B**	**A**
User information (layer 3 protocol)								
ANSI T1.607				0	0	0	1	0
Recommendation X.25 packet layer				0	0	1	1	0
Layer 3 identifier		1	1					
Extension bit								
Octet continues through the next octet	0							
Last octet	1							

The user service information parameter is used when the subscriber is requesting data transmission on the voice facility without use of a modem. In this case, the subscriber is using either ISDN or X.25 packet switching as an interface to the PSTN.

The purpose of this parameter is to send the data transmission parameters to the distant exchange, so that the called party can receive the same parameters and establish the proper connection for receipt of the data. This parameter does not address the requirements of Asynchronous Transfer Mode (ATM) or Broadband ISDN. These are addressed through another protocol, B-ISUP.

There are many notes and prerequisites listed in the ANSI standard regarding this parameter. Not all of these are listed here. The use of several of the octets is dependent on the use of other octets. For example, octet 3b is only used when octet 3 indicates ITU standardized rate adaptation V.110/X.30. Many of the notes are provided, but the best source for this parameter is the ANSI T1.113-1992 publication.

Broadband ISUP Message Types

Broadband ISUP has been developed to support the usage of ATM voice circuits in the Public Switched Telephone Network (PSTN). The ISUP protocol was developed to support the connection and teardown of digital channelized circuits, such as DS1 and DS3. The ISUP protocol provides the data necessary to connect and control these channels in digital circuits, but ATM is based on virtual paths and virtual circuits not on channels.

The B-ISUP protocol is based on ISUP, and uses the same signaling procedures when possible. It is important to understand the role of B-ISUP so as not to confuse its procedures with those used when ATM SS7 links are used. The B-ISUP protocol is intended for the setup and teardown of voice transmissions over ATM; not the management of SS7 over ATM links in the same node.

Rather than duplicate the efforts from the previous section, this section will serve to identify the message types and their parameters used for broadband ISUP signaling within the SS7 network. These message types and their parameters are published as they are currently defined, and could change as the standards continue to evolve.

Figure 9.13 reflects the message types that are defined for use in the broadband networks. They are defined in the previous section for the most part, for the exception of those that are new. The new message types are defined below.

Message Type	HGFE	DCBA
Address Complete	0 0 0 0	0 1 1 0
Answer	0 0 0 0	1 0 0 1
Blocking	0 0 0 1	0 0 1 1
Blocking Acknowledgment	0 0 0 1	0 1 0 1
Call Progress	0 0 1 0	1 1 0 0
Confusion	0 0 1 0	1 1 1 1
Consistency Check End	0 0 0 1	0 1 1 1
Consistency Check End Ack	0 0 0 1	1 0 0 0
Consistency Check Request	0 0 0 0	0 1 0 1
Consistency Check Request Ack	0 0 0 1	0 0 0 1
Forward Transfer	0 0 0 0	1 0 0 0
IAM Acknowledgment	0 0 0 0	1 0 1 0
IAM Reject	0 0 0 0	1 0 1 1
Initial Address Message	0 0 0 0	0 0 0 1
Network Resource Management	0 0 1 1	0 0 1 0
Release	0 0 0 0	1 1 0 0
Release Complete	0 0 0 1	0 0 0 0
Reset	0 0 0 1	0 0 1 0
Reset Acknowledgment	0 0 0 0	1 1 1 1
Resume	0 0 0 0	1 1 1 0
Segmentation	0 0 1 1	1 0 0 0
Subsequent Address	0 0 0 0	0 0 1 0
Suspend	0 0 0 0	1 1 0 1
Unblocking	0 0 0 1	0 1 0 0
Unblocking Acknowledgment	0 0 0 1	0 1 1 0
User Part Available	0 0 1 1	0 1 0 1
User Part Test	0 0 1 1	0 1 0 0
User-to-user Information	0 0 1 0	1 1 0 1
Reserved for Narrowband ISDN	0 0 0 0	0 0 1 1
	to	
	0 0 1 1	0 1 1 1
Reserved for Code Extension	1 1 1 1	1 1 1 1

Figure 9.61 Broadband ISUP (B-ISUP) requires some new message types, in addition to those already supported in ISUP. This figure shows all of the B-ISUP message types and their indicator values.

In addition to some changes in message types, some additional parameters have been added to support ATM and broadband ISDN circuits. These parameters are illustrated below with their respective message types, and described in the section following. The values of the various parameters are defined in the Bellcore and ANSI standards, as well as the ITU-TS Recommendation Q.2931.

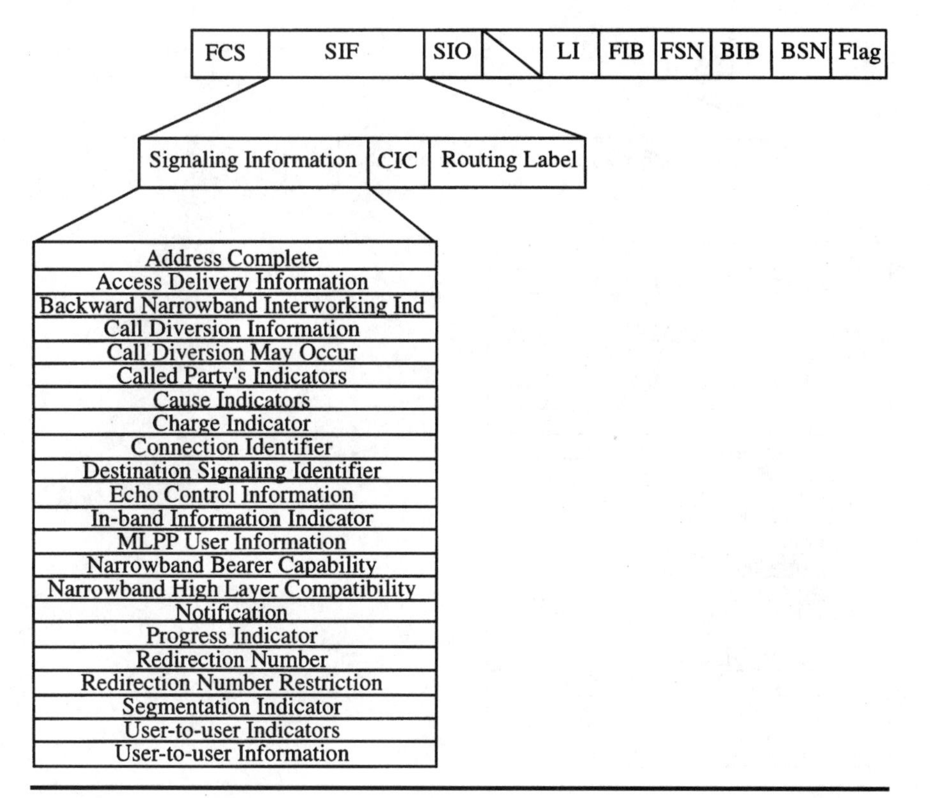

Figure 9.62

Address complete (ACM)

The address complete message has added several new parameters to support the virtual path connection identifiers and additional broadband parameters. These can be seen in the table above. The function has not changed, however.

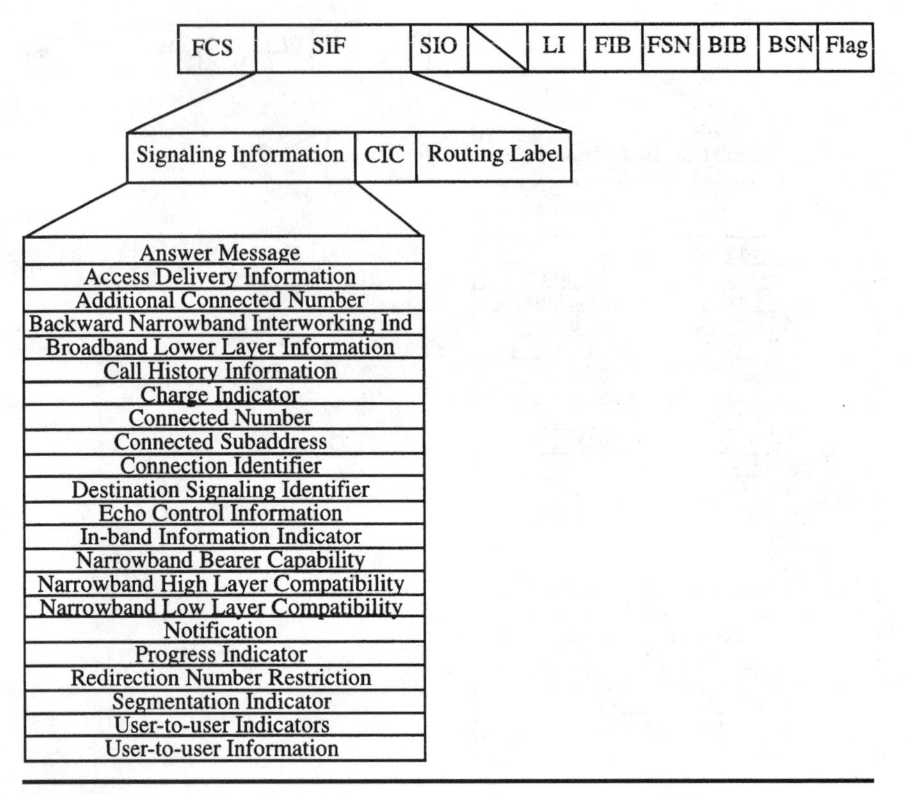

Figure 9.63

Answer (ANM)

The answer message also remains the same in function. The cut-through on a voice circuit (virtual path connection identifier in the case of broadband) does not take place in both directions in ANSI networks until the called party goes off-hook, and the ANM has been received. Tones are passed over the voice circuit in one direction only, meaning the cut-through takes place in the backward direction.

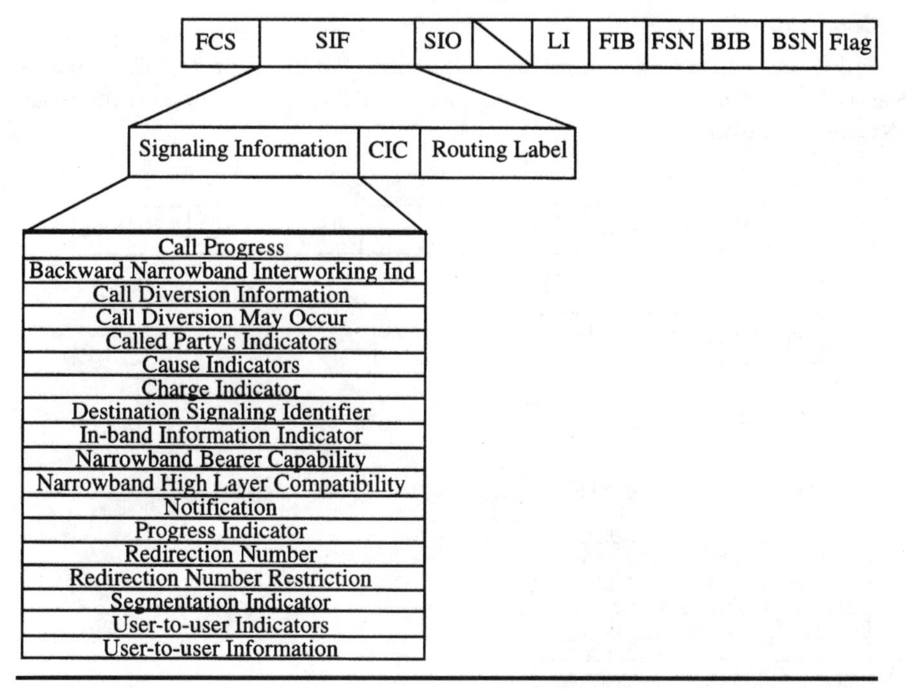

Figure 9.64

Call progress (CPG)

This message allows additional information regarding the status of a call to be sent to a remote exchange. Used only in broadband, this allows exchanges to send event information in either direction while a call is in progress.

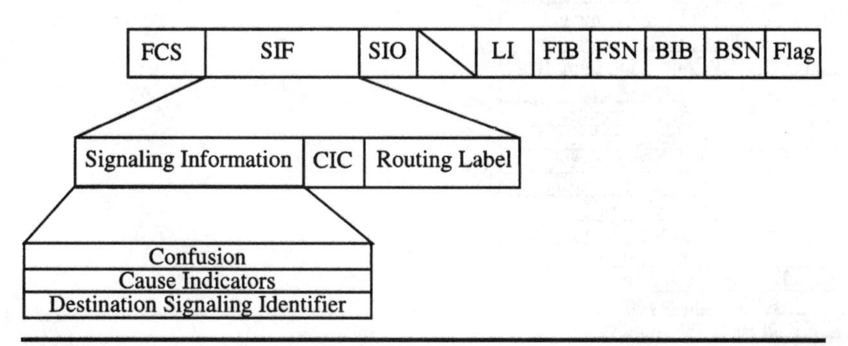

Figure 9.65

Confusion (CFN)

This message is the same as in normal ISUP. When a message is received which the exchange cannot identify, it returns the message type of confusion.

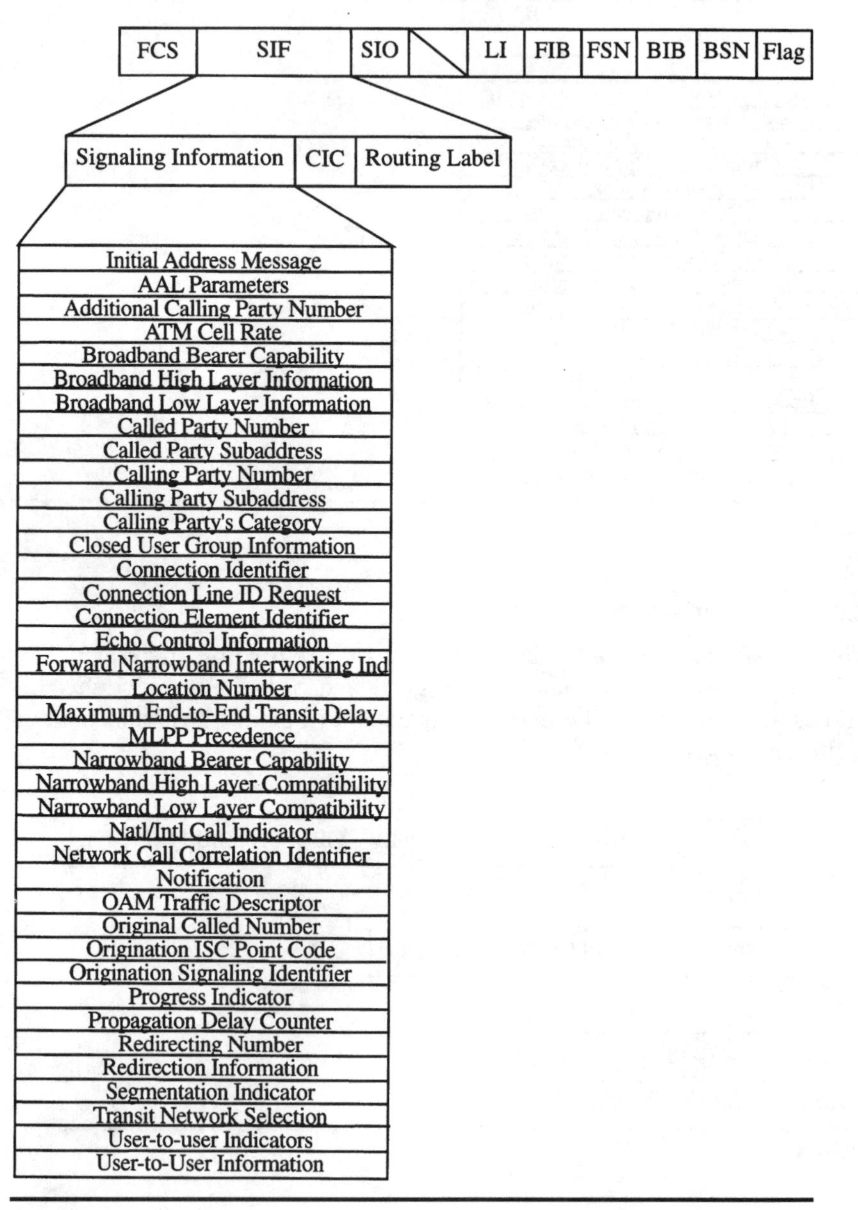

Figure 9.66

Exit (EXM)

This message is virtually the same as in ISUP. It is used when interworking with other networks to indicate a message has been successfully passed to another network.

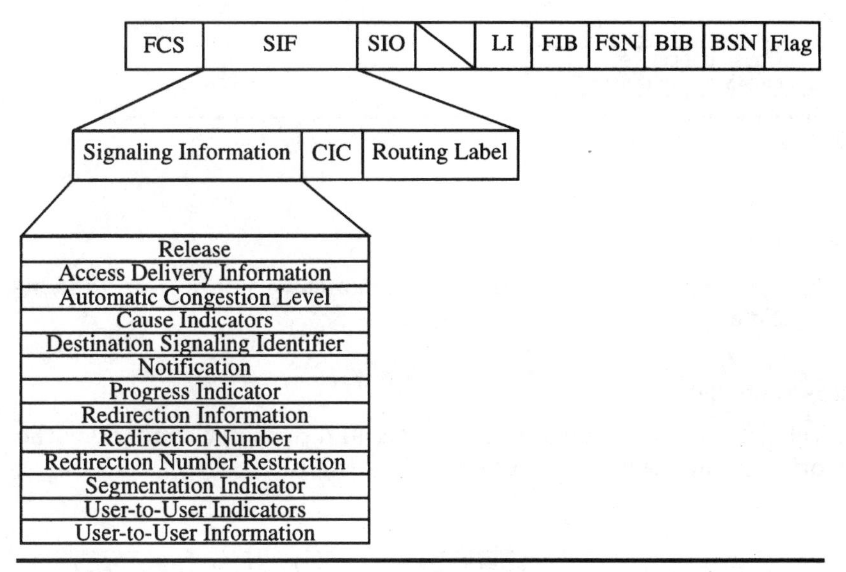

Figure 9.67

Initial address message (IAM)

This also is the same as in normal ISUP. The main difference between the two protocols and their IAMs is in the procedures. Normally, the routing label would identify which trunk circuit is to be used. In broadband, it identifies the virtual path connection identifier, if there is one available at the originating exchange that will accommodate the call being requested. Each exchange (originating and destination) is responsible for half of the virtual path connections, which prevents the possibility of glare.

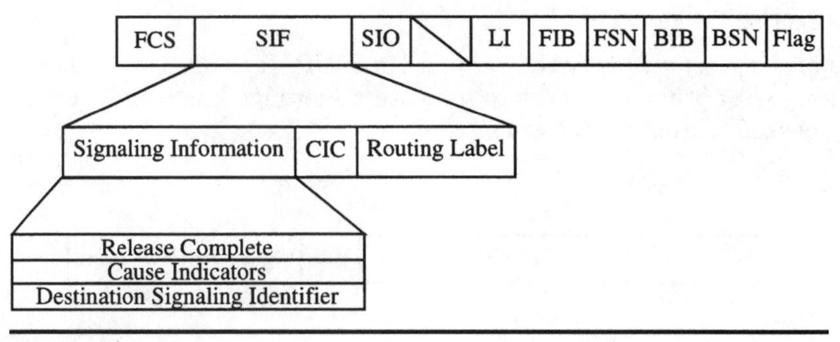

Figure 9.68

Release (REL)

The release message has several new parameters. All pertain to identifying the circuit that is to be released.

Release complete (RLC)

The release complete remains the same, and does not change for the exception of the destination signaling identifier.

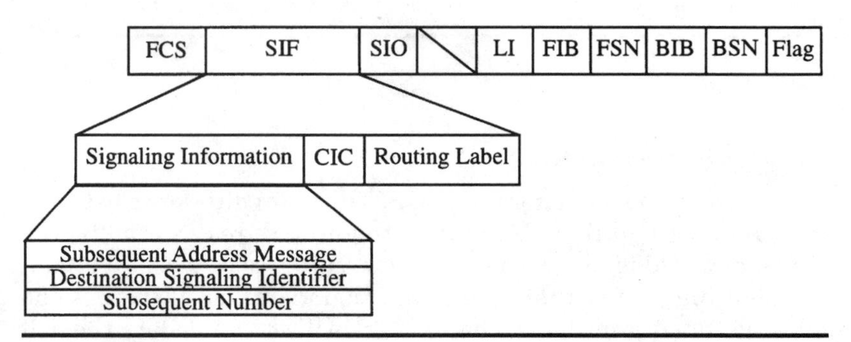

Figure 9.69

Subsequent address (SAM)

This message type is used in international networks only, and provides additional addressing information. This does not apply to U.S. networks.

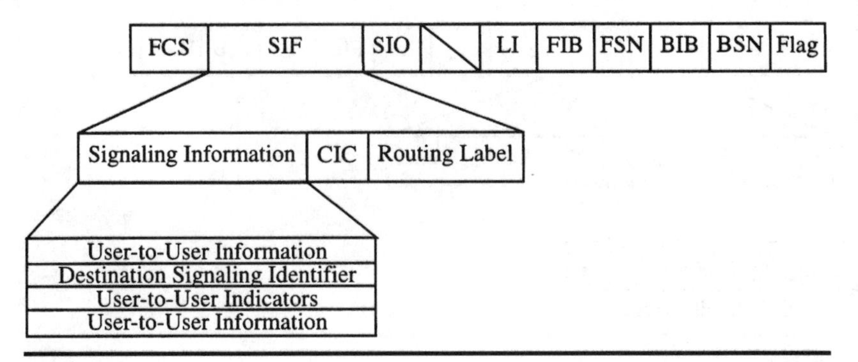

Figure 9.70

User-to-user information (USIS)

There are no procedures currently defined for this parameter in U.S. networks. The purpose of this message is to provide additional information to a user part from a remote user part in relation to a call already in progress. The term "user-to-user" refers to two user parts sending information to one another.

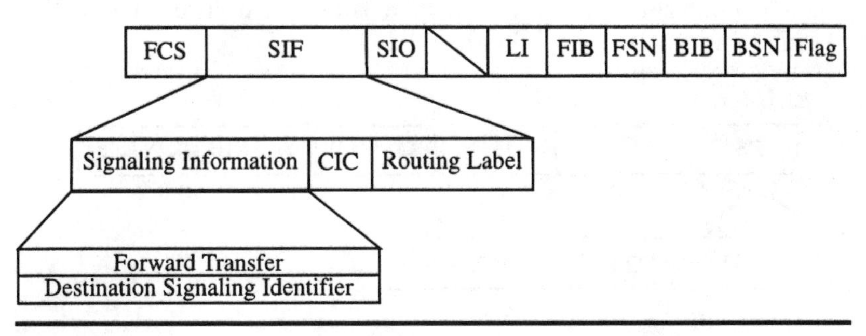

Figure 9.71

Forward transfer (FOT)

Used when an operator is requested during direct dial to an international number. The operator is also recalled when the call is terminated to provide additional assistance. Currently, this is only used in international networks.

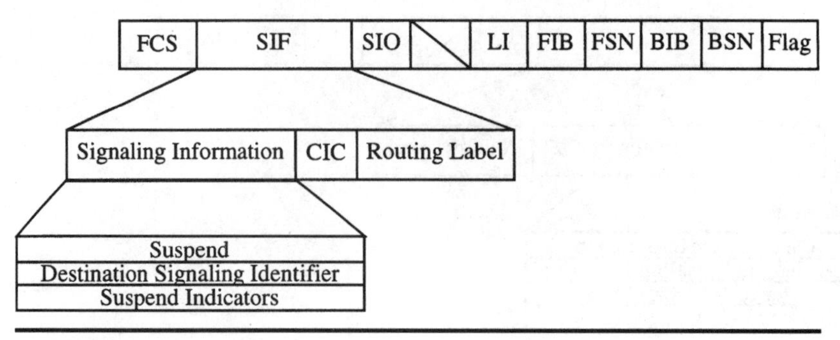

Figure 9.72

Suspend (SUS)

This message is sent in either direction to indicate that the called party has been disconnected. This would be used only if the connection had first been established, then the called party disconnected. Rather than send a release, the suspend allows the call circuit to be maintained for a period of time before initiating the release procedure. For example, if a called party is using call waiting, they may issue a flash hook to answer the other party. The suspend would be sent to the first calling exchange to hold the call circuit until a timeout, or until the called party came back to the original call.

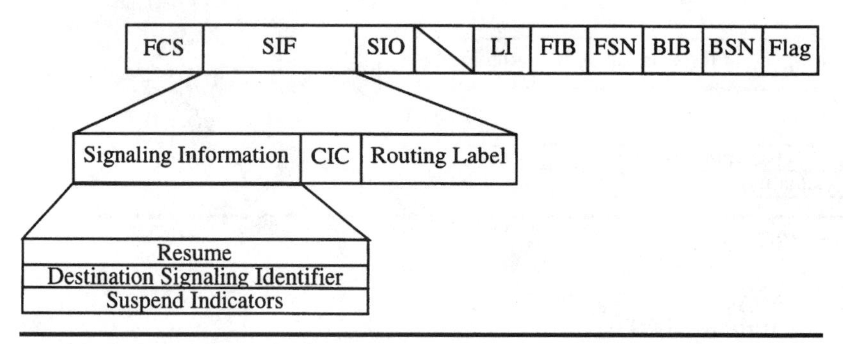

Figure 9.73

Resume (RES)

This is used along with the suspend message type. When the called party comes back to the call previously suspended, the resume message type is sent to indicate the call progress can continue as normal.

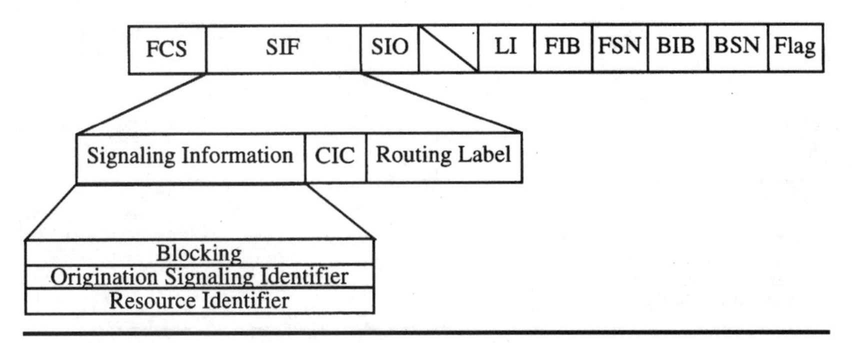

Figure 9.74

Blocking (BLO)

This is a maintenance message used for blocking a particular circuit from a connection. It can be initiated either manually or automatically by network management. Allows a circuit to be removed from service while still being able to send traffic over the circuit. Maintenance personnel may choose to block a circuit while they send test messages over that circuit.

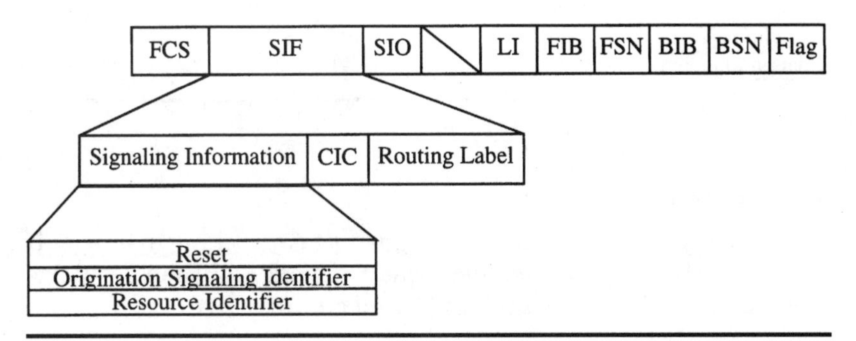

Figure 9.75

Reset (RSM)

Used when memory allocated to a specific virtual path connection identifier gets corrupted and can no longer remember the state of the connection. The reset is used to start both ends in the same known state. All resources are released and counters are reset, yet the connection is maintained.

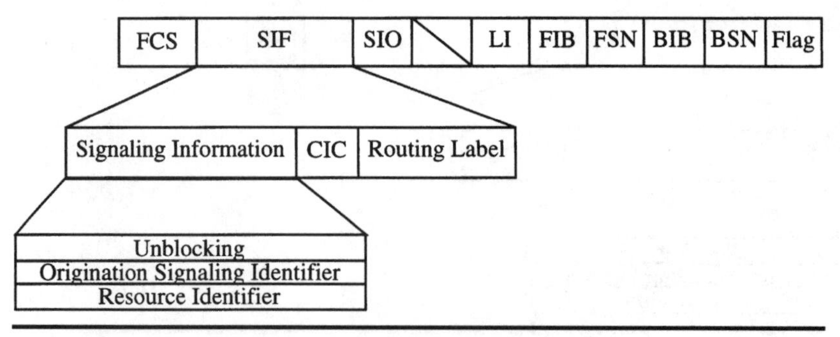

Figure 9.76

Unblocking (UBL)

This is the opposite of the blocking message, used to unblock a circuit.

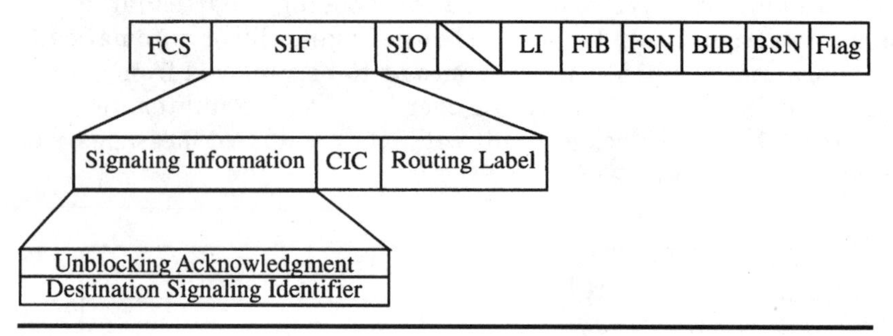

Figure 9.77

Blocking acknowledgment (BLA)

This message type is sent in acknowledgment to a blocking message. It indicates that the blocking message was received and the user parts have been blocked from using the indicated circuit.

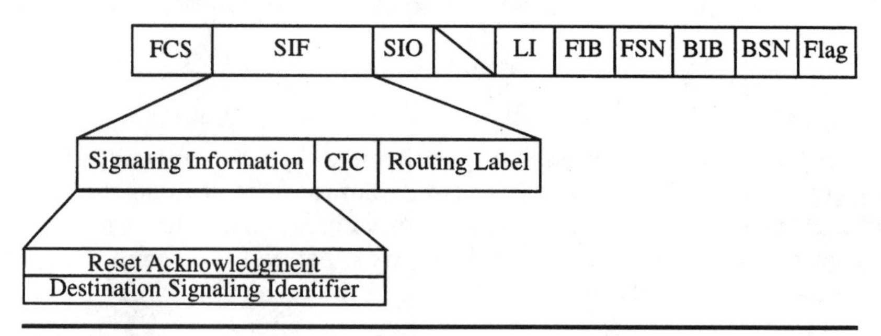

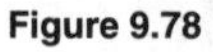
Figure 9.78

Reset acknowledgment message (RAM)

When memory at either end of a connection becomes corrupted, the signaling point may lose track of the state of a connection. In this case, a reset is requested on the specified virtual path connection identifier. This reset will release all resources associated with the connection.

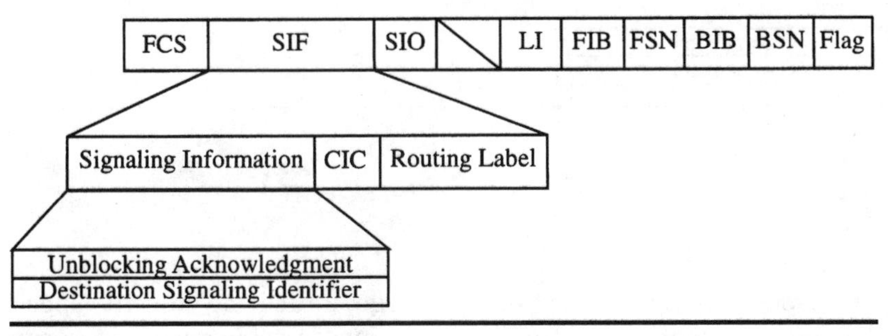

Figure 9.79

Unblocking acknowledgment (UBA)

This message indicates receipt of an unblocking message. This acknowledges receipt of the unblocking message and confirms that the circuit identified has been unblocked from the user parts.

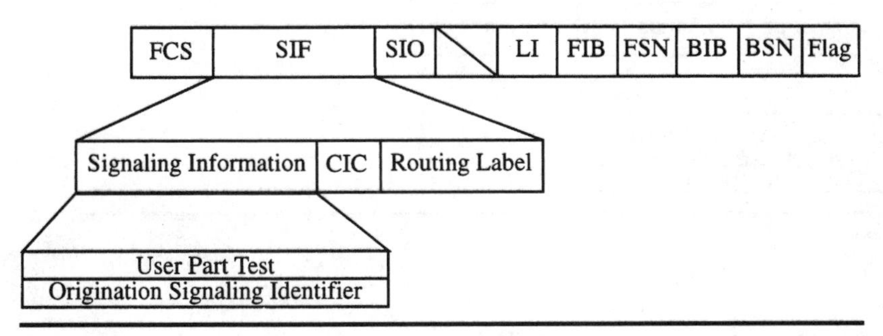

Figure 9.80

User part test (UPT)

This test message is used to verify the status of the specified user part. The purpose is to ensure that a user part marked as available or prohibited at the sending exchange is an accurate representation of the true state of the user part at the destination signaling point.

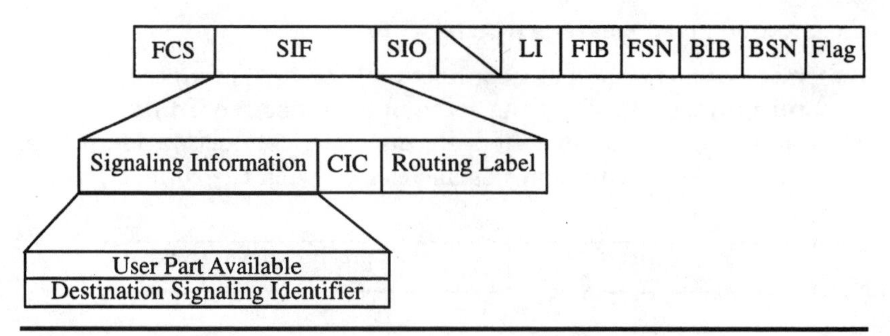

Figure 9.81

User part available (UPA)

Sent in response to a user part test message, indicating that the specified user part is available. User part test messages are used to verify the status of a given user part.

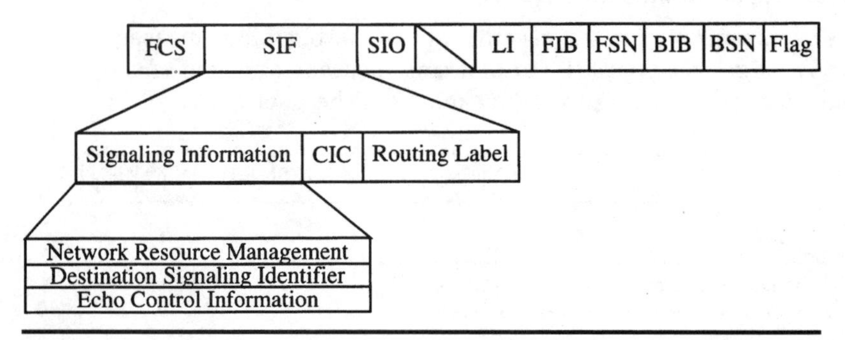

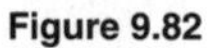
Figure 9.82

Network resource management (NRM)

This message is sent in the backward direction whenever the resources allocated to an established call need to be modified (i.e., additional resources allocated for additional bandwidth requirements).

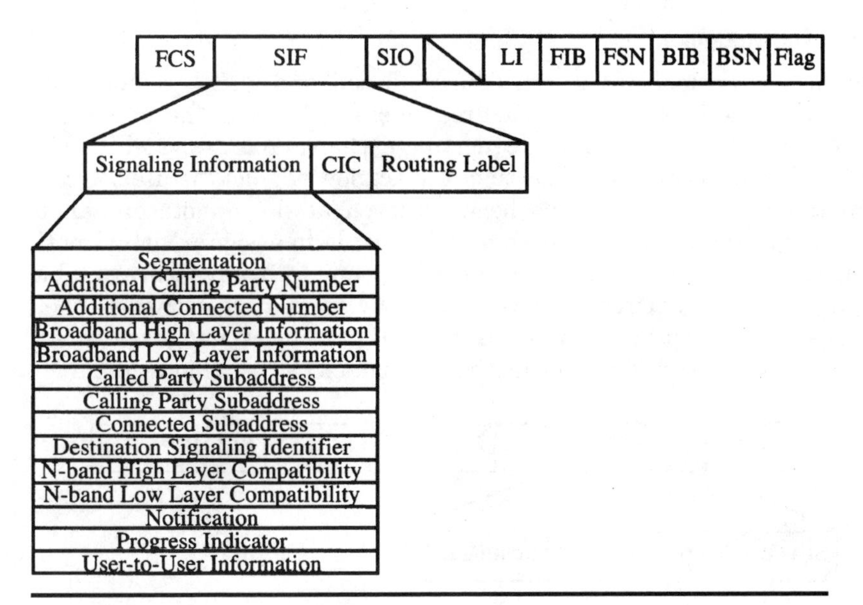

Figure 9.83

Segmentation message (SGM)

When messages exceed the maximum size of the SS7 packet (272 octets) the message must be segmented. This message type is used to send an additional segment to the destination signaling point.

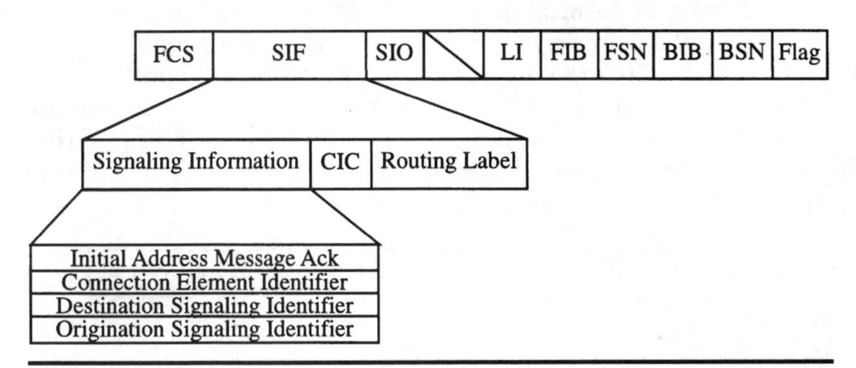

Figure 9.84

IAM acknowledgment (IAA)

Unlike the normal ISUP procedures, broadband ISUP requires an acknowledgment to an initial address message (IAM). The purpose is twofold. In the event that the requesting exchange has assigned the virtual path connection identifier, then the acknowledgment is used to confirm that the connection has been reserved at the remote exchange. However, if the originating exchange is unable to assign a virtual path connection identifier (because there is not ample bandwidth available within the range of circuits controlled by the originating exchange) then the acknowledgment is used to notify the originating exchange which virtual path connection identifier has been assigned by the remote exchange.

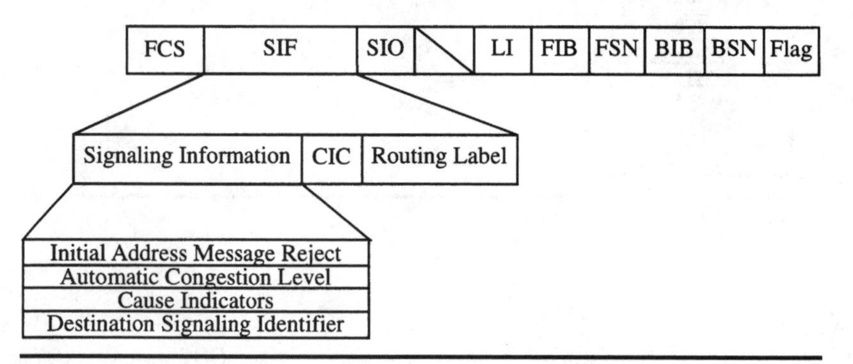

Figure 9.85

IAM reject (IAR)

This is used to notify the originator of an IAM that the requested bandwidth is not available on any of the virtual path connection identifiers within the control of the remote exchange. Therefore, a connection cannot be established. Since the originating exchange has not assigned the virtual path connection identifier, it is assumed that the originating exchange does not have ample bandwidth within its range of circuits either. The call connection attempt is aborted.

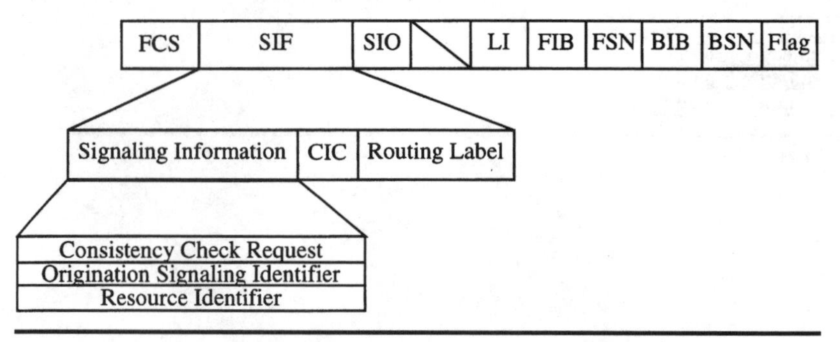

Figure 9.86

Consistency check request (CCR)

The consistency check request is sent to request the beginning of a "continuity" test on a specified virtual path connection identifier within a virtual path. This is much like the continuity check used in the ISDN User Part (ISUP) between two exchanges. The purpose is to ensure that a virtual connection can be established on the user plane between two exchanges. The test can be initiated by either exchange.

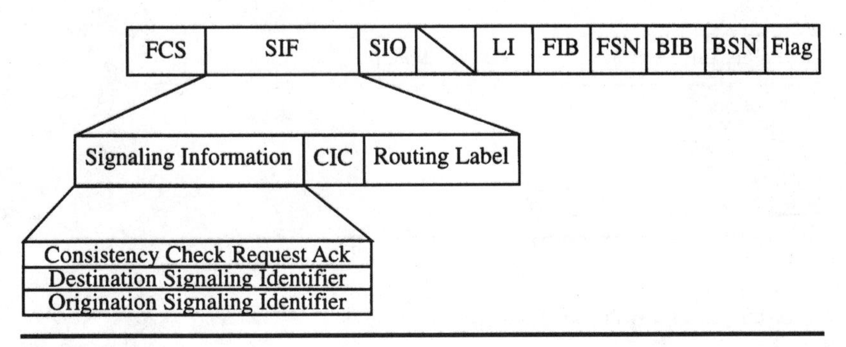

Figure 9.87

Consistency check request acknowledgment (CCRA)

This is sent in response to the consistency check request, and establishes the connections necessary to perform the test. Once the acknowledgment has been received, a test pattern may be sent over the proposed virtual path connection identifier.

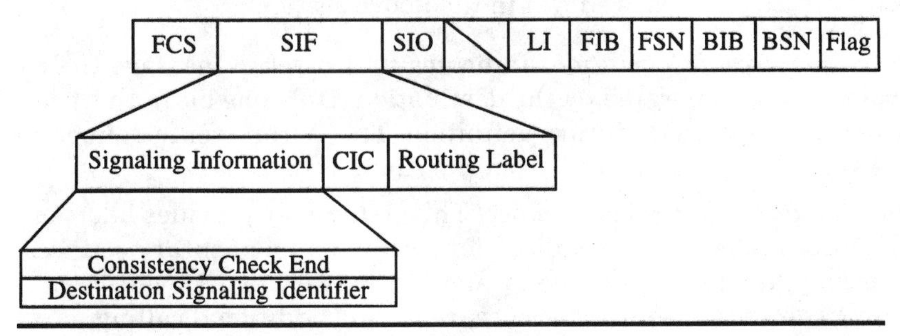

Figure 9.88

Consistency check end (CCE)

This parameter indicates the end of a consistency check. It can only be sent when the consistency check has been completed. The consistency check end is treated as an acknowledgment that the test has been completed.

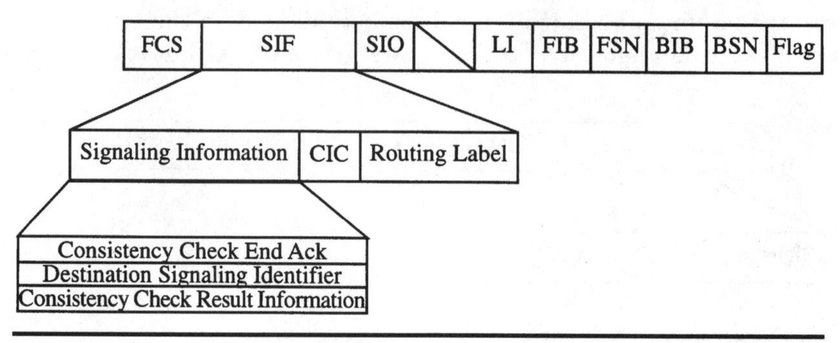

Figure 9.89

Consistency check end acknowledgment (CCE)

This is sent in response to a consistency check end message. It indicates the end of a test. The acknowledgment cannot be sent until the test flow has been stopped.

Broadband Parameters

Following is a description of the parameters that have been identified for usage in broadband ISUP. The association of these parameters to specific message types is shown in the above section.

Access delivery information. Indicates that a setup message (ISDN protocol) was generated by the destination. Only one bit in this octet is used, the rest is for future definition. This parameter is defined in the ITU-TS standards but not in the ANSI standards.

Additional calling party number. This parameter provides additional address information required for certain supplementary services. Typically, another exchange or another network entity will be providing the supplementary service, requiring additional calling party address information from the other entity. The same format as the calling party address is used. This parameter is defined in the ITU-TS standards but not in the ANSI standards.

Additional connected number. This parameter is sent in the backward direction and is used in the same manner as the additional calling party number. The difference is this parameter is associated to an additional *connected* party number. The same format as the additional calling party number is used. This parameter is defined in the ITU-TS standards but not in the ANSI standards.

ATM adaptation layer (AAL) parameters. This parameter is used to indicate which adaptation services will be required for the call. This information is of use to both the local exchanges, who will be providing the services, as well as the subscriber, who will be providing the end-point connection. The originating subscriber will be providing this information, based on the type of data that they will be sending, and depending on the type of source network being used at the originating subscriber's premise. Adaptation allows protocols such as Ethernet and Token Ring to interface to broadband ISDN or ATM circuits seamlessly and transparently. The header information is stripped and encapsulated into an ATM header, then transported across the network to its destination, where it can then be reassembled to the destination network.

ATM cell rate. This information is used by the receiving exchange in setting up the ATM circuit. The receiving exchange must know how many cells per second are needed for this particular call to be successful. The originator will specify the transmission rate through the BISDN interface using the BISDN *setup* message, which maps to the SS7 IAM message.

Automatic congestion level. When an exchange reaches a specified level of congestion, it will use this parameter to indicate which level of congestion it has reached to the opposite exchange. The message is not propagated through the network, and is addressed to the adjacent exchange with which a connection has been established.

Backward narrowband interworking indicators. This parameter identifies whether or not ISDN is used as the interface by the subscriber, and whether or not ISUP is available end to end within the SS7 network.

Broadband bearer capability. This parameter is used to indicate to the adjacent exchange how much bandwidth will be needed on the subscriber interface (broadband ISDN interface) to accommodate the call. This applies to the bearer channel only, and has no meaning to the signaling network.

Broadband low layer information. The parameters within this parameter are defined in the Q.2931 ITU-TS standards. This is used to ensure compatibility at the lower layers of the protocol stack at the subscriber interface. They are sent to the distant exchange to ensure the distant exchange is compliant.

Broadband high layer information. Like the low layer information parameter, this parameter and its contents are defined by ITU-TS Q.2931. The purpose is to ensure compatibility at the higher layers of the protocol stack at the subscriber interface at the distant end.

Call diversion information. When a call attempt is unsuccessful, this parameter is used to determine what type of treatment to provide for an attempted call. In some cases, an announcement may be requested, while in others, a busy tone or some other service tone may be used. This parameter is defined in the ITU-TS standards but not in the ANSI standards.

Call diversion may occur. Used to indicate that a call diversion may be indicated in a later message. This is necessary when interworking with narrowband ISDN (NISDN) networks. This parameter is defined in the ITU-TS standards but not in the ANSI standards.

Call history information. The purpose of this parameter is to provide a means for advising the distant exchange as to how long a call took to reach its destination. This is accomplished by providing a time value, sending it to the distant exchange, and the distant exchange comparing the time to its own real-time clock. The difference is considered propagation delay, which is sent to the originating exchange for processing. This parameter is defined in the ITU-TS standards but not in the ANSI standards.

Called party's indicators. This is used to identify the type of called party, such as normal subscriber or pay phone, and is used to determine if special treatment should be given to the call (such as in the case of a pay phone).

Called party number. This parameter provides the called party number, just as it does in normal ISUP. The called party number now includes a screening parameter, which is used to indicate whether or not the called party number is to be presented or screened from view. The telephone company is allowed access to the number for routing purposes, but if the number is to be screened, they cannot provide the number to the subscriber. This is used with certain services (such as 800 services) where the called party may not know what number the calling party dialed. The number would then be displayed to them, allowing them to determine how the call should be answered.

Called party subaddress. This information is also defined in ITU-TS Q.2931. This parameter carries the information defined in that standard through the SS7 network for delivery to the distant exchange.

Calling party number. This is sent in the forward direction during a call setup procedure to identify the calling party. The presentation parameter determines whether or not the calling party number may be displayed to the called party. If not, the calling party number cannot be transferred to the broadband ISDN interface.

Calling party's category. This parameter identifies the origin of the call, i.e., pay phone, data terminal, or an ordinary subscriber. The parameter also provides operators with a language indicator to indicate which language they should use when answering the call. Spanish, Russian, French, English, and German are presently defined within the protocol. Other languages would have to be defined by the agencies using the network and agreed upon mutually. The protocol provides additional codes to support network-defined languages.

Calling party's subaddress. Same as the called party's subaddress, but sent in the opposite direction.

Carrier identification code. This parameter is used to identify the carrier that a subscriber has selected. It is sent in the forward direction during call setup.

Carrier selection information. This identifies how the carrier selection code was selected; by the subscriber dialing digits or through preselection.

Cause indicator. Used when a call has ailed or has been cleared for any reason.

Charge indicator. Sent in the backward direction to indicate to the originating exchange whether or not a call is chargeable or not chargeable, based on the destination and the called party number.

Charge number. Provides the number to be billed for a call. Sent only in the forward direction.

Closed user group information. Currently, there are no procedures defined in ANSI or Bellcore networks for this parameter. A closed user group treats a group of numbers as if they were PBX extensions, much in the way Centrex treats a group of numbers as belonging to a business group. The user group can then be assigned specific features and privileges, allowing them to call within the group, but possibly blocking them from outside access.

Connected line identity request indicator. This is sent in the forward direction to indicate a request for identity of the connected number, rather than the called number (such as in the case of a forwarded call). Currently there are no procedures defined in ANSI or Bellcore networks.

Connected number. While there are no procedures defined in U.S. networks for this parameter, its intended use is to identify the number to which a forwarded call was actually connected.

Connected subaddress. This parameter is used along with the connected number, and provides subaddress information according to the ITU-TS Recommendation I.330. As mentioned in the connected number description, these two parameters are used to identify which number a call was eventually terminated to such as in the case of a forwarded or a transferred call.

Connection element identifier. This information is sent in the forward direction to identify the ATM virtual connection. In the event that the originating exchange does not have any virtual connections within its control available, this parameter would be sent by the destination exchange indicating which virtual circuit it has assigned for the requested call.

Consistency check request information. Used to indicate the result of the consistency check. The consistency check is like the continuity check used in normal ISUP procedures, but because of the nature of broadband, continuity tests can no longer apply. The consistency check is used instead, and is capable of testing the assignment of the virtual connection as well as the ability to transmit data through that virtual connection.

Destination signaling identifier. This parameter is used to associate the signaling connection with a virtual connection. This will most likely be used when B-ISUP is used in a fully associated signaling configuration.

Echo control information. As in the normal ISUP procedures, this parameter indicates whether or not echo control is required for half of the circuit or all of it.

Egress service. This parameter provides information about the network of the terminating exchange.

Forward narrowband interworking indicator. When interworking with a narrowband ISDN interface, this parameter provides information in the forward direction regarding the signaling abilities on the connection.

Generic address. This is used in supplementary services, and can identify a destination number (dialed number).

Generic digits. These digits can be account codes, authorization codes, or any type of number used in supplementary services.

Generic name. Provides specific name-related information used in supplementary services.

In-band information indicator. This parameter is used to indicate that in-band signaling information or an appropriate pattern is available on the connection specified.

Jurisdiction information. Same as the JIP parameter used in ISUP. Provides the LRN of the originating party to be used by billing systems when the calling party number has been ported to another service provider.

Location number. This parameter provides the same information as the called or calling party number, but is associated with a user within a narrowband ISDN connection. This is only used when interworking with narrowband ISDN networks.

Maximum end-to-end transit delay. This parameter indicates the maximum transit delay allowed for a message traveling through the network. The delay maximum is for an end-to-end transmission.

MLPP precedence. Used primarily with military networks, this parameter indicates the level of precedence supported for the connection. The Multilevel Precedence Preemption supplementary service is used in military installations to allow officers of higher rank to seize a trunk in use based on rank. This feature used to be limited to large AUTOVON systems installed on military bases, but is now offered through telephone service providers.

MLPP user information. This is a one-octet parameter, which uses only one bit in the entire octet. That one bit is used to indicate whether or not the called party is an MLPP user.

Narrowband bearer capability. During the setup phase of a connection, the originator may request for a specific bandwidth. In some cases, the originator may allow a lower bandwidth for the connection, only if the requested bandwidth is not available. This is referred to as the *fallback* bandwidth. This parameter is used to indicate the fallback bandwidth that has been allocated for the call connection.

Narrowband high layer compatibility. This parameter is used to ensure compatibility between two exchanges. When an exchange requests service to a specific exchange, it may request a specific level of service, or allow fallback to an available service. In the event the requested service is not available, and the fallback service is assigned, this parameter notifies the distant exchange of the fallback service assignment so that it may set up its end to be compatible.

Narrowband lower layer compatibility. This parameter is like the above parameter, but is concerned with compatibility at the lower layers rather than the upper layers. This also ensures compatibility at both ends of the exchange.

National/international call indicator. Used to indicate to a national exchange that the origin of the call is from a national network or an international network. The call handling procedures for an international call are somewhat different than they are for a national call.

Notification. This parameter may be sent in either direction, and is used to notify the other exchange regarding supplementary services such as call diversion procedures. Call diversion procedures encompass the use of service tones and announcements if a call cannot be completed (such as in the case of a cellular subscriber being away from their car, and the cellular phone being turned off).

Notification indicator. Used with supplementary services to send notification to the user.

OAM traffic descriptor. This parameter indicates the cell rate that is required by Operations Administration and Maintenance (OAM) traffic on a specified virtual connection.

Original called number. When a call has been redirected, whether through forwarding or a transfer, this parameter identifies the original called party number. The connected called number will also be provided through that parameter. This is used when interworking with narrowband ISDN connections.

Originating line information. Provides the toll class of service for a call. Sent only in the forward direction.

Origination ISC point code. This information is provided in an Initial Address Message (IAM) to indicate the originating point code of an international ISC.

Origination signaling identifier. Sent by the originator of a signaling or control message to identify the association with the distant signaling connection.

Originating facility identifier. Only used when intranetworking. Identifies the outgoing facility selected to reach the adjacent network.

Parameter compatibility information parameter. This information is used to inform the receiver how to interact in the event it receives an incorrect parameter it cannot recognize.

Progress indicator. This is used to indicate an event which has taken place during the lifetime of a call. This is defined in Q.9231.

Propagation delay counter. While being transferred through the network, this parameter is accumulating timer information. Each time it passes through a network entity, the timer information is updated in 1 ms increments. By the time it reaches its final destination, the destination exchange can determine the amount of propagation delay experienced by the message.

Redirecting number. When a call is forwarded to another number, this parameter identifies the number that forwarded the call. For example, a number may be forwarded to another number, which in turn has diverted the call to yet another number. The number that provided the diversion would be indicated in this parameter.

Redirection information. This parameter is also related to call diverting, or call forwarding. When a call is redirected to another number, this parameter indicates the nature of the redirection. For example, if a mobile subscriber is not available, then the call is diverted to an announcement. This parameter would then provide information as to why the call was diverted and the nature of the call.

Redirection number. When a call is being diverted to another number, this parameter identifies the number to which the call has been diverted. This parameter is sent in the backward direction.

Redirection number restriction. Also related to diverted calls, this parameter identifies whether or not the diverted user allows presentation of the telephone number. Because of legislature in many states, telephone companies must provide the option of screening the calling party number identification to the called party. This parameter indicates whether or not that is the case.

Resource identifier. When a resource has been blocked or unblocked (such as an announcement), this parameter indicates which resource it applies to.

Signaling identifier. The signaling identifier allows each exchange to assign independently of one another a signaling association, and correlate messages with each signaling association. A signaling association is received over a signaling identifier.

Segmentation indication. When a message is segmented into several messages, this parameter is used to indicate that the message is segmented and that there is additional segmentation information.

Special processing request. This parameter is sent only in the forward direction when special processing is required at the terminating exchange. Special processing includes verifying authorization codes or translating private network numbers.

Subsequent number. When the called party requires additional information this parameter is used to provide the additional numbers, or subsequent addresses.

Suspend/resume indicators. This parameter is sent in either the suspend or resume message to indicate whether or not the suspend or resume message was initiated by an ISDN subscriber or by the network. The suspend is used to temporarily place a circuit into a hold state, such as when the subscriber invokes another feature such as call waiting by performing a hook flash. The resume is used to indicate that the circuit is now "reconnected" and transmission can resume.

Transit network selection. When an Initial Address Message (IAM) is sent, this parameter indicates which transit network, if any, is to be used to carry the message. A transit network is one that is used to access another carrier's network when network boundaries must be crossed.

User-network interaction indicator. This parameter is sent in the backward direction when additional information is needed from the calling party to process a call.

User-to-user indicators. This parameter carries the information defined in ITU-TS Q.2931, when the users at each end of the connection have information to share. The indicators are used to indicate that such information exists, while the information parameter (below) provides the actual information. The indicators also specify whether or not a reply is to be sent.

User-to-user information. This parameter is used to send information from a user to the user at the remote exchange. The information is passed transparently through all intermediate exchanges.

Chapter

10

Local Number Portability

By far the biggest project to ever hit the telephone industry has been Local Number Portability (LNP). Never before has anything this big been attempted in such a short time frame. There are a lot of questions and issues surrounding LNP, as well as a lot of uncertainties.

Local Number Portability was defined in the Telecommuni-cations Act of 1996 as the "ability of users of telecommunications services to retain, at the same location, existing telecommunications numbers without impairment of quality, reliability, or convenience when switching from one telecommunications carrier to another."

The Telecommunications Act mandated that all telecommunications service providers provide, to the extent technically feasible, number portability in accordance with the requirements prescribed by the Commission. LNP got little attention until the FCC issued a mandate in June of 1996, requiring the implementation of LNP according to a very aggressive schedule. Rather than cite the specifics of the FCC mandate (FCC Docket 95-116), I will leave it to the reader to review the mandate itself, which is being published in three different phases. However, it should be understood that LNP affects everyone involved in wireline *and* wireless industries.

Following are some highlights from the FCC docket:

The solution must support existing services and features.

LNP must use the existing numbering resources efficiently.

LNP cannot require subscribers to change their telephone numbers.

There can be no unreasonable degradation in service (such as call setup delays) or network reliability degradation when subscribers switch carriers.

No carrier can have a proprietary interest.

The LNP solution must be able to accommodate location and service portability in the future.

There can be no significant adverse impact outside areas where number portability is deployed.

The intent of LNP is to open up local telephone service to competition. The authors of the Telecommunications Act feel that the biggest roadblock to competition is the ownership of telephone numbers. Subscribers are reluctant to switch to a new service provider because they have to give up their telephone numbers when they switch to a new service provider. LNP allows subscribers to switch to a new provider while keeping their existing telephone numbers.

This presents a huge challenge to the telephone industry. Until now, routing of telephone calls has been based on the first six digits of the telephone number (NPA NXX, or area code and office code). If a subscriber moves to a new area, or elects to change service providers, this is no longer possible. The telephone switches once identified by the old numbering plan are suddenly faced with servicing numbers from other service providers.

If a subscriber moves across the country, the problem becomes more complex. Telephone equipment and software throughout the network have been designed to use the telephone number to determine the geographical location of subscribers. For example, everyone knows that the 212 area code is Manhattan. However, with Local Number Portability, a subscriber with a 212 area code could live in California. This problem is compounded when one looks at billing systems, operations support systems (OSSs), and other network subsystems which all rely on the numbering plan for determining a caller's geographic location and service provider.

This chapter will outline the impact of Local Number Portability (LNP) on the telecommunications industry, how it works, and how SS7 is used to implement LNP for both wireline and wireless networks.

Introduction

There are three phases to Local Number Portability. The first phase, *service provider portability,* is being implemented now. This allows a subscriber to select a new local service provider while keeping their existing telephone number. It also allows a subscriber to move within their rate center while maintaining the same telephone number.

The next phase of Local Number Portability is *service portability.* This allows subscribers to change the type of service they have while keeping their telephone numbers. For example, if a subscriber changes from a POTS line to an ISDN service, he or she has to obtain a new telephone number, because the switching equipment used to provide the ISDN service supports a different block of numbers. With LNP, the subscriber does not have to give up the telephone number when changing the type of service.

The most difficult phase of Local Number Portability is *location portability.* This will allow a subscriber to move from city to city, or even state to state, while maintaining the same telephone number. This has a much more global impact. Even subscribers are accustomed to associating geographical areas with telephone numbers. It will be difficult for anyone to determine where he or she is calling once location portability is implemented.

Currently, porting is only supported within a rate center. A rate center is a geographic area, usually within a LATA, which is used for determining the time and distance used in billing of phone calls. Rate centers are determined by using (V)ertical and (H)orizontal coordinates. Porting a number outside of a rate center will present many technological challenges in itself, and is being addressed in later implementations of number portability.

The Telecommunications Act of 1996 and the FCC mandate (docket 95-116) do not specify exactly how LNP is to be implemented. They simply outline the ground rules to be used when implementing it in the network. The first trial of LNP took place in Illinois, under the direction of the Illinois Commerce Commission (ICC). The ICC specified some rules regarding the actual implementation, such as using a Location Routing Number (LRN). These have been published in the AT&T specification FSD 30-12-0001. These implementation requirements have become the de facto standard for LNP and have been widely accepted throughout the industry.

LNP Impact

There have been several proposals for providing LNP, without implementing a database solution. The first solution was call forwarding. This was quickly rejected by the FCC as an interim solution because of the delay imposed on the calling party while the carriers tried to route the call. The FCC did not want subscribers punished for changing providers and it has stringent requirements regarding the level of service provided to subscribers when they switch carriers.

Another approach was Query-on-Release (QoR). When a call is routed to a number that has been ported, the receiving switch identifies the number as being vacant, and returns an SS7 RELease message with an appropriate cause code. The originating switch would then initiate a database query to determine if the number had been ported. This approach certainly reduces the traffic across the SS7 network, and lessens the impact of the database queries, but again it places unnecessary delays on setting up of telephone calls to subscribers who have changed carriers. QoR was also rejected as an interim method.

The solution that was chosen was the LRN method. The end office switches in the rate center have a table identifying all NPA-NXXs, which have numbers in them that have been ported. The specific number is not provided in the database, so the switch must initiate a query if it is determined that the number dialed was to an NPA-NXX considered as ported.

The database provides a Location Routing Number (LRN), which is explained later when we discuss call flows. The LRN method places a higher demand on the SS7 network, but ensures there is no degradation of quality or service for the subscriber who changes carriers. Unfortunately, LRN also imposes huge unrecoverable costs on telephone companies.

There has to be some event that causes a query to take place. The industry has agreed that IN/AIN triggers should be used to initiate queries. A trigger expands the call processing capabilities of switches by *triggering* defined events to take place (like initiating an LNP query) when specific criteria are met. For example, if received dialed digits equal a specific value a query is sent to obtain additional routing instructions. This will require software upgrades in all switching equipment to support IN/AIN triggering.

To understand the impact that LNP has on the telephone network routing, you must first understand how networks treat telephone numbers. The first usage of telephone numbers came with the early manual switchboards. Local town operators were finding it difficult keeping up with all of the town's citizens by name. In those days, if you wanted to call someone, the operator had to connect the call for you. There were no telephone numbers; everyone was known by name. It was a doctor who first suggested using a numbering system instead of names, using his own practice as an example (doctors maintain an elaborate and numbered filing system to maintain patient records).

The early telephone numbers were sometimes given names for the "exchange" they served. The first three numbers after the area code were given letters; those letters identifying the exchange the telephone number resided in. The exchange was the area that a particular telephone switch (or group of telephone switches) served. For example, in my study is an antique telephone from an office in Los Angeles, with the exchange "MAD" for Madison exchange still on its dial.

Over time this changed, and the names were lost, but numbers continue on. Telephone switches are assigned blocks of numbers, with the first three digits (office code) identifying the particular switch or central office the subscriber number is served by. When calls are routed, only the first six digits are used (area code/office code, or NPA/NXX). When the call is delivered to the correct destination central office, the switches recognize the office code as their own and route the call by the last four digits (the subscriber number).

Likewise, billing systems use the telephone numbers in much the same way. Telephone companies established rate centers by dividing the "exchanges" into geographical areas. These areas could be measured for distance by using (H)orizontal and (V)ertical coordinates. When a call is placed, the area/office code is used to determine if the calling party is making a local or a long-distance call. If it is long distance, the charges are applied according to a complex formula using the H/V coordinates. What happens if you take that telephone number and use it in another city? You can begin to see the complexity now of Local Number Portability.

This complexity is even more severe when you examine the wireless network. Cellular providers face a greater challenge ahead because of the very nature of their networks. Subscribers are already "portable," moving from cell site to cell site. Their billing is determined by a completely different plan and does not match the same system used by wireline providers.

Each subscriber is assigned to a "home" Mobile Switching Center (MSC), which falls within a specific wireline rate center. If a mobile subscriber calls a wireline number, the billing is determined by the distance from the MSC to the wireline number. However, that same subscriber could be roaming (outside their home area), making the billing more difficult. Previous to LNP, the cellular subscriber's mobile identification number (MIN) has been used for determining the home MSC and how they would be billed for calls when roaming. If they take that MIN to another cellular provider, it becomes even more difficult to determine what charges should be applied. The MIN is also used for identifying the carrier providing the wireless services. The first six digits of the MIN identify the service provider for that subscriber. This means the MIN can no longer be used for call processing, because the MIN cannot be ported.

Wireless networks rely on the mobile identification number (MIN) for call processing, billing, and virtually every transaction related to a mobile subscriber. However, use of the MIN to address portability would require too many database queries, and impacts global title databases. The wireless industry has elected to change the identification of mobile subscribers by assigning two numbers; the mobile directory number (MDN) and the mobile station identifier (MSID). The MSID can use the MIN format, but the MSID is not portable. The mobile directory number (MDN) is portable. This changes the way call processing takes place in the wireless networks, impacting virtually every element in the wireless network.

In international cellular networks, the mobile identification number (MIN) is not recognized. Instead, networks use what is called the international mobile station identifier (IMSI), which is recognized in any network overseas. These networks are usually based on GSM technology. Cellular providers in the United States have been preparing to change from MIN format to ISMI format, so they could support international roaming. LNP offers an opportunity to use the IMSI format when assigning MSIDs to mobile subscribers.

Another impact on wireless networks will be the method used to query databases. Wireline networks have agreed on the IN/AIN triggers for querying databases; however, wireless networks do not necessarily support IN/AIN. The industry is looking at IS-41 and GSM protocols for querying the LNP database. This will place a burden on developers of LNP products, who must be able to support all solutions for both wireline and wireless. Both the IS-41 and GSM protocols are being modified to support additional parameters for LNP.

Suffice it to say there are a number of issues surrounding the implementation of LNP in both the wireline networks and the wireless networks. A lot of these issues have been resolved; some are still being investigated. What is clear is that the networks can no longer rely on the telephone number for routing, billing, and network operations. This means that millions of dollars will be spent changing the software in existing equipment and adding new hardware (such as databases, and data links to access those databases) to make LNP work.

In addition to changes in the network equipment, SS7 has been modified as well. The solution to LNP has required new parameters to the protocol, mostly in the area of ISUP. Additional information is now required to route calls. The solution agreed upon by the industry was the assignment of a new number, called the Location Routing Number (LRN). The LRN is assigned to each end office, using the same existing numbering plan in place today. In most cases, if an end office serves, say, the 919-460 area, NPA NXX; its LRN would be 919-460-0000. This preserves the concept of associating routing with geography, but does require changes to routing tables and software. We'll discuss how this all works a little later.

Not all networks support SS7. There are still some networks that rely on multifrequency (MF) signaling. There are interworking procedures defined when SS7 networks must interwork with MF networks, however, there is an efficiency problem introduced with LNP.

If an MF network becomes part of the call path, the information contained in the ISUP IAM message cannot be transferred across the network. This means all LNP data will be lost. If there is an SS7 network on the other side (in other words, the MF network is in between two SS7 networks), then the SS7 network receiving calls routed from the

MF network must initiate another query to an LNP database to determine how the call is to be routed.

This introduces additional delay on processing the call, and introduces the risk that the call may be handled improperly if for some reason the query cannot be performed at the receiving SS7 network. These conditions are still under study, and procedures are being established for handling calls with MF signaling networks.

Databases and operations support systems (OSSs) used throughout the SS7 network face changes. Many of these systems rely on specific fields within the SS7 ISUP messages, which have now been modified to support LNP. These systems will have to be upgraded to understand the new message format. For example, calling number delivery depends on the ISUP IAM calling party field. This may no longer be the actual directory number of the person calling, but the LRN of the calling party's exchange. There are many similar scenarios, all of which have been defined in the Bellcore specification GR-2936-CORE.

Congestion within an SCP is a real threat, especially with so many queries being made for LNP. Automatic call gapping (ACG) has been identified as the method for preventing SCP congestion. ACG defines how long (duration) and how often (intervals) transactions should be placed under control. ACG will be supported in both wireline and wireless networks.

The FCC mandate provided a schedule for the implementation of LNP. This schedule has been extremely aggressive, and by the time this book has been published, most of the LNP milestones will have been met. The mandate required LNP to be demonstrated within the top 100 MSAs by December 1998. However, the FCC issued a deployment schedule starting as early as October 1997. LNP must be fully operational in the top 100 MSAs by June 1999.

Wireless service providers were issued a reprieve, and have until December 1998 to be able to route calls to ported numbers, and June 1999 to have LNP implemented in the top 100 MSAs. They have 9 months to implement LNP when they are requested by another carrier to provide number portability. This means that if a carrier loses a subscriber to another carrier, and that carrier requests the number to be ported, the wireless provider has 9 months to make that happen. Wireline service providers were given 6 months after a request to implement LNP.

Smaller service providers are allowed to file for exemption from LNP under certain conditions. If the carrier has fewer than 50,000 access lines, they can be exempt from the LNP requirements. There are other conditions that must be met as well, all published in the FCC mandate.

Of major concern to all carriers is the responsibility of sending a query to an LNP database. A carrier could have a subscriber place a call to a number that has been ported, and if they do not have LNP in their network, route the call as normal. The receiving network would be unable to connect the call, because the number would be found as vacant in the switch's database. The receiving network would then be left with the responsibility of sending a query to an LNP database to see if the number had been ported.

This impacts other networks and shifts the fiscal responsibility on other carriers, which the FCC mandate clearly cited as a violation. To prevent this type of "dumping," the industry has agreed on an N-1 scheme. The next to the last network in a call route is responsible for making the database query. For example, if there are only two carriers involved in routing a call, the originating network is responsible for sending the database query.

If there are several networks involved in routing and connecting the call, the next to the last network (which would be a long-distance carrier) is responsible for sending the database query and routing the call to the appropriate network. This approach is the most equitable way to prevent carriers from dumping calls on other carriers and makes everyone responsible for implementing LNP.

The LNP Elements

There are several elements required in the network to support LNP. Figure 10.1 shows the general architecture of the LNP network. Each of the elements is described below.

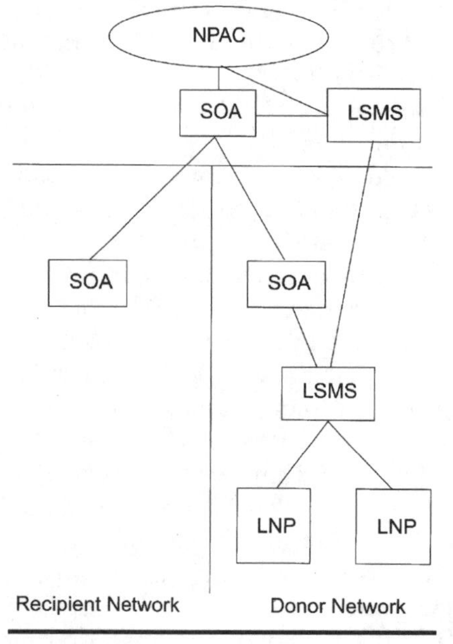

Figure 10.1

Number portability administration centers (NPACs)

The NPAC is managed by a third-party company with no interest in the telephone business. There are presently two corporations responsible for the management of NPACs, Martin Marietta and Perot Systems. The NPAC is responsible for receiving requests from carriers (recipient carriers) for the porting of telephone numbers (from donor carriers). They then coordinate the porting of the number by sending the data to the donor network, confirming the request has been accepted, and then downloading the ported number data (which is the new LRN for the telephone number, and other routing information) to all of the other networks connected to that NPAC.

There are seven NPAC regions, roughly aligned with the regions designated for the Bell Operating Companies. Perot Systems manages the Southeast, West, and West Coast regions. Martin Marietta manages the Midwest, Southwest, Mid-Atlantic, and New England regions. The NPAC network includes a service order administration (SOA) system as well as a service management system (SMS). Their functions are the same as described below.

Local service order administration (LSOA)

All order information regarding a number being ported is sent to the service order administration (SOA) system. The SOA is used to process a subscriber's order and track the order through completion. It provides all departments a single record location regarding a service order, and is used to coordinate and track service order activities.

In the case of LNP, the subscriber data includes the location routing number (LRN) of the serving (donor) carrier, the date/time the number is to be ported to the new carrier (recipient), and other pertinent information. The SOA communicates this information to the SOA in the NPAC, which in turn passes the information to the SOA in the donor network.

It is important to understand that the only purpose of the SOA is to track the activities of an order, and in the case of LNP, provide the specifics about when a number is to be ported, and who the donor network is. Some of this information will be used by the billing center, while some of it is passed to provisioning systems to be configured in switches and databases in the new network.

Local service management system (LSMS)

Each carrier has an LSMS, which serves as the interface between the carrier networks and the NPAC. The LSMS is responsible for collecting porting data and downloading it to the LNP databases. The LSMS is usually a computer system with database storage, and must be able to verify the data within the database with the data stored at the NPAC. This is accomplished through periodic audits between the LSMS and the NPAC.

Likewise, the LSMS also audits the LNP databases within its own network. It is crucial that the data is accurate. If there are discrepancies in the data, then telephone calls cannot be completed to ported numbers properly. The computer systems used for LSMS must have high reliability, and in many networks, they are deployed as mirror systems.

A mirrored system is one where a duplicate LSMS is placed in a different geographical location, but contains a mirrored image of the data at its mate system. This configuration provides diversity within the network. Diversity is as crucial with LNP as it is within other parts of the SS7 network.

The industry standard interface used between the LSMS and the NPAC is a Q3 interface, using the OSI stack for communications. The protocol used to communicate between the two entities is CMISE and ROSE. The communications protocol is used at the network layers of Q3 and TCP/IP.

The interface between the LSMS and the LNP database has not yet been standardized, but Bellcore is actively working on defining Q3 as the interface between these two entities as well. Most systems being implemented today are using this Q3 interface between the LSMS and the LNP database.

LNP database

The actual database used to maintain LNP data can be one of two types; a Service Control Point (SCP) or an integrated Signal Transfer Point (STP). The integrated STP solution is more favorable in many cases because of the throughput modern STPs can provide in comparison to SCPs. The top of the line SCP is only capable of 850 transactions per second, while some newer STPs are capable of 20,000 transactions per second and higher.

The actual throughput requirement of the database itself will depend on the calculated number of database queries to be performed. To calculate the number of queries (transactions) per second, remember that every call made to an NPA-NXX with a directory number that has been ported will require a database query. This can be significant in some areas.

You also have to factor in the average message size of an LNP query. LNP queries average around 120 bytes in length, so if you are using 56-kbps links, you can transmit approximately 58 messages per second, if your link is transmitting at 100% of its capacity (56,000/[120 × 8]). This of course is never the case, since links are engineered to transmit at 40% capacity. Actual throughput of a 56-kbps link with LNP message size of 120 bytes would be more like 23 LNP transactions per second.

Summary

These are the basic elements used in an LNP network. Notice that nothing was mentioned about STPs, or SCPs for that matter. The database function itself can reside in either an STP or an SCP. The only connection to the SS7 network provided is the database itself. The LSMS does not connect to the SS7 network, and neither does the NPAC. All communications between these entities is through a private communications link, using Ethernet and TCP/IP protocols.

The SS7 network uses the information provided by the LNP database to route calls through the network. This function is much like the database function used in 800 services. Figure 10.2 illustrates the relationship between the SS7 network elements and LNP elements.

Now that we understand the elements needed to implement LNP, let's look at how LNP works, first in the wireline network and then in the wireless network. We will review the role of each of the elements in these discussions as well.

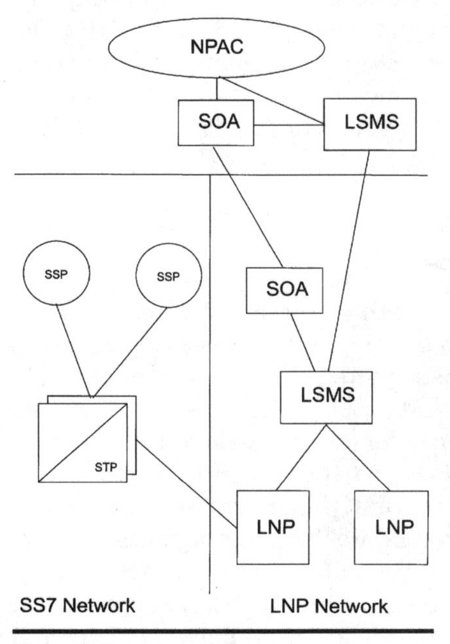

Figure 10.2

The Wireline Solution

In the wireline network, LNP is simpler. This is because the subscribers are all fixed in location. In the wireless network, the subscribers are mobile, making billing applications and OSSs more complicated. Before discussing the specifics about how a number is ported and then routed, let's first look at the changes made within the SS7 ISUP protocol to support LNP.

The ISUP IAM message has been modified to include information regarding a ported number. The most significant change is the placement of the location routing number (LRN). The called party number field is no longer the directory number of the called party, but is the LRN of the called party if the called party directory number has been ported. The directory number of the called party is moved to the generic access parameter (GAP) of the IAM message. The type of address parameter in the IAM message is "ported number."

The forward call indicators (FCI) field of the IAM now includes a "translated called number field." This 1-bit parameter in the 8-bit field is used to indicate whether or not a number has been translated. This is used by switches to determine if a query is needed, and whether or not routing should be based on the GAP parameter or the called party number field. Even if a number has not been ported, the forward call indicators will still indicate "number translated."

The jurisdiction information parameter (JIP) is used for billing purposes. The LRN of the originating end office is placed in the JIP field. Billing systems can then determine how the call is to be processed. Only the first six digits of the LRN are needed.

Now that we understand the modifications to the SS7 protocol, let's look at how telephone numbers are ported and how calls are routed to ported numbers.

Porting a number in the wireline network

When a subscriber places an order with a competitive local service provider (recipient carrier), it is up to the new carrier to send a request to have the subscriber's telephone number ported. The request is sent from the recipient network LSOA to the NPAC SOA. The NPAC SOA is then responsible for notifying the original carrier (donor carrier) of the request.

The NPAC sends the request down to the donor carriers LSOA, which then verifies the number, is currently assigned in the network, and then confirms receipt of the porting data. The porting data consists of the subscriber's telephone number, the date and time the number is to be ported, and the location routing number (LRN) of the recipient carrier. The NPAC then forwards the confirmation back to the recipient carrier's LSOA.

The NPAC then sends the porting information to its own LSMS, which then distributes the porting information immediately down to all of the LNP databases within its region. The number has now been ported and the actual switch from one carrier's switching equipment to the other takes place according to the porting data. Obviously, coordination between the two carriers is critical in order to prevent the subscriber from being disconnected by his or her original carrier before the recipient carrier can active service.

The donor carrier is not responsible for notifying the recipient carrier of the services in the subscriber's profile. It is still the responsibility of the subscriber to notify the recipient carrier which services they want to have when they port. If the subscriber disconnects his or her service at any time, the telephone number is returned to the donor carrier (until the LNP service matures and number assignment becomes more universal).

Routing a call with LNP in the wireline network

Each end office switch is assigned a 10-digit location routing number (LRN). Each switch is required to support at least two LRNs (in the event the switch previously serviced more than one NPA-NXX). This LRN consists of the same area/office code previously assigned to the end office. The last four digits are typically all zeros. By using an LRN, the existing numbering plan can be utilized. Routing is based on the LRN instead of the dialed telephone number when a number has been ported.

When a number is dialed, the originating switch looks up the NPA-NXX of the number in a table. This table will indicate whether or not the NPA-NXX is considered as ported or not. If even one telephone number within an NPA-NXX is ported to another carrier, then the entire NPA-NXX is considered ported. This means every phone call to that NPA-NXX requires a database query, to determine whether or not the dialed number has actually been ported to another carrier, and if so what the LRN of the new carrier is.

Before we continue discussing routing, we need to understand why the end-office switch only knows about the NPA-NXX and not the subscriber number. It would be easy to store routing information about every number in the end-office switch, alleviating the requirement to access a database. However, propagating this information to many switches and maintaining that data is a different story altogether.

The alternative is to query a database on each and every call, but the impact to SS7 is too large, and the capacities of almost every network would seriously be taxed. It was for these reasons the industry elected to use the LRN method for routing, and propagating only the NPA-

NXX information in the individual switches. The impact on networks is still high, but the solution is a compromise to more costly methods.

If an end-office switch determines that a number belongs to a ported NPA-NXX, it originates a query to the LNP database (provided the switch has been equipped with that capability). In the event the switch has not been equipped to perform LNP queries, the call is routed as normal through the network to the donor carrier. When the call reaches the donor network's end office, the switch will determine that the call cannot be completed, because that switch no longer services the dialed number. The directory number will appear as vacant in the end-office switches database.

The question is then raised as to who is responsible for originating an LNP query. It would be very easy for carriers to simply route calls to neighboring networks without performing queries, thereby alleviating the need to implement any method of LNP in their own networks. Simply let the neighbors bear the cost of implementation and pass all your traffic to them. The FCC recognized the potential of call dumping and passed a ruling, that the next to the last carrier (or network) was responsible for originating the LNP query. This is known as the N-1 method. If there are only two carriers involved in the call setup, the original carrier is responsible for originating the LNP query.

The LNP database will provide the LRN of the servicing carrier if a number has been ported. This information is placed in the SS7 ISUP message using the called party parameter. The actual called party directory number is moved to the generic access parameter (GAP). This is one of the reasons upgrades to virtually every network node in the SS7 network is required to support LNP; ISUP messages have been modified to support this feature. The ISUP parameters that have been modified are discussed above.

Let's look at a typical call to a number that has not been ported, but is in an NPA-NXX considered "portable." Figure 10.3 shows an example of the call flow in the case of a nonported number.

In Fig. 10.3, the calling party dials the number (919) 460-2100. The end-office switch servicing the called party identifies that particular NPA-NXX as one that has been ported. The end-office switch then originates a database query to the LNP database.

The query is sent to the STP, which then examines the address information in the SCCP portion of the message to determine how to handle the query. The SCCP portion of the message identifies the type of database to direct the query to, if that is known. If the end-office switch does not know the network address of the database, the STP can perform a global title on the SCCP portion to determine the type of query and the database to send the query to.

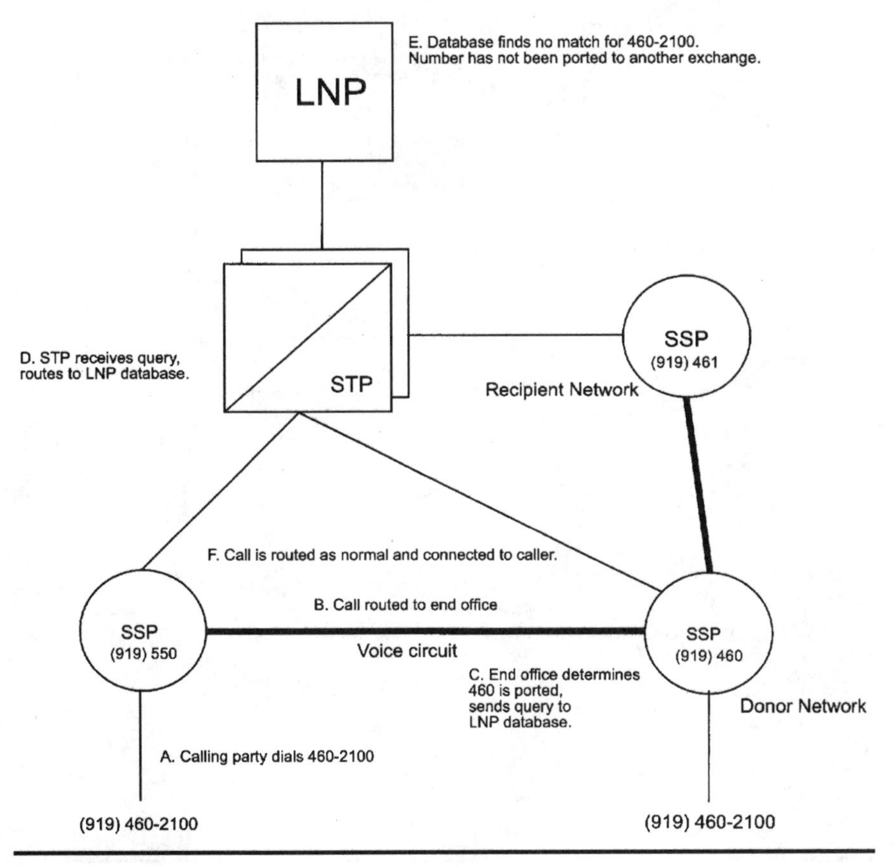

Figure 10.3

The LNP database checks the database, and determines that the called number (919) 460-2100 has not been ported to another carrier, and returns the query results to the end-office switch. The end-office switch then routes the call based on the called party address field in the ISUP message. The call is connected.

This is an example of how a call is connected to a nonported number. Now let's look at how a call gets connected to a ported number. In Fig. 10.4, a call is made to (919) 460-2100, but this time the called party has switched carriers, and their number has been ported to the new carrier.

Again, the end office has determined that 919-460 is a ported NPA-NXX, and initiates a query to the LNP database. The LNP database looks for the dialed number in its database. A match is found, and the LNP returns the results back to the end office.

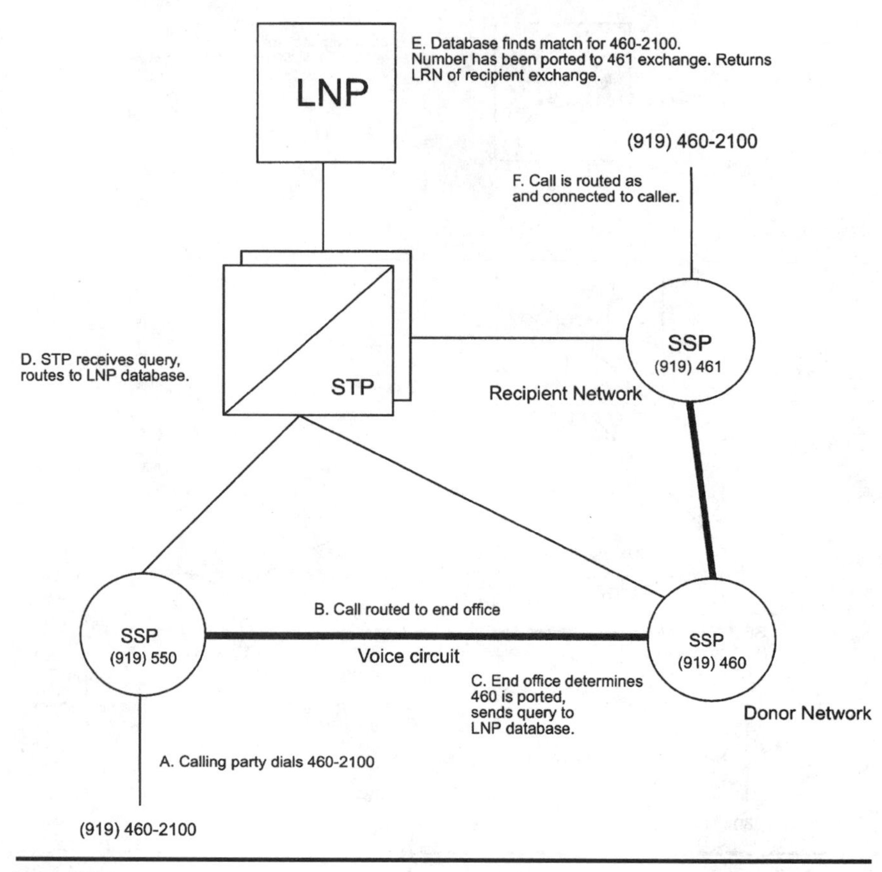

Figure 10.4

The end office now has the information needed to route the call. The call is routed based on the location routing number (LRN) instead of the called party number. The originating end office generates an ISUP IAM message with the LRN placed in the called party address field, and the dialed number in the GAP parameter. The JIP parameter identifies the NPA-NXX to be used for billing reference.

The call is routed to the recipient network, which routes the call to the subscriber using the dialed digits in the GAP parameter. Billing systems to determine the origin of the call use the LRN in the JIP parameter. The calling party number was used to determine the geographic-specific digits needed by billing systems previous to LNP.

This explains the basics of routing a call in the wireline network, using SS7 to deliver the necessary information to the network elements. The wireless network operates differently than the wireline network, and presents challenges of its own.

The Wireless Solution

In the wireless network, the information needed to route calls to mobile subscribers is somewhat different than what is required in the wireline network. For beginners, the number itself does not render enough information about the location of the called party. They can be anywhere in the network.

Each mobile subscriber has a mobile identification number (MIN) which is used to identify the individual subscriber to the network. The wireless network relies on databases to locate the location of mobile subscribers, which changes as they move through the network. We won't go into details about the routing of calls in the wireless network, as that was discussed in the beginning of this book. We will look at how calls are routed using LRN in the wireless network, and the impact LNP has on the wireless network.

A cellular telephone is a radio device. The phone must be compatible with the cell site equipment in the areas where it operates. There are cases where a subscriber will not be able to "port" his or her cellular phone. In other words, subscribers may change wireless service providers and maintain their existing telephone numbers, but they will have to buy a new cellular telephone.

Another fundamental difference between wireline and wireless is in the billing and network boundaries. In the wireline network, boundaries are set by LATAs. The wireless industry does not recognize LATAs, and the service provider determines the boundaries. This can cause difficulty in determining how LNP is to be supported, because the ruling says that LNP must be supported within a rate center, which is not recognized by the wireless industry. The Cellular Telecommunications Industry Association (CTIA) is currently defining this. Some of the fundamentals of porting a number in the wireless network are discussed in the next section.

Porting a number in the wireless network

When a subscriber ports to a new wireless service provider, the new carrier must provide service in the same area as the donor network. This is one of the principal requirements, since rate centers are not recognized in wireless networks. In the wireline network, the service provider had to be providing local service within the same rate center as the subscriber.

There are other requirements before porting can be supported. The wireline rate center associated with the directory number assigned to the mobile subscriber must be in the same geographic area as the home service area of the subscriber. This is also true of wireline subscribers wishing to port their numbers to a wireless service provider.

Routing a call with LNP in the wireless network

Routing in wireless networks is different than in wireline networks, because the subscriber is mobile. Databases must be used to locate the mobile subscriber, which results in delays in the call routing. With LNP, we are now introducing yet another database query prior to even locating the proper network.

There are several call scenarios in wireless; land-to-mobile, mobile-to-mobile, and mobile-to-land. All are handled somewhat differently. We will start by looking at a simple mobile-to-land call.

Figure 10.5 shows the call flow for a mobile-to-land call. When a mobile subscriber dials a number, the receiving MSC establishes a trunk connection with the PSTN. The receiving switch in the PSTN examines the dialed digits and determines that the dialed number is in a portable NPA-NXX. The switch sends a query to the LNP database using IN/AIN triggers to obtain additional routing instructions. The MSC in the serving cellular network can just as easily send this query, if the wireless provider has the capability.

The query to the LNP database results in a response, providing the LRN to be used in routing the call. This is no different than the routing procedures used in the wireline network. The actual IN/AIN triggers have been added (Info_Analyze and Analyze_Response) in Fig. 10.5.

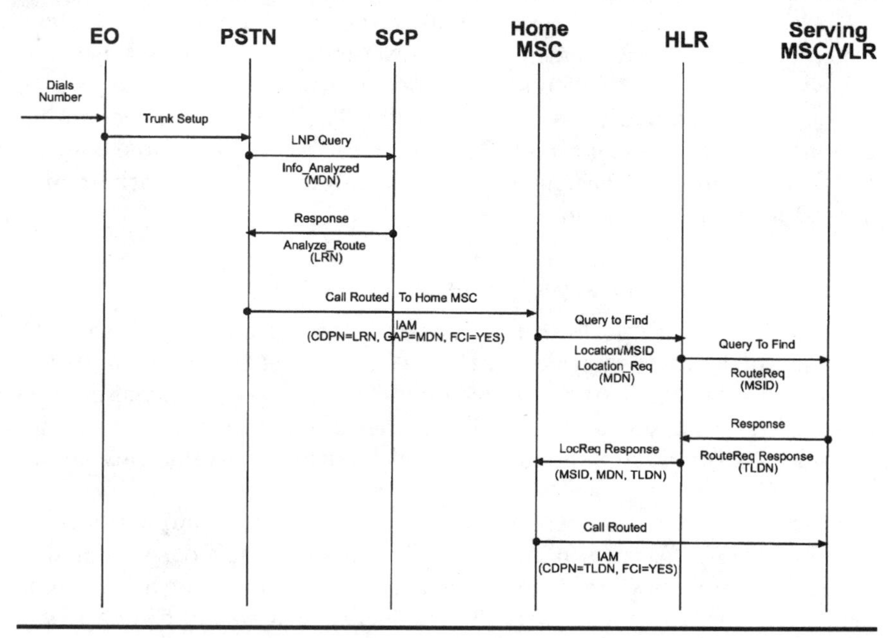

Figure 10.5

Once the switch has the routing information it needs, it generates an ISUP IAM message containing the LRN, dialed number, and the mobile directory number that originated the call. This call flow is not the same when a call is made land-to-mobile, as you are about to see. The location of the subscriber must be obtained first.

In Fig. 10.6, we see the events that take place in a land-to-mobile call. The end-office switch initiates a query to the LNP database, sending the mobile directory number (MDN). The LRN is returned (if the mobile number has been ported), and the end office establishes a trunk connection with the mobile's home MSC. Notice the IAM Fig. 10.6 provides, and the MDN, but not the MSID.

The home MSC searches its own HLR to determine the location of the mobile station. The HLR shows the mobile station is in another network and provides the identity of the serving MSC. The HLR then sends a query to the serving MSC's VLR, which will provide location information.

The serving VLR returns a response to the HLR, providing the temporary local number (TLDN) used in roaming. The TLDN is a temporary number assigned to the mobile station as long as it is served in the visited network. This number is then assigned to a new mobile station when the previously assigned mobile leaves the network.

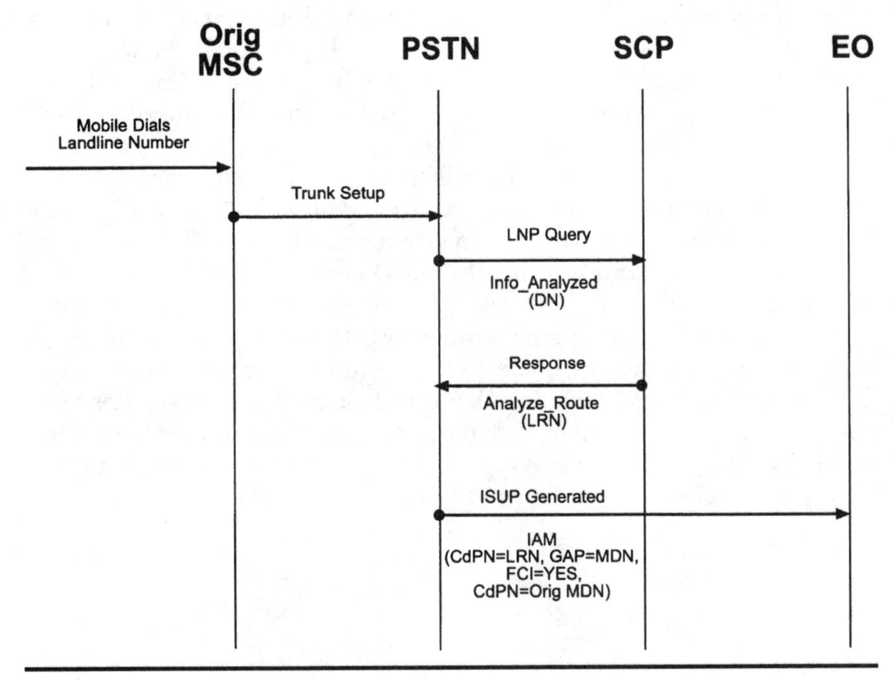

Figure 10.6

The HLR then sends a response to the home MSC providing the MSID, MDN, and TLDN. The home MSC can now generate the ISUP IAM. In this case, the called party number contains the TLDN, not the MDN. If the mobile station were still in its home network, this parameter would contain the MDN.

These call flows have been simplified considerably for ease of reading. There are many other events that must take place to route calls in the wireless network. This should give you an idea how the protocol works in obtaining the necessary routing information when the numbers have been ported in wireless networks.

Summary

Local number portability will impact everyone. The conventional methods of routing calls are gone, and new procedures will require many modifications to our nation's networks. These changes are not just impacting U.S. networks. The international telephone companies are also looking at implementing LNP in their networks. Many international carriers have already begun choosing solutions for their networks, and have identified deployment schedules.

Developers of telecommunications products will also be impacted, but in a positive way. Their products must be modified to support LNP solutions. As seen in our previous discussions about wireline and wireless portability, these solutions will vary. Equipment vendors must support all solutions to market products in both the wireline and wireless markets. This means supporting IN/AIN triggers as well as IS-41 and GSM protocol changes. As LNP moves forward, the subscriber will begin to feel the impact as well, but hopefully in a positive way. As subscribers, we will have many more choices available to us. We will now be able to choose whom we want to provide our local telephone service, as well as long-distance service. In many cases, service providers will provide bundled services. These bundled services will include local, long distance, wireless, paging, and even Internet access. Hopefully, as more competition enters the market, we will also see reduced prices for these services. Don't count on substantial savings, however; these network changes cost millions of dollars, and LNP alone is costing some telephone companies billions of dollars to deploy. These costs have to be recouped somehow, and always trickle back to the subscriber.

Appendix

A

ITU-TSS SS7 Publications

The following listing includes all SS7 standards published by the ITU-TSS as of 1994. These standards define the protocol and the network at the international plane. The standards in the United States and in all other countries are based on these ITU standards.

Message Transfer Part (MTP)

Q.701 Functional Description of the Message Transfer Part
Q.702 Signaling Data Link
Q.703 Signaling Link
Q.704 Signaling Network Functions and Messages
Q.705 Signaling Network Structure
Q.706 MTP Signaling Performance
Q.707 Testing and Maintenance
Q.708 Numbering of International Signaling Point Codes
Q.709 Hypothetical Signaling Reference Connection
Q.710 Simplified MTP Version for Small Systems

Signaling Connection Control Part (SCCP)

Q.711 Functional Description of SCCP
Q.712 Definition and Functions of SCCP Messages
Q.713 SCCP Formats and Codes

Q.714 SCCP Procedures

Q.716 SCCP Performance

Telephone User Part (TUP)

Q.721 Functional Description of TUP

Q.722 General Function of Telephone Messages and Signals

Q.723 Formats and Codes

Q.724 Signaling Procedures

Q.725 Signaling Performance in the Telephone Applications

ISDN Supplementary Services

Q.730 ISDN Supplementary Services

Q.731 Stage 3 Description for Number ID Supplementary Services Using SS7

Q.733 Stage 3 Description for Call Completion Supplementary Services

Data User Part (DUP)

Q.741 SS7 Data User Part

ISDN User Part (ISUP)

Q.761 Functional Description of ISUP

Q.762 General Function of Messages and Signals

Q.763 Formats and Codes

Q.764 Signaling Procedures

Q.766 Performance Objectives in the ISDN Application

Q.767 Application of the ISUP for International ISDN Interconnections

Transaction Capabilities Application Part (TCAP)

Q.771 Functional Description of TCAP

Q.772 Transaction Capabilities Information Element Definitions

Q.773 TCAP Formats and Encoding

Q.774 TCAP Procedures

Q.775 SS7 Test Specification General Description

Test Specification

Q.780 SS7 Test Specification—General Description

Q.781 MTP Level 2 Test Specification

Q.782 MTP Level 3 Test Specification

Q.783 TUP Test Specification

Q.784 ISUP Basic Call Test Specification

Q.785 ISUP Protocol Test Specification for Supplementary Services

Monitoring and Measurements

Q.791 Monitoring and Measurements for SS7 Networks

Operations, Administration, and Maintenance (OAM)

Q.795 Operations, Maintenance, and Administration Part

Appendix

B

American National Standards Institute (ANSI)

The following listing is of SS7 standards published by the ANSI. These standards define the protocol and the network at the national plane.

ANSI T1.110 Telecommunications Signaling System No. 7 (SS7)—General Information

ANSI T1.111 Telecommunications Signaling System No. 7 (SS7)—Functional Description of the Signaling System Message Transfer Part (MTP)

ANSI T1-112 Telecommunications Signaling System No. 7 (SS7)—Signaling Connection Control Part (SCCP)

ANSI T1.113 Telecommunications Signaling System No. 7 (SS7)—Integrated Services Digital Network (ISDN) User Part (ISUP)

ANSI T1.113 Telecommunications Signaling System No. 7 (SS7)—Integrated Services Digital Network (ISDN) User Part (NxDS0 Multi-Rate Connection)(supplement to ANSI T1.113-1992)

ANSI 1.114 Telecommunications Signaling System No. 7 (SS7)—Transaction Capabilities Application Part (TCAP)

ANSI T1.115 Telecommunications—Monitoring and Measurements for Signaling System No. 7 Networks

ANSI T1.116 Telecommunications Signaling System No. 7 (SS7)—Operations, Maintenance, and Administration Part (OMAP)

ANSI T1.118 Telecommunications Signaling System No. 7 (SS7)—Intermediate Signaling Network Identification (ISNI)

ANSI T1.226 Telecommunications—Operations, Administration, Maintenance, and Provisioning (OAMP)—Management of Functions for Signaling System No. 7 (SS7) Network Interconnections

ANSI T1.609 Telecommunications—Interworking between the ISDN User-to-Network Interwork Interface Protocol and the Signaling System No. 7 ISDN User Part (ISUP)

ANSI T1.611 Telecommunications—Signaling System No. 7 (SS7)—Supplementary Services for non-ISDN Subscribers

ANSI T1.631 Telecommunications—Signaling System No. 7 (SS7)—High Probability of Completion (HPC) Network Capability

Bell Communications Research (Bellcore)

The following publications are Bellcore publications, which are directly related to the SS7 network and its protocol. There are many other related publications that define the requirements of various network components not listed here.

TR-TSY-000024 Service Switching Points (SSPs) Generic Requirements

TR-NWT-000082 Signaling Transfer Point (STP) Generic Requirements

TR-NWT-000246 Bell Communications Research Specification of Signaling System Number 7

TA-TSY-000372 CCS Switching System Maintenance Requirements

TR-NWT-000394 Switching System Generic Requirements for Interexchange Carrier Interconnection Using the Integrated Services Digital Network User Part (ISDNUP)

TR-TSV-000905 Common Channel Signaling (CCS) Network Interface Specification

Bibliography

ANSI, *Standards Committee T1—Telecommunications Procedures Manual,* June 1991.

Bell Laboratories, *Engineering and Operations in the Bell System,* 2d ed., 1984.

Bellcore, *GR-246-CORE—Bell Communications Research Specification of Signaling System Number 7*, December 1994.

Bellcore, *GR-2936-CORE—Bell Communications Research Local Number Portability Capability Specifications*, Issue 1, July 1996, Revision 1, October 1996.

Bellcore, *TR-NWT-000082—Signaling Transfer Point Generic Requirements,* Issue 4, December 1992.

Bellcore, *TR-NWT-000246—Bell Communications Research Specification of Signaling System Number 7,* Issue 2, June 1991, Revision 3, December 1993.

Bellcore, *TA-TSY-000387—Generic Requirements for Interim Defined Central Office Interface (IDCI)*, Issue 2, July 1990.

Bellcore, *TA-NPL-000458—Digital Signal Zero "A" (DS-0A 64kb/s) Systems Interconnection,* Issue 1, February 1988.

Bellcore, *FA-NWT-001109—Broadband ISDN Transport Network Elements Framework Generic Criteria,* Issue 1, December 1990.

Bellcore, *TA-NWT-001111—B-ISDN Access Signaling Generic Requirements,* Issue 1, August 1993.

Bellcore, *TA-NWT-001114—Generic Requirements for Operations Interfaces Using OSI Tools: ATM/Broadband Network Management,* Issue 2, October 1993.

Berkman, Roger, and John H. Brewster, *Perspectives on the AIN Architecture,* Datapro Research, August 1992: #8205.

Costello, Rich, *Central Office Switches: Overview,* Datapro Research Corp., April 1993.

Cellular Telecommunications Industry Association, *CTIA Report on Wireless Number Portability*, Revision 1.0, April 11, 1997.

Datapro, *SS7—Good for More Than Telephone Companies,* Datapro Research Corp., May 1987.

Federal Communications Commission, *FCC in Brief,* Public Services Division, FCC, June 1993.

Glowacz, Dave, "AIN Services Get New Life in 1993," *Telephony Magazine,* January 11, 1993, pp. 27–32.

Jeppason, Richard, *Common Channel Signaling #7,* Datapro Research Corp., May 1992.

Kearns, Timothy, and Maureen C. Mellon, "The Role of ISDN Signaling in Global Networks," *IEEE Communications Magazine,* July 1990, pp. 36–43.

Langer, Mark R., *Advanced Intelligent Networks (AIN)*, Datapro Research Corp., January 1993: #3070.

Martin, James, *Telecommunications and the Computer,* 3d ed., Prentice Hall, New Jersey, 1990.

Mason, Charles F., "Bell Atlantic Cuts Over Seamless Cellular," *Telephony Magazine*, January 11, 1993, p. 1.

Modaressi, Abdi R., and Ronald A. Skoog, "Signaling System No. 7: A Tutorial," *IEEE Communications Magazine,* July 1990, pp. 19–35.

Petersen, Phil, "Positioning PCS on the Telecom Landscape," *Telephony Magazine*, Dec. 13, 1993, pp. 26–32.

Schriftgiesser, Dave, "Key Trends in Broadband Communications: The Next Five Years," *Telecommunications Magazine,* January 1994, pp. 101–105.

Stallings, William, *Handbook of Computer Communications Standards—The Open Systems Interconnection (OSI) Model and OSI-Related Standards,* Macmillan, New York, 1987.

Valenta, Rose A., *Integrated Services Digital Network (ISDN) Standards,* Datapro Research Corp., October, 1991: #8430.

Index

ABOUT THE AUTHOR

Travis Russell has over 15 years experience in the telecommunications industry. His experience in R&D projects encompasses many technologies, including SS7, Frame Relay, FDDI, ISDN and GSM. He has delivered a number of technology seminars and lectures in many countries, for both industry and universities. He also provided switch training for a number of companies from the Middle East, Pacific Rim, South America and China.

Travis has worked as field engineer, writer and instructor for the switching divisions of Pacific Telephone, AT&T, Memorex Telex, and now Tekelec. He is currently manager for the switching division of Tekelec, a leading provider of SS7 solutions.